Tissue World Americas 2012

Miami, Florida, USA
20-23 March 2012

ISBN: 978-1-5108-2171-2

Printed from e-media with permission by:

Curran Associates, Inc.
57 Morehouse Lane
Red Hook, NY 12571

Some format issues inherent in the e-media version may also appear in this print version.

Printed by Curran Associates, Inc. (2016)

For permission requests, please contact UBM Exhibition Singapore PTE LTD
at the address below.

UBM Exhibition Singapore PTE LTD
10 Hoe Chiang Road
#20-05 Keppel Towers
Singapore 089315

Phone: +65 6592 0888
Fax: +65 6221 2282

www.ubmasia.com

Additional copies of this publication are available from:

Curran Associates, Inc.
57 Morehouse Lane
Red Hook, NY 12571 USA
Phone: 845-758-0400
Fax: 845-758-2633
Email: curran@proceedings.com
Web: www.proceedings.com

TABLE OF CONTENTS

FORUM

SESSION 1

SESSION 2

SESSION 3

SESSION 7

WORKSHOP

TOSCOTEC

AMERICAS TISSUE WORLD

STEEL YANKEE DRYER TT SYD

Practical applications and field experiences

INTRODUCTION

AMERICAS TISSUE WORLD

- ✓ Yankee Dryers represent key elements of tissue machines
- ✓ Steel Yankee Dryers are quite a recent technological application
- ✓ Significant field data are today available to support design and process troubleshooting

2

TOSCOTEC

INTRODUCTION

- The combined action of Yankee and Hood determines the drying of the sheet, so it has a great influence on a plant performance
- There are substantial field data up to date about applications with Steel Yankee Dryers

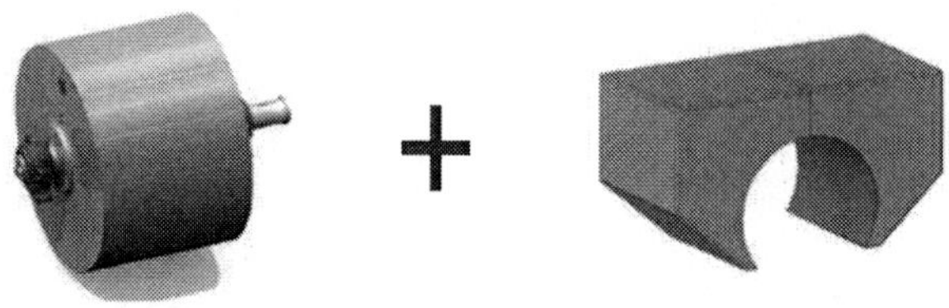

3

TOSCOTEC

THE HISTORY

- **1960** – Starts industrial application of Steel Drying Cylinders

- **Today** – Field data available over **1200** operating pressure vessels manufactured in steel

4

TOSCOTEC

THE HISTORY

- ✓ **1999** – Design and test for utilization of steel on **Yankee Dryers**

- ✓ **2000** – First Steel Yankee Dryer with a diameter of **2500** mm manufactured and in operation
- ✓ **2006** – **15FT** (4572 mm)
- ✓ **2010** – **16FT** (4876 mm)

TOSCOTEC

THE HISTORY

Today

- ✓ Technologically available diameters up to 18FT (**5486 mm)** and with **7200 mm** maximum face length. Manufacturing codes ASME - PED - CSEI (PRC) – JIS

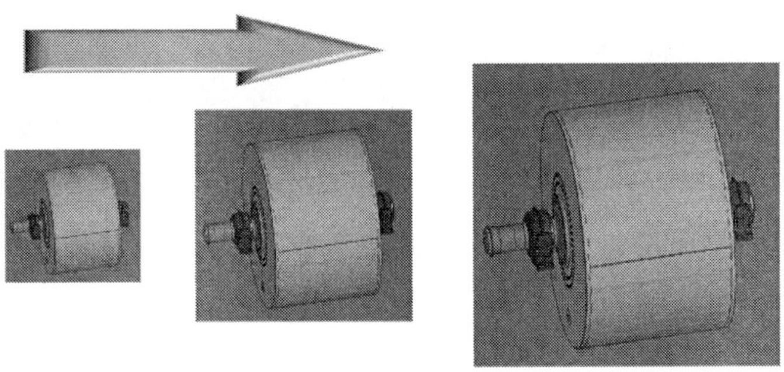

TOSCOTEC

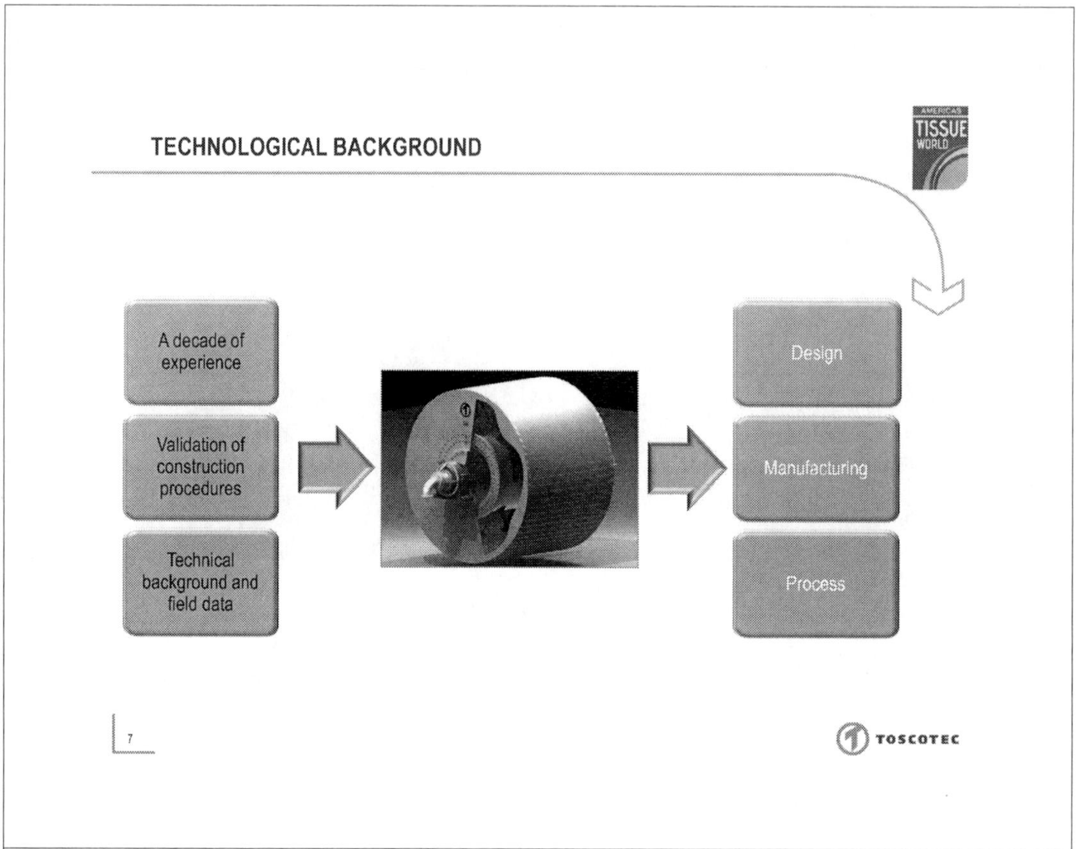
TECHNOLOGICAL BACKGROUND
AMERICAS TISSUE WORLD
A decade of experience
Validation of construction procedures
Technical background and field data
Design
Manufacturing
Process
7
TOSCOTEC

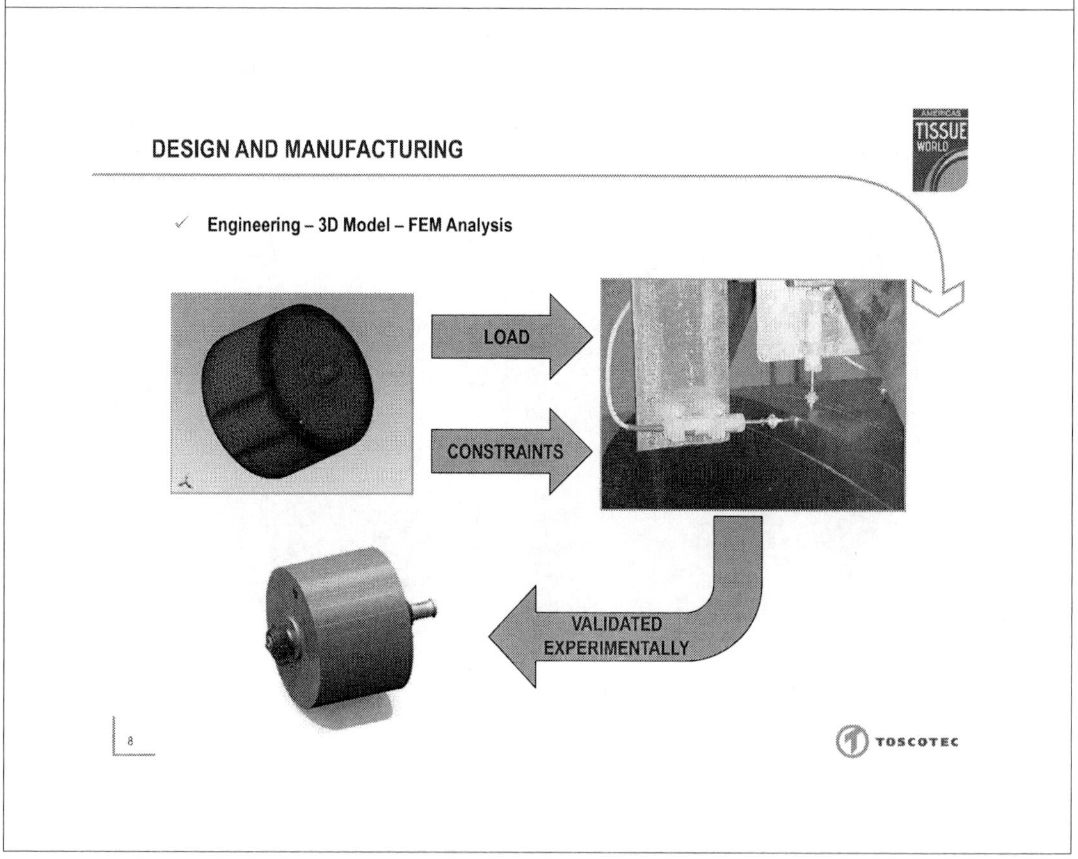
DESIGN AND MANUFACTURING
AMERICAS TISSUE WORLD
✓ Engineering – 3D Model – FEM Analysis
LOAD
CONSTRAINTS
VALIDATED EXPERIMENTALLY
8
TOSCOTEC

AMERICAS TISSUE WORLD

FINITE ELEMENTS ANALYSIS

- ✓ **Working condition:** to evaluate Yankee stress and displacements in working condition
- ✓ **Broken paper condition:** in this configuration the Yankee is not in contact with the wet paper
- ✓ **Hydro-Test condition:** to evaluate Yankee stress and displacements in Hydro-Test condition

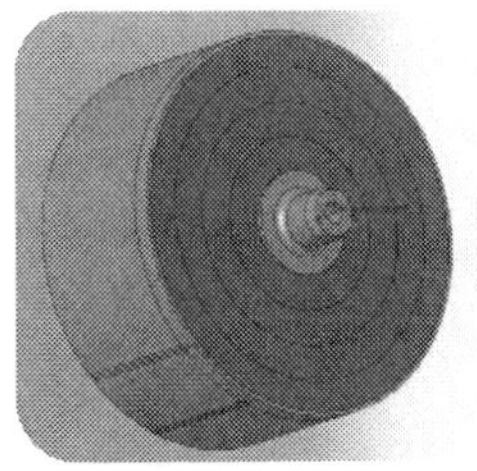

9

TOSCOTEC

AMERICAS TISSUE WORLD

FINITE ELEMENTS ANALYSIS

- ✓ **Output data:** temperature and stress distribution

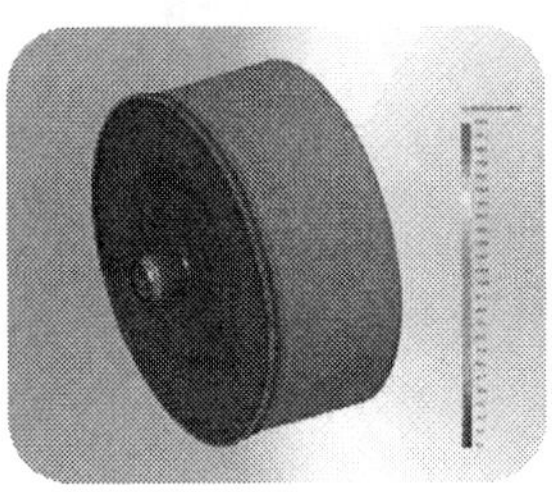

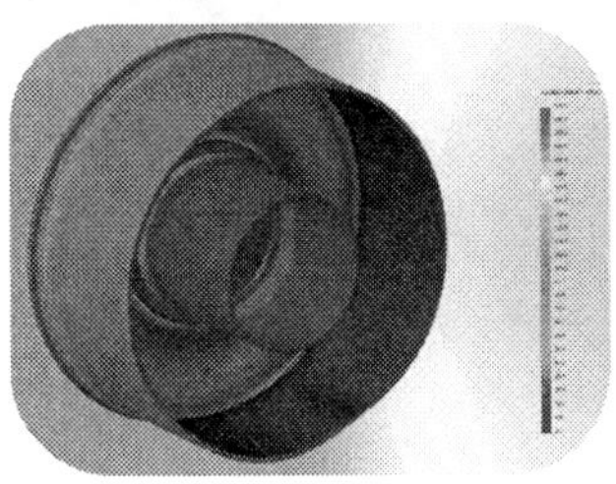

10

TOSCOTEC

VALIDATION OF THE MODEL

- **Laboratory test:** strain gauge and displacement measurements

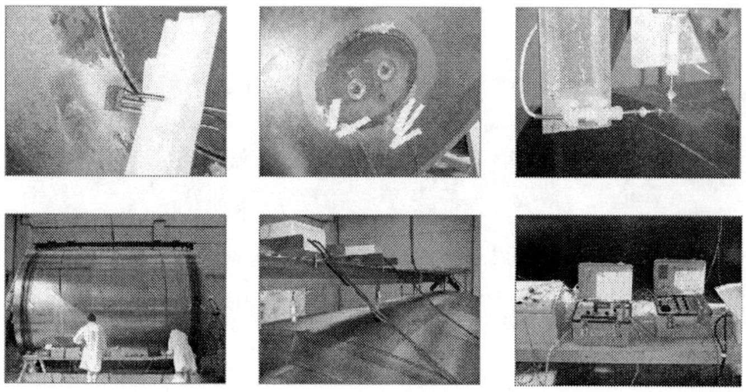

11

TOSCOTEC

MANUFACTURING PROCESS

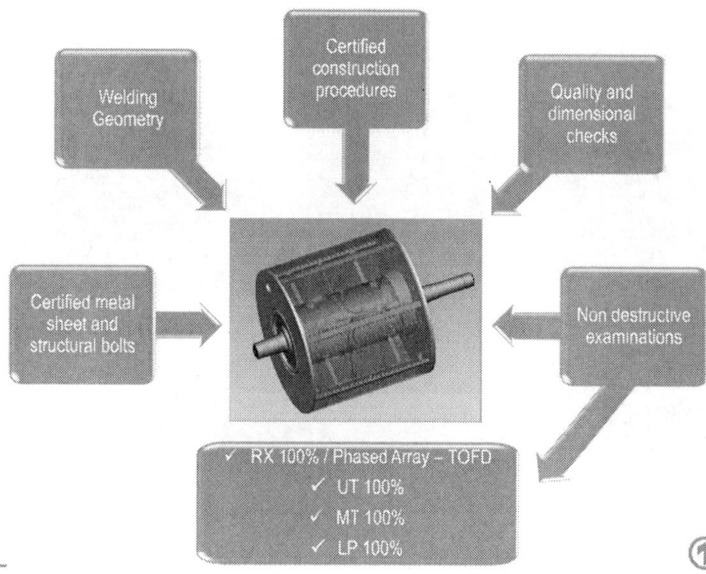

12

TOSCOTEC

PERIODICAL CHECKS AFTER START-UP

- STRUCTURAL WELDS AND BOLTS / MOST STRESSED AREAS → PHASED ARRAY; UT, MT, NDE
- CONDENSATE REMOVAL SYSTEM
- TEMPERATURE PROFILE → WITH/WITHOUT PAPER
- RUN-OUT → AMBIENT TEMPERATURE + HOT CONDITION
- METALLIZATION

13

TOSCOTEC

OVERVIEW OF FIELD DATA AND EXPERIENCES

- ✓ Shell heat transfer
- ✓ Head insulation
- ✓ Soda straw plugging
- ✓ Accidental damage

14

TOSCOTEC

FIELD DATA – HEAT TRANSFER

Yankee replacement
Comparison at 5,5 barg

15FT RIBBED YANKEE			
	Steel Yankee Dryer	Cast Iron Yankee Dryer	
Test steam pressure	5,5	5,5	barg
Sheet width at Yankee Dryer	2850	2850	mm
Yankee diameter	15FT	15FT	mm
Drying rate	**81,8**	**58,5**	kg H_2O/m^2hr

15

TOSCOTEC

FIELD DATA – HEAT TRANSFER

Yankee replacement
Comparison at 7,8 barg

15FT RIBBED YANKEE			
	Steel Yankee Dryer	Cast Iron Yankee Dryer	
Test steam pressure	7,8	7,8	barg
Sheet width at Yankee Dryer	2850	2850	mm
Yankee diameter	15FT	15FT	mm
Drying rate	**98**	**70,2**	kg H_2O/m^2hr

16

TOSCOTEC

FIELD DATA – HEAT TRANSFER

✓ **Increased heat transfer:** validation by field measures

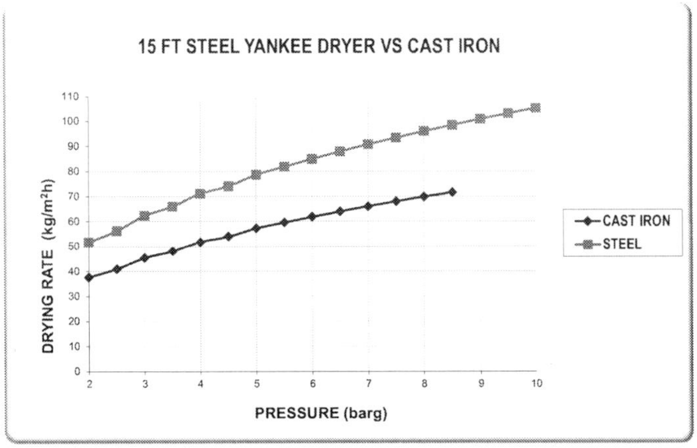

TOSCOTEC

FIELD DATA – HEAT LOSS REDUCTION

✓ **Head insulation:** reduction of heat flow to the ambient

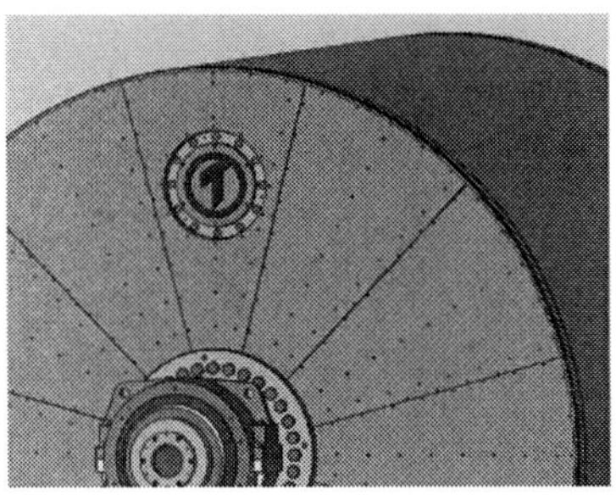

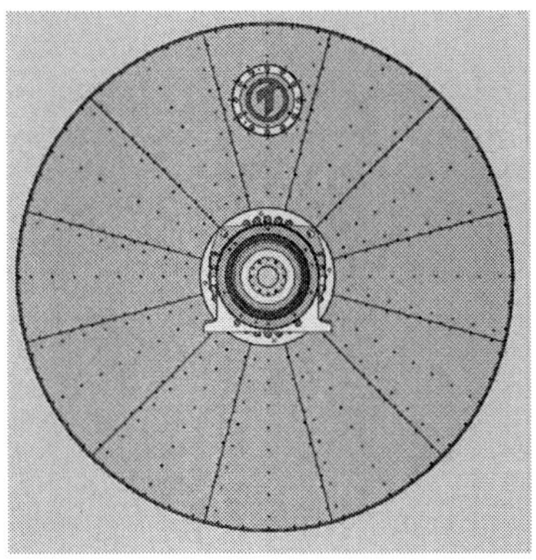

TOSCOTEC

ENERGY LOSS REDUCTION - HEAD INSULATION

- **Heat loss reduction:** the energy flow is a function of the insulation thickness and of the yankee dimensions

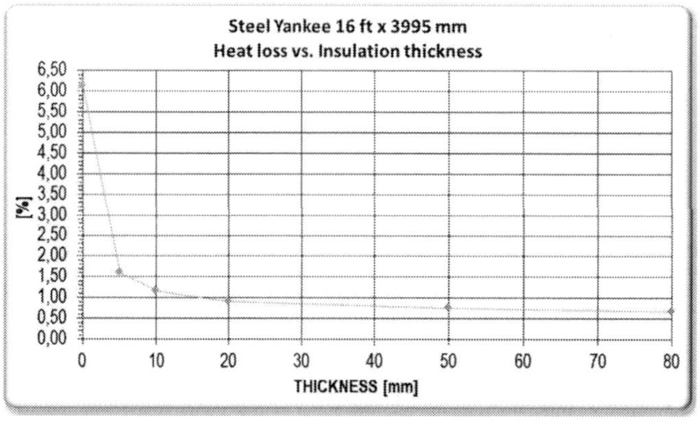

19

TOSCOTEC

ENERGY LOSS REDUCTION - HEAD INSULATION

- Thermo scan measures
- Heat flow reduction Yankee size dependent
- Head surface temperature significantly reduced

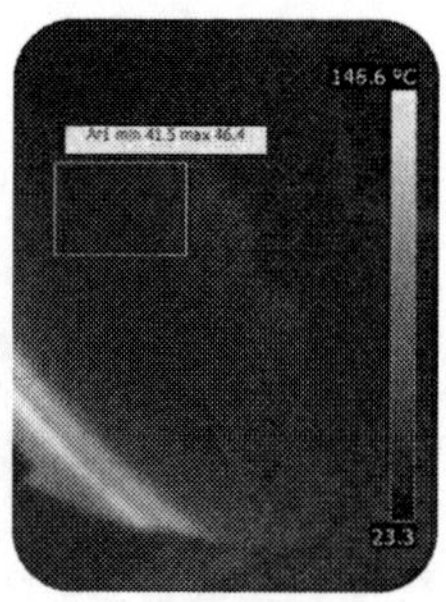

20

TOSCOTEC

OPERATION OF TT SYD – TROUBLESHOOTING

- ✓ Poor feed water quality to boiler: possible plugging of soda straws
- ✓ Basic guidelines for boiler waters

Parameters	Min./Max. recommended	Expected
Total Hardness (ppm = mg $CaCO_3$/kg)	0-4	0-1
pH	8,0 - 9,5	8,7 - 9,2
Oxygen (ppb)	20	<10
Carbon Dioxide (ppm)	<5	0
Iron (ppm)	0,1	<0,05
Copper (ppm)	0,05	<0,025
Conductivity (μmhos/cm)	<7000	2000-5000
TDS - Total Dissolved Solids (ppm)	<3500	<2000

21 TOSCOTEC

TROUBLESHOOTING – SODA STRAW PLUGGING

- ✓ Low quality of steam → Excessive formation and detachment of black oxide
- ✓ Black oxide (Magnetite) homogeneous, thin and tightly fixed layer
- ✓ Black oxide → Creates a barrier that stops the corrosion phenomena
- ✓ The reaction is self-limiting !

22 TOSCOTEC

TROUBLESHOOTING – SODA STRAW PLUGGING

- Steel Yankee Dryer in operation from year 2000 – Picture made in May 2011

23

TOSCOTEC

TROUBLESHOOTING – WET STREAKS

- Wet streaks on the paper

Bad

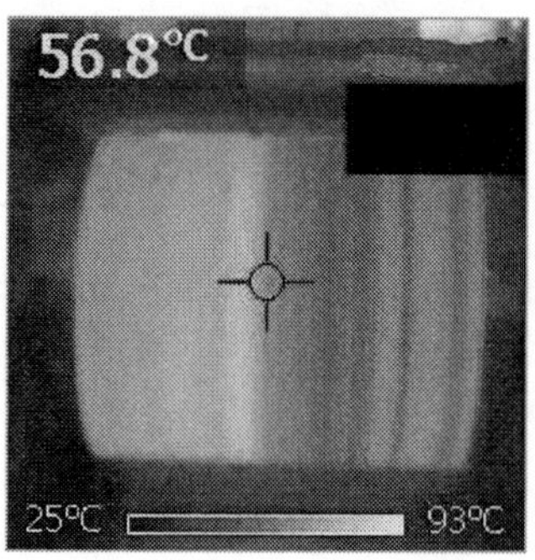

Good

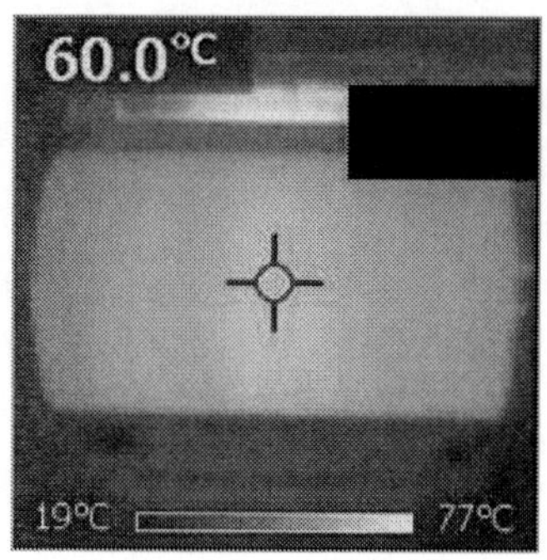

24

TOSCOTEC

TROUBLESHOOTING – SODA STRAW PLUGGING

- ✓ Knowledge of product **+** ✓ Various cases observed in 10 years **+** ✓ Infrared Technology

- ✓ In 92 % of cases → ✓ The cause was found on events non related to the Yankee Dryer

- ✓ In 8% of cases → ✓ Soda straw plugging → ✓ Low quality of the steam

25

TOSCOTEC

TROUBLESHOOTING – ACCIDENTAL DAMAGE

- ✓ Shock stress in the nip
- ✓ Intensive nip pressure

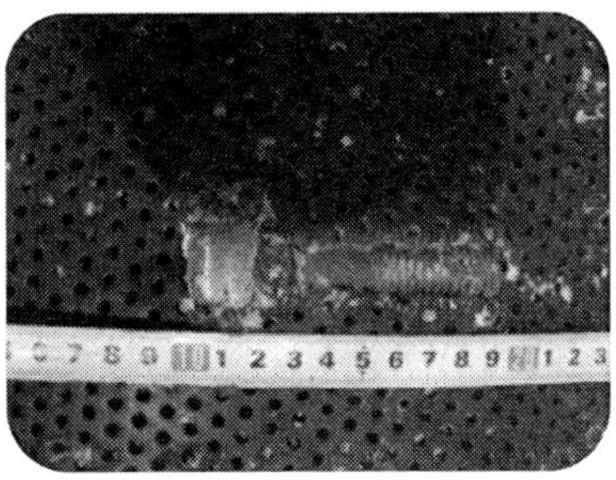

26

TOSCOTEC

TROUBLESHOOTING – ACCIDENTAL DAMAGE

AMERICAS TISSUE WORLD

- ✓ Impact on yankee surface
- ✓ Non destructive examination

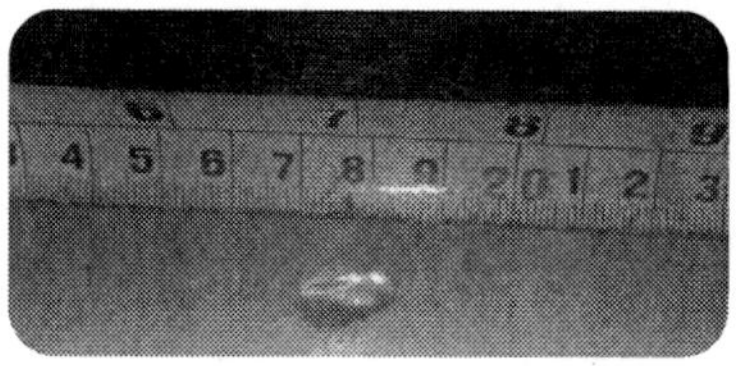

27

TOSCOTEC

TROUBLESHOOTING – ACCIDENTAL DAMAGE

AMERICAS TISSUE WORLD

- ✓ Yankee inspection (NDE)
- ✓ Metallization repaired

28

TOSCOTEC

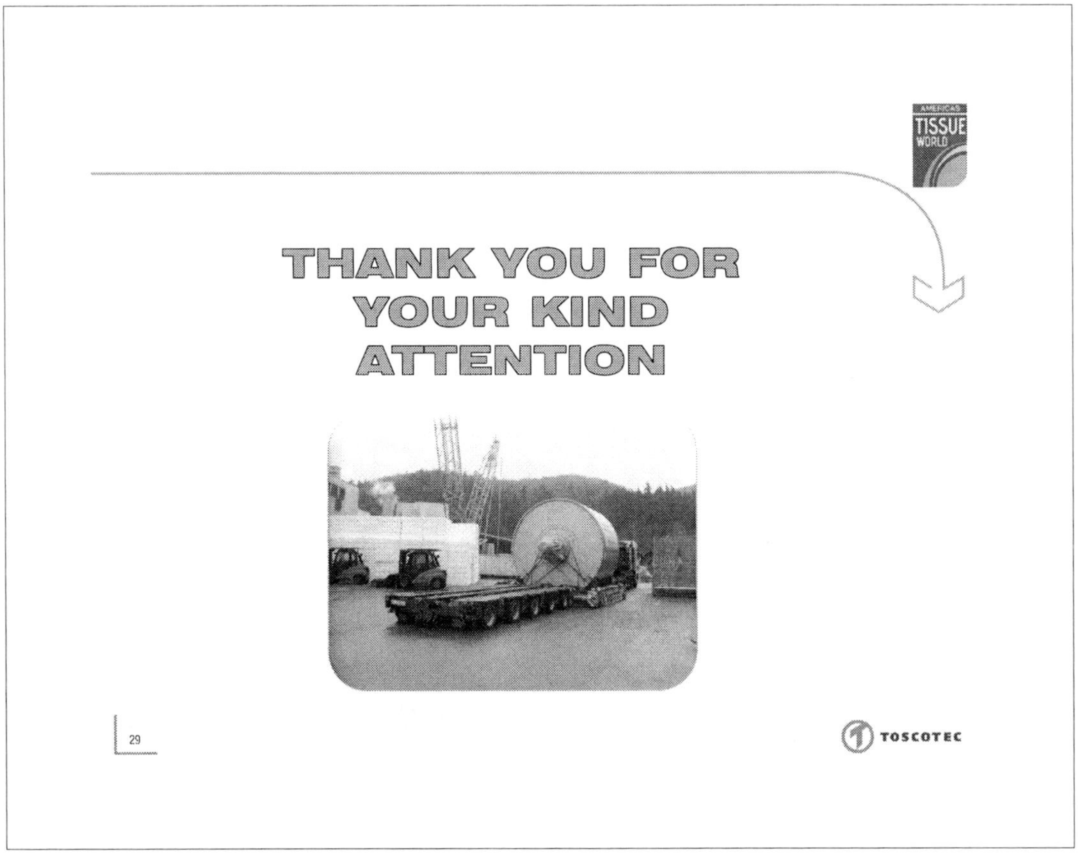
AMERICAS
TISSUE
WORLD
THANK YOU FOR YOUR KIND ATTENTION
29
TOSCOTEC

ENERQUIN AIR

Improvement of the drying and heat recovery

AMERICAS TISSUE WORLD

How to keep an eye on your machine drying and energy performance!

By

Dominique Thifault, Eng.

Director, Yankee and Technical Services,

Enerquin Air

Antoine Hofer, Eng.

Yankee Application Engineer,

Enerquin Air

From solid engineering to total client satisfaction

www.enerquin.com

ENERQUIN AIR

Improvement of the drying and heat recovery

AMERICAS TISSUE WORLD

OVER 700 MACHINE AUDITS

On-going audits on hundreds of Yankee hoods have revealed interesting findings...

...Data collected typically shows **energy saving potentials** in the **15-20%** range... and more!

...by proceeding with

- Process air **fine tuning** strategy
- Improving **heat recovery**
- Hood drying **equipment upgrade**

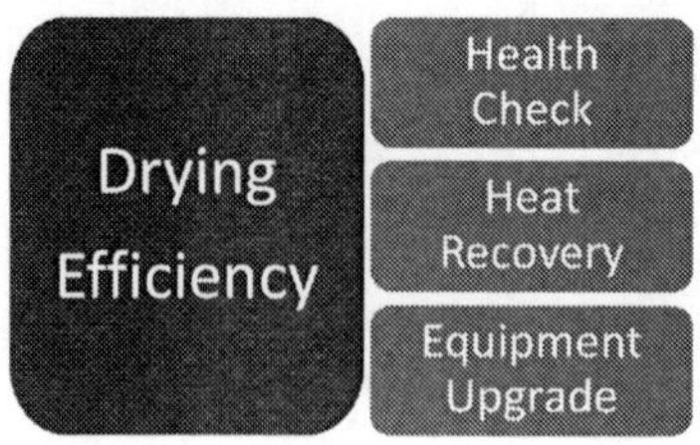

Tissue World Americas, Miami 2012 / Improving Yankee Hood Efficiency and Sustainability / Enerquin Air

ENERQUIN AIR

Improvement of the drying and heat recovery

AMERICAS TISSUE WORLD

What is a Health Check?

Health Check
Heat Recovery
Equipment Upgrade

Expert intervention after a comprehensive Yankee Hood and Process Air System audit has been conducted.

Objective:
Provide mill assistance when process deviation occurs, without field support.

…. Experts do not always need to be on site to provide technical assistance!

Typical process deviations:

-Gas flow increase
-Production decrease
-Sheet quality problems
-Wet felt lifetime reduction
-Safety hazards
-Etc…

Tissue World Americas, Miami 2012 / Improving Yankee Hood Efficiency and Sustainability / Enerquin Air

ENERQUIN AIR

Improvement of the drying and heat recovery

AMERICAS TISSUE WORLD

Typical "Health Check" Flow sheet

Health Check
Heat Recovery
Equipment Upgrade

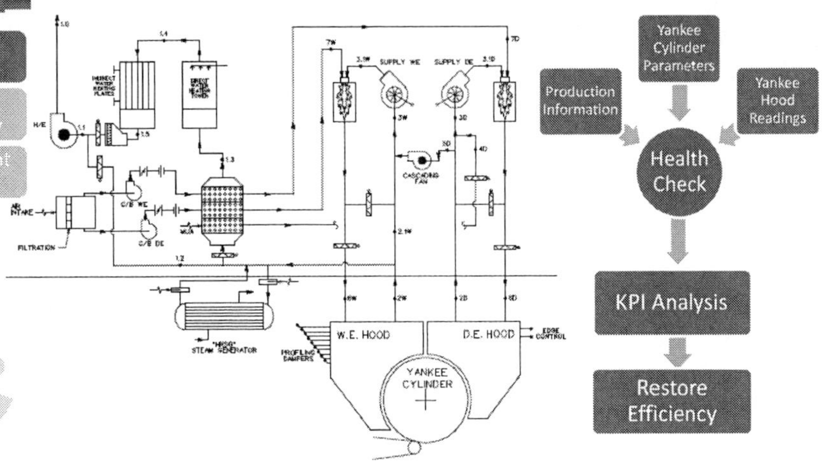

Tissue World Americas, Miami 2012 / Improving Yankee Hood Efficiency and Sustainability / Enerquin Air

Improvement of the drying and heat recovery

ENERQUIN AIR

AMERICAS TISSUE WORLD

Health Check

Heat Recovery

Equipment Upgrade

Typical "Health Check" guidelines

Yankee Hood Optimization: Centerlines and "Health" Check

Focus Area	"Optimized"	"Actual"	Units of Measure	How To Measured
Production Data				
Date	2011		---	---
Time	12:00		---	---
Grade	Towel		---	---
Basis Weight @ Reel	15.5		g / m²	DCS reading/display
Yankee Speed	[illegible]		m / min	DCS reading/display
Reel Speed	1305		m / min	DCS reading/display
Sheet Moisture @ Reel	4%		% Moisture	DCS reading (Measurex System)
Yankee Cylinder Pressure	8		kg/cm²	DCS reading/display
Crepe Ratio	20%		---	DCS reading/display
Sheet Width	3540		mm	Field measurement or DCS access
Steam Flow	4295		kg steam / hr	DCS reading/display
Gas Flow	[illegible]		m³/hr	DCS reading/display
Hood Temperatures				
By Paper Machine/By Grade "Type"				
Supply Wet End	400 °C		°C	Temperature Transmitter - DCS
Supply Dry End	400 °C		°C	Temperature Transmitter - DCS
Return Wet End	= 75% of Supply T°		°C	Temperature Transmitter - DCS
Return Dry End	= 75% of Supply T°		°C	Temperature Transmitter - DCS
Hood Controls				
WE/Main Exhaust Fan Speed	[illegible]		%	Field and/or DCS display
Cascading Fan Speed	100		%	Field and/or DCS display
DE Make-Up Air Damper	100		%	Field and/or DCS display
WE Supply Damper	100		%	Field and/or DCS display
DE Supply Damper	100		%	Field and/or DCS display
Hood Controls				
WE/Main Exhaust Fan Amperage			A	Field and/or DCS display
Cascading Fan Amperage			A	Field and/or DCS display
WE Supply Amperage			A	Field and/or DCS display
DE Supply Amperage			A	Field and/or DCS display
Wet End/Main Exhaust Humidity				
Common Grades	0.50 - 0.60		kg H₂O/kg D.A	Humidity Transmitter
Air : Fuel Ratio				
Wet End Combustion	12-14 : 1		---	Automated Burner system
Dry End Combustion	12-14 : 1		---	Automated Burner system
Hood Air Balance				
Wet End	100%		%	Requires Field Testing or Based on hood control setting (refer to Enerquin's survey)
Dry End	[illegible]		%	
Hood Thermal Efficiency [illegible]	4500-4500		kJ / kg H₂O	Requires field testing
Yankee Drying Rate "Conditions"				
Condensate Flow	4300		kg H₂O/hr	Flow meter (orifice plate or electronic device) or condensate rise test
RW Steam	50-60		kg H₂O/hr/m²	N/A : Would require external expertise
Drying Efficiency	---		%	
Hood Drying Rate "Conditions"				
RW Air	85-95		kg H₂O/hr/m²	N/A : Would require external expertise
Drying Efficiency			%	
Hood Pressures				
Wet End	760		mm w.g.	Pressure Transmitter - DCS access
Dry End	760		mm w.g	Pressure Transmitter - DCS access
Hood/Cylinder Gap	20-30		mm	Field visual check

Production Info

Energy Usage: Gas , Steam and Electricity

Hood Settings

Other Settings & KPI

Tissue World Americas, Miami 2012 / Improving Yankee Hood Efficiency and Sustainability / Enerquin Air

Improvement of the drying and heat recovery

ENERQUIN AIR

AMERICAS TISSUE WORLD

Health Check

Heat Recovery

Equipment Upgrade

Key Performance Indicators (KPI)

	Can it be monitored?		
	Yes	No	
Hood Pressures	✓		
Hood Temperatures	✓		
Steam flow / Condensate Flow	✓		
Hood/Cylinder Clearance	✓		
Stoichiometric Air/gas Ratio	✓		
Hood Humidity	✓		
Moisture entering the cylinder		✓	Punctual test
Hood Control settings /Damper / Fan Speed	✓		
Yankee Cylinder Drying Rate		✓	Calculated
Hood Drying Rate		✓	Calculated
Hood Thermal Efficiency		✓	Calculated

Tissue World Americas, Miami 2012 / Improving Yankee Hood Efficiency and Sustainability / Enerquin Air

Improvement of the drying and heat recovery

ENERQUIN AIR

AMERICAS TISSUE WORLD

Health Check
Heat Recovery
Equipment Upgrade

Monitoring and Maintaining Conditions

Sustainability

- To remain efficient, optimized operation conditions must be maintained.
- **Keeping a good eye on your KPI is the KEY!!!**

Monitoring

HEALTH CHECK DATA		
1st Level (Easily accessible)	**2nd Level**	**3rd Level**
Hood Supply Temp. (DCS) Fans Speed Damper Settings Yankee Cylinder Pressure Hood pressures	Hood Humidity Sensors Entering Sheet Dryness Condensate Flow	Air Flows Fully Automated Systems

Improvement of the drying and heat recovery

ENERQUIN AIR

AMERICAS TISSUE WORLD

Health Check
Heat Recovery
Equipment Upgrade

The Health Check loop!

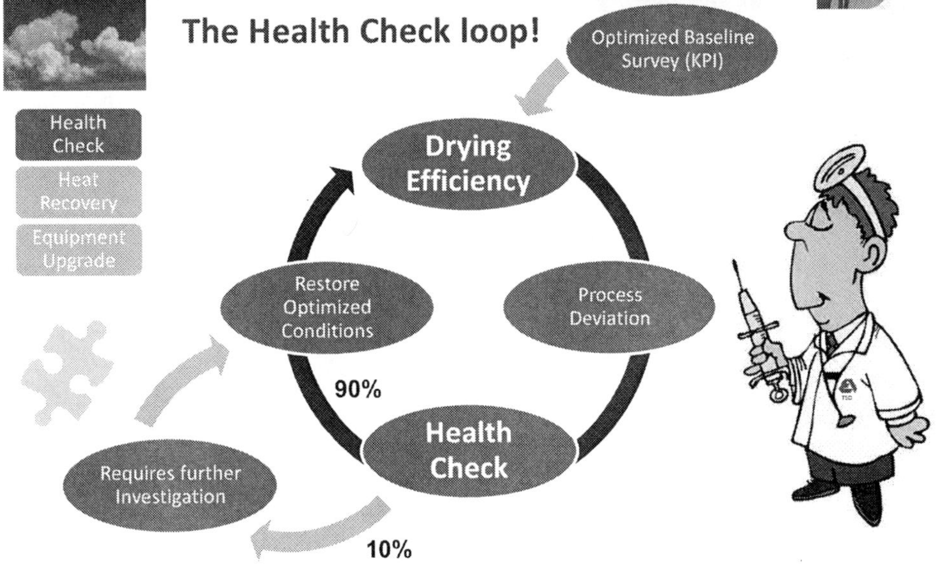

Improvement of the drying and heat recovery

ENERQUIN AIR

AMERICAS TISSUE WORLD

Health Check
Heat Recovery
Equipment Upgrade

Improving Heat Recovery

- Heat Recovery is now recognized as an **essential part of tissue making**.
- Total energy introduced into the drying process:
  - Burners (natural gas) : 50 to 65%
  - Yankee cylinder (steam) : 50 to 35%
  - } 75% can easily be recovered
- Heat recovery potential is determined by…
  - Hood exhaust flow, temperature and humidity
  - Availability of heat sinks

… Optimized heat recovery should not be limited to seasonal heat demand.

Tissue World Americas, Miami 2012 / Improving Yankee Hood Efficiency and Sustainability / Enerquin Air

Improvement of the drying and heat recovery

ENERQUIN AIR

AMERICAS TISSUE WORLD

Health Check
Heat Recovery
Equipment Upgrade

Typical Optimized Heat Recovery System

Sensible Heat (30% typically.)

Hood Exhaust → HRSG → MUA A/A HR → C/B A/A HR → Indirect A/W HR → Direct A/W HR → Roof Stack

Latent Heat (70% typically)

Tissue World Americas, Miami 2012 / Improving Yankee Hood Efficiency and Sustainability / Enerquin Air

Improvement of the drying and heat recovery

ENERQUIN AIR

AMERICAS TISSUE WORLD

Health Check

Heat Recovery

Equipment Upgrade

Typical Heat Recovery Breakdown for a 18 ft Yankee cylinder

Burner capacity:	36 MMBTU/h (2 x 18 MMBTU/h)	53.5 MMBTU/h
Steam flow:	18,000 lb/hr or 17.5 MMBTU/h	
Exhaust flow:	1,000 lb D.A./min @ Humidity: 0.45 lb H_2O/lb D.A 75% recoverable: 40 MMBTU/h	

	Heat Recovery Unit	Heat Recovered		ROI	Comments
Sensible Heat	Steam Generator LP Make-up Air Pre-heating Combustion Air Pre-heating	7.0 MMBTU/h 1.5 MMBTU/h 1.5 MMBTU/h **10.0 MMBTU/h**	18% 4% 4%	1.5 years 2.1 years 2.1 years	Can be used for Steambox 600,000 $/yr (can vary)
Latent Heat	Indirect fluid heating Direct Water heating	15.0 MMBTU/h 15.0 MMBTU/h **30.0 MMBTU/h**	37% 37%	3 months 2 months	Seasonal variation 1,800,000 $/yr
	Total	**40.0 MMBTU/h**	100%		75% of total energy input

Savings are based on unit cost at 7 $ / decatherms = 1000000 Btu

Improvement of the drying and heat recovery

ENERQUIN AIR

AMERICAS TISSUE WORLD

Health Check

Heat Recovery

Equipment Upgrade

Recovery Opportunity 1st Stage: (HRSG)

Recovers **sensible heat** from high temperature Yankee hoods (upstream of the air/air Heat exchanger)

HP HRSG:

- Same principle as a boiler
- Generates High Pressure steam
- Can be re-injected in the Yankee cylinder (motive steam)

LP HRSG:

- Generates Low Pressure steam
- More steam generated
- Can be directed to steam box, etc

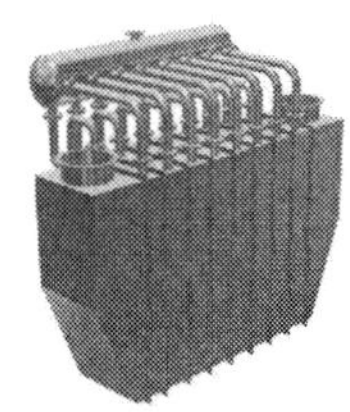

Recovery Opportunity 2nd Stage: (Air/Air)

Health Check
Heat Recovery
Equipment Upgrade

Recovers additional **sensible heat**

Pre-heating of **combustion air**:
- Always running, very good recovery potential
- Temperature rise limited (Burner components)

Pre-heating of **process make-up air**:
- Not always required (depends on Hood Balance)
- No temperature rise limitations
- Make-up Air is the first cut-back by the operators

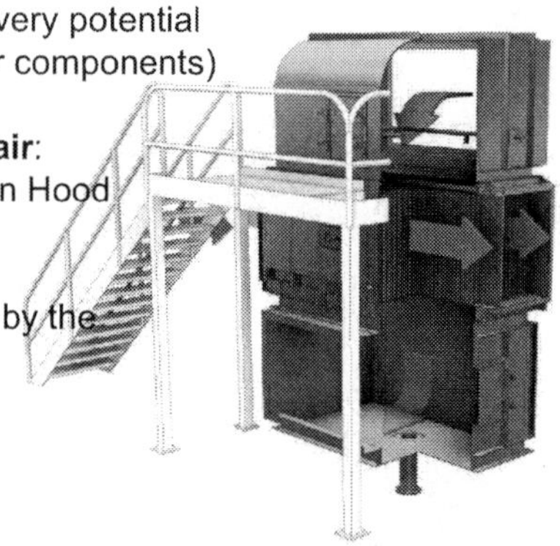

Recovery Opportunity 3rd Stage: (Air/Water)

Health Check
Heat Recovery
Equipment Upgrade

Recovers **latent heat**

Indirect Air/Water (or Glycol):
- Use for clean water or thermo fluids
- Useful for building ventilation (seasonal demand)

Direct Air/Water (Process Water):
- Use for unfiltered water needs
- Baffle plates and packing tower to increase the heat transfer efficiency

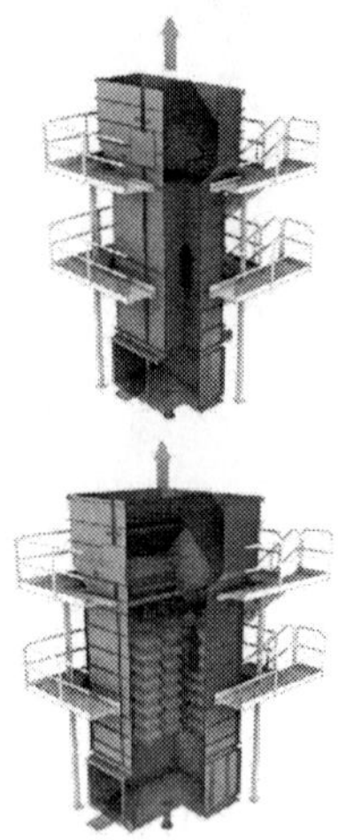

Improvement of the drying and heat recovery

ENERQUIN AIR

AMERICAS TISSUE WORLD

Health Check
Heat Recovery
Equipment Upgrade

Typical System Layout

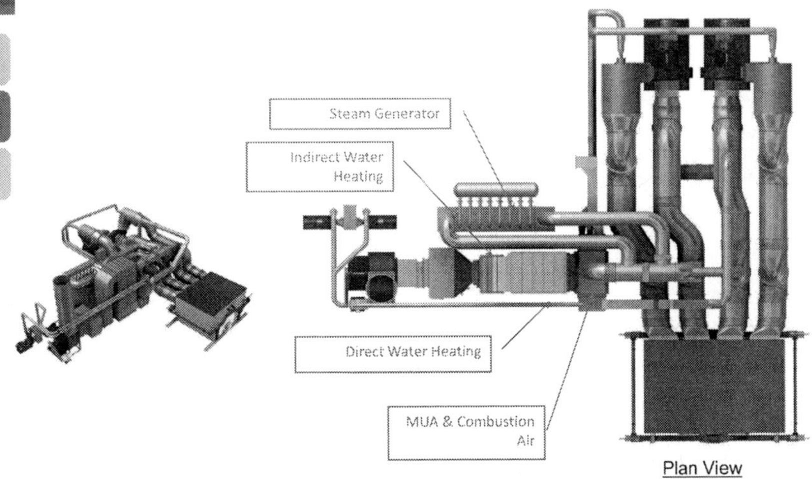

Plan View

Improvement of the drying and heat recovery

ENERQUIN AIR

AMERICAS TISSUE WORLD

Health Check
Heat Recovery
Equipment Upgrade

Equipment Limitation

After hood optimization is completed, **equipment in place can often represent your bottlenecks**.

Typical obstacles:

- Old and Obsolete hood technology (inefficient hood design)
- Damaged or deteriorated equipment
- Low supply fan pressure (low impingement velocity)
- Absence of heat recovery

Improvement of the drying and heat recovery

ENERQUIN AIR

AMERICAS TISSUE WORLD

Equipment Upgrade Benefits

Health Check
Heat Recovery
Equipment Upgrade

TE Hood Btu / lb H2O

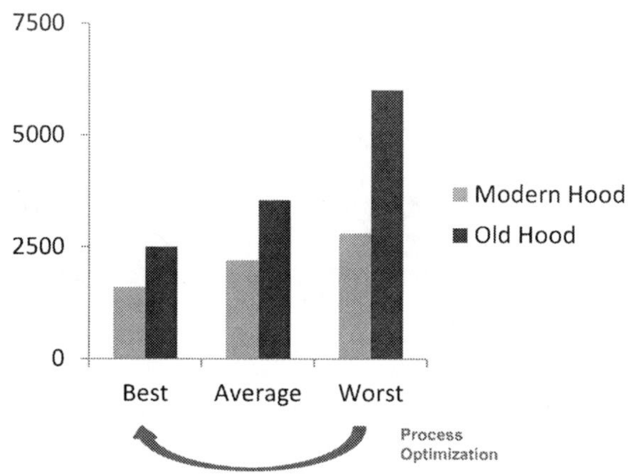

Tissue World Americas, Miami 2012 / Improving Yankee Hood Efficiency and Sustainability / Enerquin Air

Improvement of the drying and heat recovery

ENERQUIN AIR

AMERICAS TISSUE WORLD

Equipment Upgrade ROI

Health Check
Heat Recovery
Equipment Upgrade

- Typical payback… **12 to 24 months.**
- **New Health Check program** after equipment upgrade.
- **Keeping operating conditions optimized** reduces payback time.

… Failure to maintain new equipment at optimized operating conditions has been found to be a serious issue.

Tissue World Americas, Miami 2012 / Improving Yankee Hood Efficiency and Sustainability / Enerquin Air

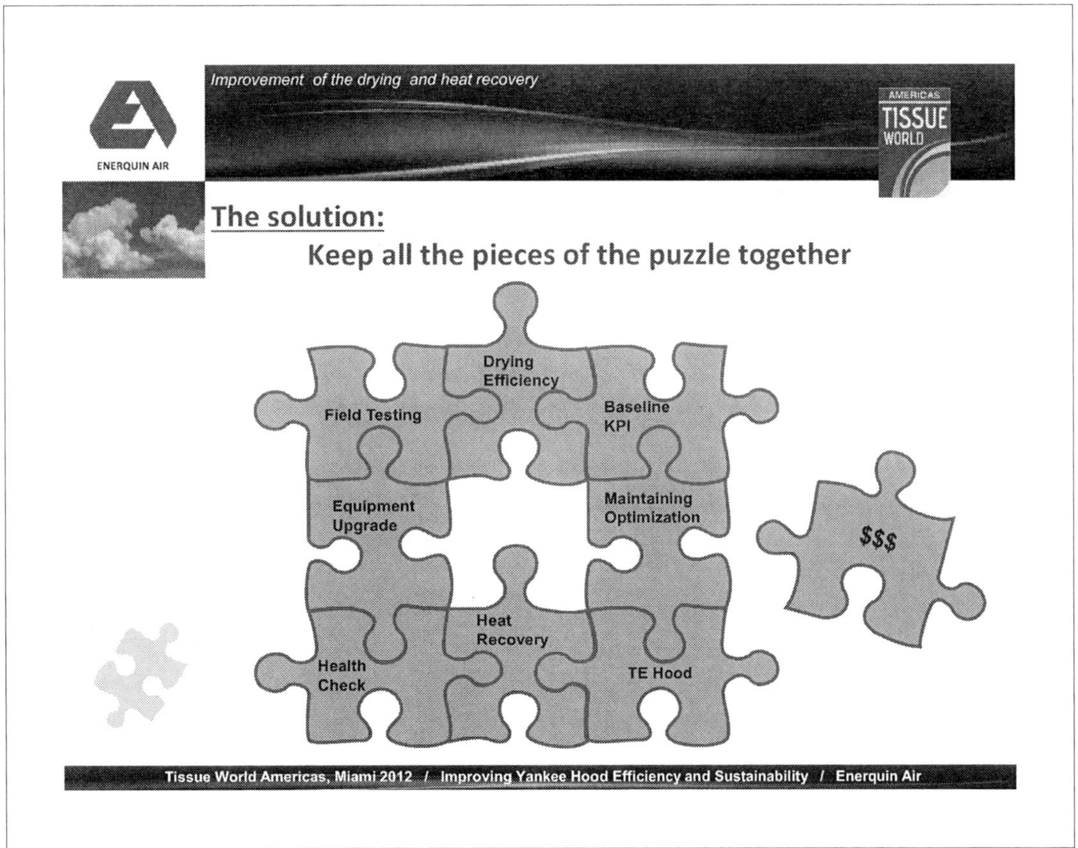
Improvement of the drying and heat recovery
ENERQUIN AIR
AMERICAS
TISSUE
WORLD
The solution:
Keep all the pieces of the puzzle together
Drying Efficiency
Field Testing
Baseline KPI
Equipment Upgrade
Maintaining Optimization
$$$
Heat Recovery
Health Check
TE Hood
Tissue World Americas, Miami 2012 / Improving Yankee Hood Efficiency and Sustainability / Enerquin Air

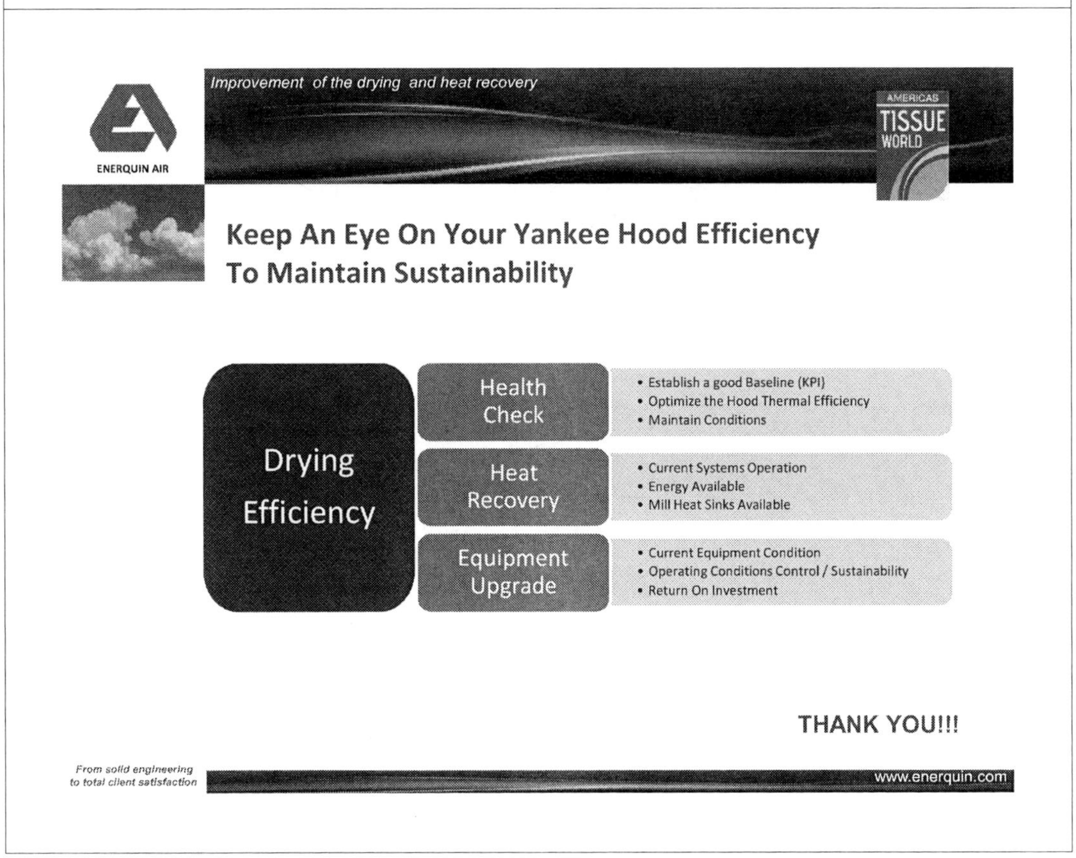
Improvement of the drying and heat recovery
ENERQUIN AIR
AMERICAS
TISSUE
WORLD
Keep An Eye On Your Yankee Hood Efficiency
To Maintain Sustainability
Drying Efficiency
Health Check
• Establish a good Baseline (KPI)
• Optimize the Hood Thermal Efficiency
• Maintain Conditions
Heat Recovery
• Current Systems Operation
• Energy Available
• Mill Heat Sinks Available
Equipment Upgrade
• Current Equipment Condition
• Operating Conditions Control / Sustainability
• Return On Investment
THANK YOU!!!
From solid engineering
to total client satisfaction
www.enerquin.com

Buckman
Commitment makes the best chemistry.

Sustainable Yankee Release Chemistry

Tissue World Americas

March 20, 2012

Jackie Allen & Daniel Glover
Jian Tan - Chemist

Problems with Mineral Oil Based Releases

- Crude oil based chemistries are portrayed as "Unsustainable to the Environment"
- Old technology in the tissue business
- Limited applications for higher speeds
- Oil supply and pricing is volatile
- Oil based releases may be high in VOC
- Buckman identified a need for a greener replacement of oil releases– and this project was launched

Formulation Process

- R & D research turned up some interesting work done previously in developing food oil release products.
- This work indicated that a "Greener" release was plausible for tissue.
- The initial formulation included vegetable based oils, a modified food oil and some surfactant material.
- Lab samples were created for comparison.

Lab –Hot Plate

- Finding your favorite
  - Various adhesives
- Hotplates are videoed
- Comparison of:
  - Known programs
  - Speed of reaction
  - Film type
  - Film thickness
  - Wet tack
  - Dry tack
  - Rewet

Lab-Tack Testing

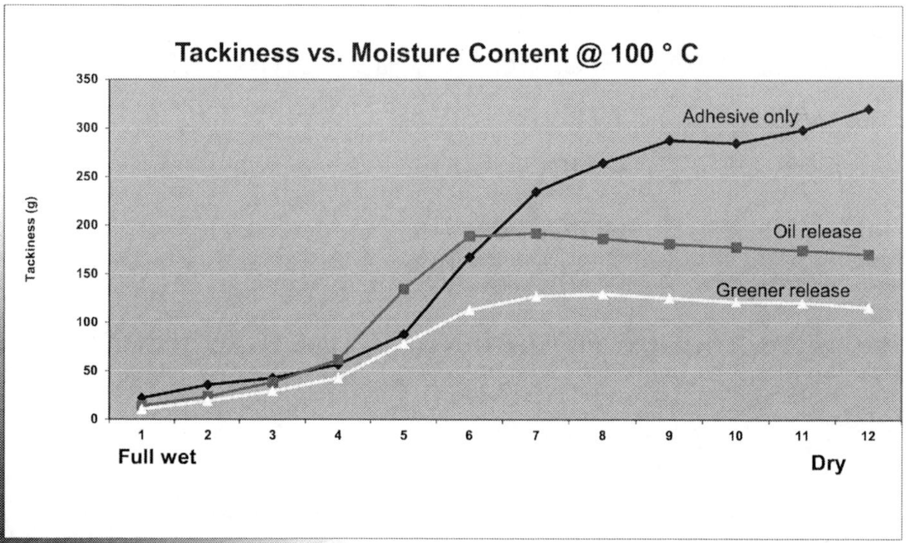

Initial Field Trials

- We found a willing partner and a gutsy salesperson
- Find the correct market– recycled tissue where release usage is high
- Can it compete with a mineral oil based product?
- Data collection and analysis
- Can we continue or do we kill it?

Formulation Optimization

- Stability
- Performance—which oil source?
- More hotplates and tack testing
- Costs
- Can we improve the product?
- Supply and availability
- Dial in on the best overall formulation

Busperse® 2888 Initial Evaluation

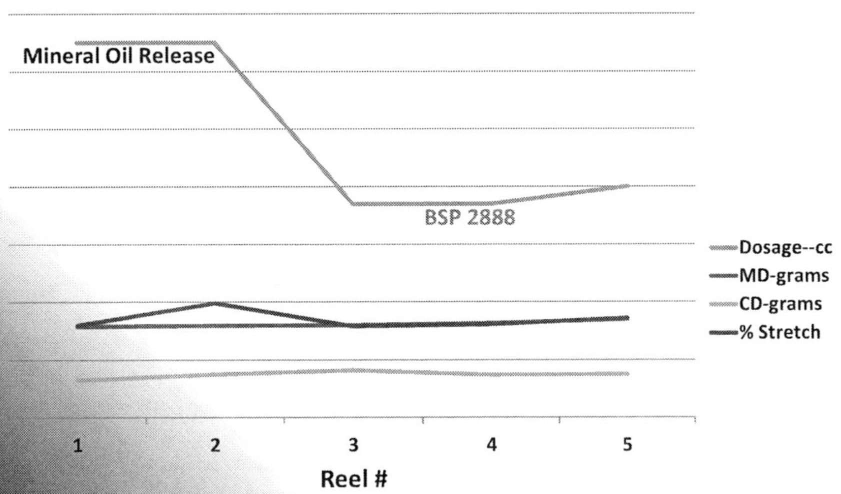

Greener Release Benefits

- Sustainable and Renewable
- Replaces high VOC products
- Very competitive against oil releases
- Faster reacting than oil programs
- The Yankee films will resemble modifier programs
- CBI are improved
- Patent pending
- BfR allowed and FDA status

What's Next ?

- The early work has proven successful in the AfH market when competing with an oil release.
- Will the market buy it?
- Other formulations can be tailored to the mill's Yankee program needs. Various oil supplies are possible.
- Buckman is listening
- Thanks for being here in Miami!

ASHLAND®

Yankee Reliability Inside and Out

Norris Johnston
Applications Team Manager- Utilities

Yankee Reliability

- Outline of mill incidents
- Diagnosing the source of the problem
- What causes corrosion inside the Yankee
- Water treatment for Asset Protection
- Summary

2

Mill Incidents

- **Case 1**
  - Mill started up a cast iron Yankee, within three months it began to lose drying capacity
  - Boiler treatment in place but no condensate sampling
  - Examination of Yankee internals on a shut found blocked straws and deposits in the Yankee
  - Causes of problem found to be drop in condensate pH during period when sampling was not available, resulting in corrosion of iron internals

3

Mill Incidents

- **Case 2**
  - Mill started up new crescent former, cast iron surface 6100 fpm.
  - Within 2 months after mechanical issues were resolved, mill put on more speed and production
  - But, streaking in Yankee coating and beginnings of chatter in thinner areas and interface between more heavily coated areas.
  - Mill chased felts, stickies, coating etc without luck
  - Internal inspection found many of the soda straws plugged
  - Issue traced to improper boiler treatment and testing.

4

Diagnosing the Source of the Problem

- Diagnosing problems is often difficult since the symptoms build-up gradually and are often attributed to other causes such as:
  - Poor felt condition
  - Press loading
  - Furnish changes
  - Yankee coating issues
  - Creping blade set up
  - Yankee hood profiling

5

Diagnosing the Source of the Problem

- A systematic approach needs to be taken involving mill personnel, clothing , machinery and chemical suppliers:
  - Felt permeability profiling
  - Press loading checks
  - Hood operation checks
  - Creping blade holder profiling
  - Creping chemical system and system chemistry surveys
  - Thermal imaging

6

Thermal Imaging of Reel

Indicates large variations in temperature profile that 'could' be due to blocked straws

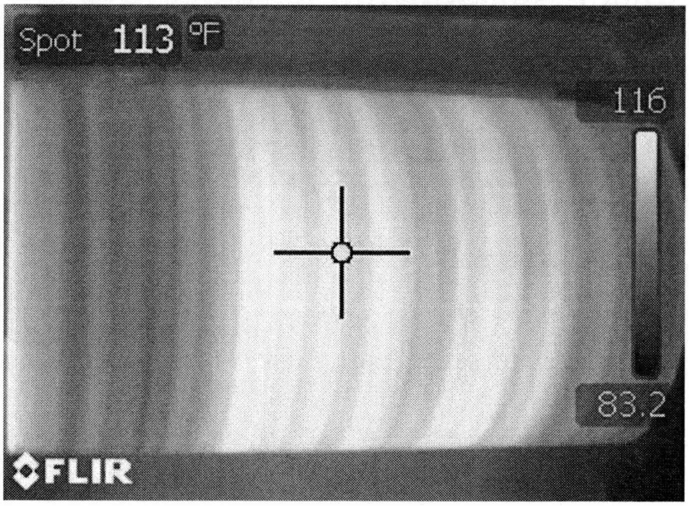

7

What can cause
Blocked Straws in the Yankee?

CORROSION

8

ACID

and

OXYGEN

9

ACID Source

- **Breakdown of alkalinity and resulting corrosion**
  - $2HCO_3^- + \text{Heat} \rightarrow CO_3^{-2} + CO_2\uparrow + H_2O$
  - $CO_2 + H_2O \rightarrow H_2CO_3$ (carbonic acid)
  - $Fe^\circ + 2H_2CO_3 \rightarrow Fe\,(HCO_3)_2 + H_2\uparrow$
- **Condensate treatment failure**

Carbonic Acid Corrosion

10

OXYGEN Source

- **Mechanical Deaeration failure**
- **Chemical Deaeration failure**
- **System Ingress**

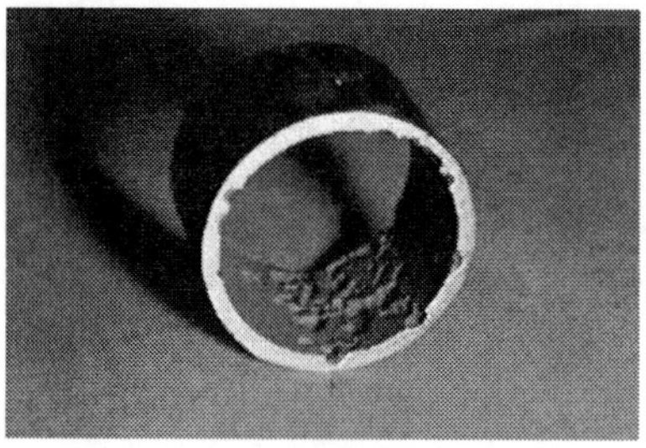

Oxygen Corrosion

11

Objective of Treatment

passivated surface

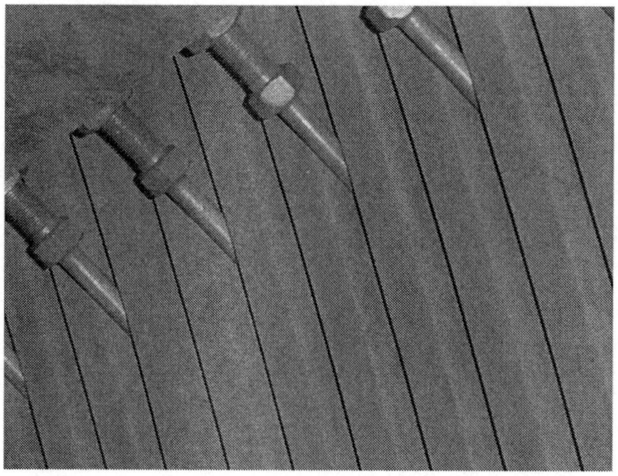

12

Water Treatment for Asset Protection

13

Goals of Boiler System Treatment

- Corrosion Control
- Deposition Control
- Carryover Control

14

Selecting a Boiler System Treatment

- Boiler pressure
- Feedwater quality
- Monitoring and Control issues
- Boiler design
- Downstream users

15

System Monitoring Sampling

- Maintain adequate velocity/minimize line length
- Run sample continuously
- Cool sample
- Stainless steel sample line
- Critical equipment
- Composite streams prior to flash tanks/receivers
- Bulk condensate return streams
- TAPPI TIP: 0416-03 and 0416-14

16

System Monitoring Analytical

- Conductivity, pH, Hardness, Silica
  - Continuous
  - Grab
- Oxygen
  - Continuous
  - Grab
- Corrosion
  - Metals analysis
  - Corrosion coupons
  - Turbidity
- Data Management

17

Monitoring a Yankee Dryer System

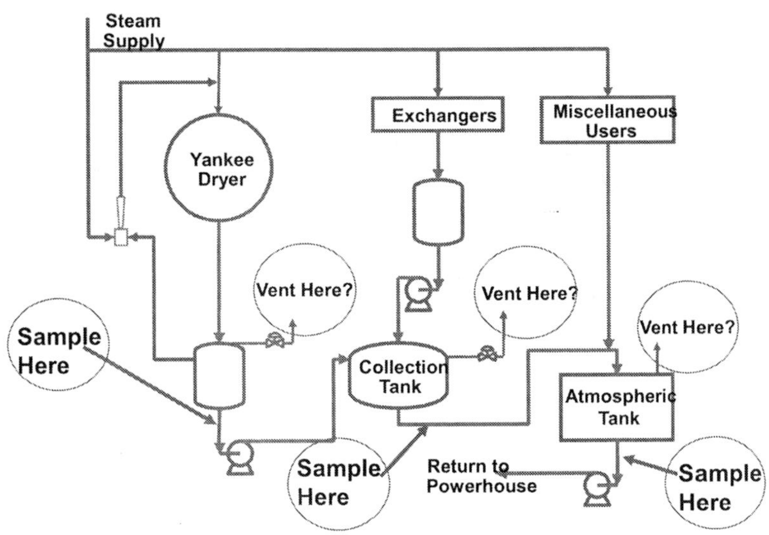

18

Summary

- Yankee internals can be have a significant impact on Yankee surface reliability
- Condensate treatment must not be overlooked
- Insure program and monitoring are a focus, especially during startups
- Consider advanced monitoring

19

ASHLAND®

With good chemistry great things happen.™

20

Crepe Blade Vibration Monitoring for Improved Efficiency and Asset Protection

By William Von Drasek, Warren Rand, Sam Archer, and Gary Furman,
Nalco Company, 1601 West Diehl Road, Naperville, IL 60563, USA

ABSTRACT

Tissue manufacturers are continually challenged to achieve higher levels of productivity and quality. As a result, operators often work with conditions such as higher machine speeds and lower sheet moisture levels that increase the probability for developing chatter. A machine experiencing chatter often leads to lost production, increased maintenance, and capital expenditures. Currently, operators rely on detecting the presence of chatter when sheet defects occur or by observing the characteristic periodic marks that form in the Yankee coating. In severe cases where the vibration energy is high enough, the doctor blade can penetrate through the coating making direct contact on the Yankee dryer surface resulting in costly damage. Avoiding this state requires detecting the onset of chatter and alerting operators of the impending negative condition. To address this, Nalco partnered with SKF®, a supplier of condition monitoring equipment, to develop an early chatter warning system. The technology is based on indirect doctor blade vibration monitoring using a customized user-friendly intuitive software module. An overview of the monitoring technology is presented along with case studies that illustrate benefits and knowledge building this tool offers for improved asset management.

INTRODUCTION

The occurrence of chatter on Yankee dryers is a well-known problem that tissue manufacturers strive to avoid, but often find themselves struggling with, due to the demand for higher quality products and optimum machine efficiency. This combination tends to stress the creping operation, thereby placing the process on an operating trajectory favorable for developing chatter. Several review papers by Archer et al.[1,2], Alessandrini et al.[3], and Corby[4] describe the different mechanical and operational conditions that promote chatter formation. Identifying the conditions favorable for chatter is the first step in developing best practices to minimize the problem. However, complete elimination of chatter is not easily achieved even when best practices are followed, because the creping operation is complex and dynamic. Consequently, attaining optimum machine runnability and asset protection requires the combination of best practices coupled with an online monitoring system that provides continuous real-time feedback to detect the onset of chatter.

Chatter formation is a vibration phenomenon caused by the crepe blade tip oscillating relative to the dryer surface plane. Under normal operating conditions, i.e., no chatter present, the creping doctor blade tip rides in the coating and experiences minimal out of plane movement, as illustrated in Figure 1A. Critical forces acting on the blade tip are tangential resistance and normal (loading) force. As the tangential resistance increases, e.g., coating becomes hard; the amplitude of the blade tip movement can increase because of the "stick-slip"[1] phenomenon. Continuous operation at this condition increases the risk to develop chatter marks that appear as cross directional (CD) defects in the coating and/or dryer surface, as illustrated in Figure 1B. Sheet defects from chatter also occur and appear as multiple holes in the machine direction, or develop a lace-like appearance, thus negatively impacting product quality. Under severe chatter conditions, when the doctor blade penetrates through the Yankee coating and makes direct contact with the dryer surface, horizontal, CD, grooves in the metal can form. Chatter marks formed in the metal is an irreversible process and can only be repaired by conducting a costly regrinding operation. Therefore, it's imperative to avoid this condition by detecting the onset of chatter early enough to perform the necessary corrective actions.

Despite the impact chatter can have on production, asset maintenance, and depreciation, tissue manufacturers have limited options for chatter detection. As a result, machine operators often rely on audible sound changes or visual inspection of the sheet quality and/or Yankee dryer surface as a first indication that chatter is present. Taking this hands-on approach is both, subjective and inconsistent, making reliable and early detection of chatter difficult. Thus, identifying chatter using these manual approaches lacks sensitivity and detection in a real-time manner. Consequently, the effectiveness of corrective actions taken is reduced, thereby increasing the risk of forming irreversible chatter damage on the dryer surface.

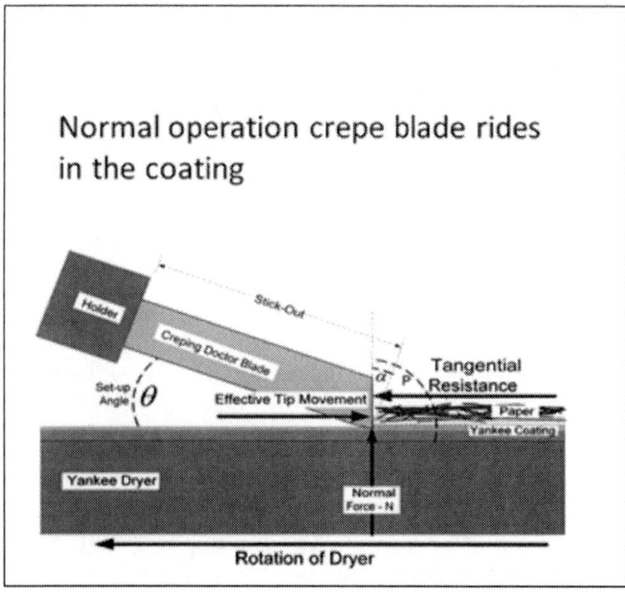

Figure 1A – *Detailed view of the creping operation showing the different forces and important setup conditions affecting crepe blade vibration.*

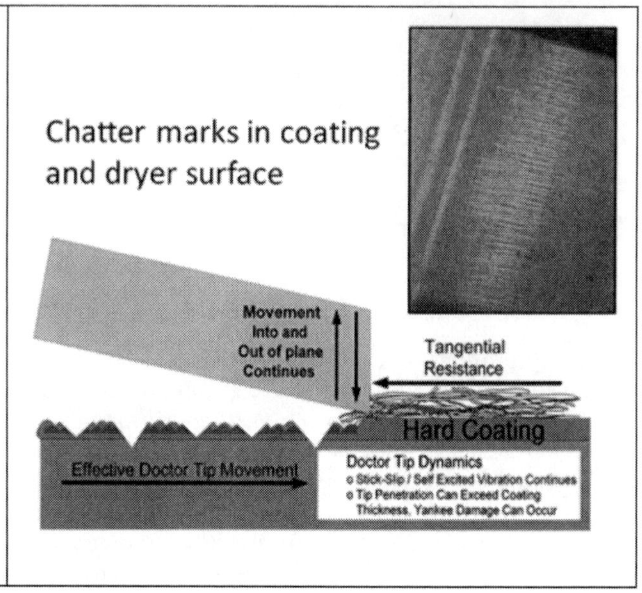

Figure 1B – *Illustration showing development of chatter by the out of plane movement of the creping blade.*

Nalco Early Warning Chatter Detection System

To address the need for an Early Warning Chatter Detection (EWCD) system, Nalco partnered with SKF to combine expertise in both the Yankee coating and process with conditioning monitoring, maintenance and asset management. At the core of the EWCD platform is the SKF hardware and software technology that is industry proven for conditioning monitoring applications covering a diverse range of industrial markets such as pulp and paper, energy, aerospace, and transportation. These applications typically focus on bearing fault detection or other lower speed rotating elements, whereas the EWCD system is tailored for crepe blade vibration monitoring. This distinction is critical for the following reasons:

1. Crepe blade vibration is dynamic compared to bearing monitoring due to the multitude of process conditions such as blade age, basis weight, furnish, coating chemistry, crepe blade setup, production rate, moisture level, etc., that will impact vibration energy transferred to the creping blade.
2. Crepe blade vibration exhibits a cyclic behavior, i.e., the vibration energy can vary over a minimum and maximum range for short or long time spans.
3. In cases where chatter develops the vibration frequencies of interest are higher than other rotating elements on a paper machine.

The functionality of the EWCD platform addresses these points using a time and intensity based alarming strategy. Additionally, EWCD is designed to be user-friendly requiring minimum training for machine operators to use. Data analysis and reporting is streamlined to eliminate the need for consulting vibration experts unless specific problems or troubleshooting is required.

The basic setup for crepe blade vibration monitoring requires installing piezoelectric accelerometers in close proximity to the crepe blade, as shown in Figure 2. Accelerometers have a small footprint (26 mm dia. x 30 mm height) allowing retrofit installation on different creping system designs. The sensors are well suited for crepe blade monitoring because they cover a wide frequency range (from 1 to 2 Hz up to 20 kHz[5]) and are immune to dusty and moist environments. In addition, the sensors are operational at temperatures > 250°F using standard commercial devices. A minimum of two accelerometers is recommended to cover front and back side zones of the dryer to identify side-to-side differences in both vibration frequency and amplitude. For larger Yankee dryer widths, additional sensors can be added for better zone coverage. Signal collection from the accelerometers and tachometer is made through water tight cabling and connections between the sensors and the SKF Multilog IMx-S conditioning monitoring hardware. The SKF Multilog® IMx-S can be mounted on the plant floor near the machine to minimize sensor cable lengths. Data transfer from the SKF Multilog IMx-S hardware to a dedicated computer running the EWCD software for alarming and reporting uses a standard CAT5e cable. Complete system installation is typically achieved during a single maintenance shutdown.

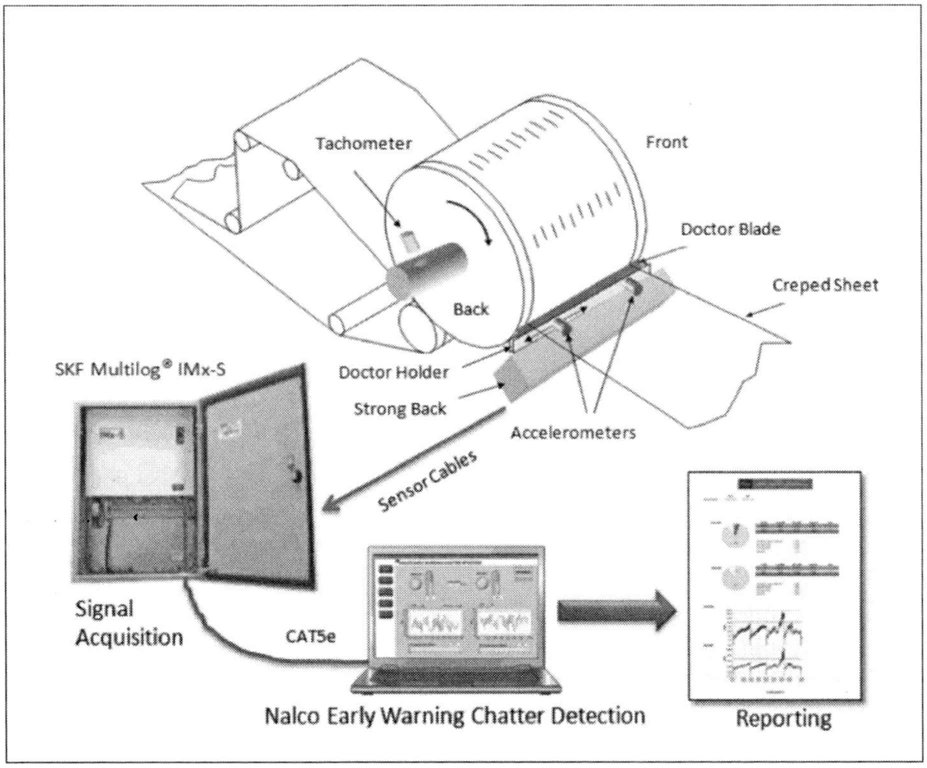

Figure 2 – *Basic setup for the early warning chatter detection system.*

EWCD HMI KEY FEATURES

The EWCD HMI (Human Machine Interface) software is composed of different layers (or screens) used in collecting and transferring information, as shown in Figure 3, with the arrows indicating the directional flow of information. Central to the HMI is the Alarm layer which is the primary viewing screen for operators to observe the current vibration state of the process. The peripheral layers labeled Configurator, Alarm Setup, and Band Analysis are used to define sampling conditions such as sensor channel, display trend length, sampling rate, etc., as well as setting the alarming parameters. Vibration data collected in the Alarm layer is stored in a MS SQL database for later use in setting the alarm parameters utilizing the Alarm Setup and Band Analysis layers, or to generate a summary report.

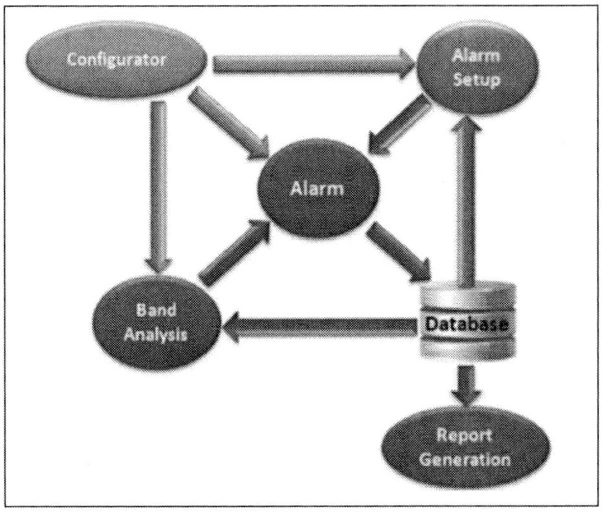

Figure 3 – *Flow chart for the early warning chatter detection software showing the different layers and direction of data collection and transfer.*

The HMI layers are intuitively designed, allowing the user to quickly navigate through the different layers to analyze data or adjust alarming parameters. Before setting the alarm parameters, baseline data collection over an extended period of time, e.g., life of a felt, production run for a specific grade, etc., is required. Baseline data collection is critical to establish the characteristic vibration signature (frequency spectrum) produced for a given machine. Factors affecting the vibration signatures from different Yankee dryers include mechanical design differences in the creping hardware, mechanical and process operating conditions, and vibration sensor mounting location. Nevertheless, common features in the vibration signature are observed from different machines.

A useful tool in the EWCD HMI is the Band Analysis layer for vibration frequency spectra analysis. This layer is used for both baseline data analysis and general data analysis to examine changes in the vibration frequency spectra and identify correlations to process operating conditions. The frequency spectrum collected from the creping process is complex containing broad spectral bands. If chatter is present the overall vibration energy will increase as well as the appearance of new broad spectral bands at high frequencies. Machine vibration sources such as pumps, felts, and rollers are at lower frequencies and have minimal contribution in the frequency range where chatter appears.

Alarming is not made on the frequency spectra but rather the overall gRMS (root mean square) value from the vibration spectra or a band gRMS value from a selected frequency range. The BAND ANALYSIS layer allows the user to recreate trends corresponding to user selected spectral regions of interest (ROI). Figure 4 shows an instantaneous vibration frequency spectrum in the top plot with four spectral regions labeled A, B, C, and D. The lower plot shows the reconstructed trends for each ROI over a 12.5 hour period. Results from this analysis show how the trend varies depending on the spectral regions with differing degrees of sensitivity to process changes. From the simulated trends ROI C exhibits the highest sensitivity to process changes, e.g., at 07:30 ROI C rapidly increases to a higher value relative to the other ROI trends. During this time period the machine experienced a sheet moisture level drop, which increased the tangential resistance on the crepe blade due to the harder coating. This example illustrates the importance of selecting spectral ROI for trending to enhance the sensitivity and response of the monitoring system for early detection of changes that stress the creping process.

Despite these numerous factors affecting the vibration signature for a given machine, common trends are still observed. For example, the ROI trends C, D, and to a lesser degree B shown in Figure 4, and the baseline overall gRMS trend shown in Figure 5, display a characteristic saw tooth pattern resulting from the crepe blade aging. When a new crepe blade is installed, some of the coating is removed, thus reducing the tangential resistance and the observed gRMS. As the crepe blade ages its efficiency to remove coating from the surface decreases due to blade tip wear, resulting in the tangential resistance increasing, which in turn increases the observed gRMS. With periodic changes in either the creping or cleaning blade, a macro cyclic behavior in the gRMS trend is produced forming the characteristic saw tooth pattern. The cyclic behavior observed is one key feature that differentiates crepe blade vibration monitoring from traditional bearing fault detection.

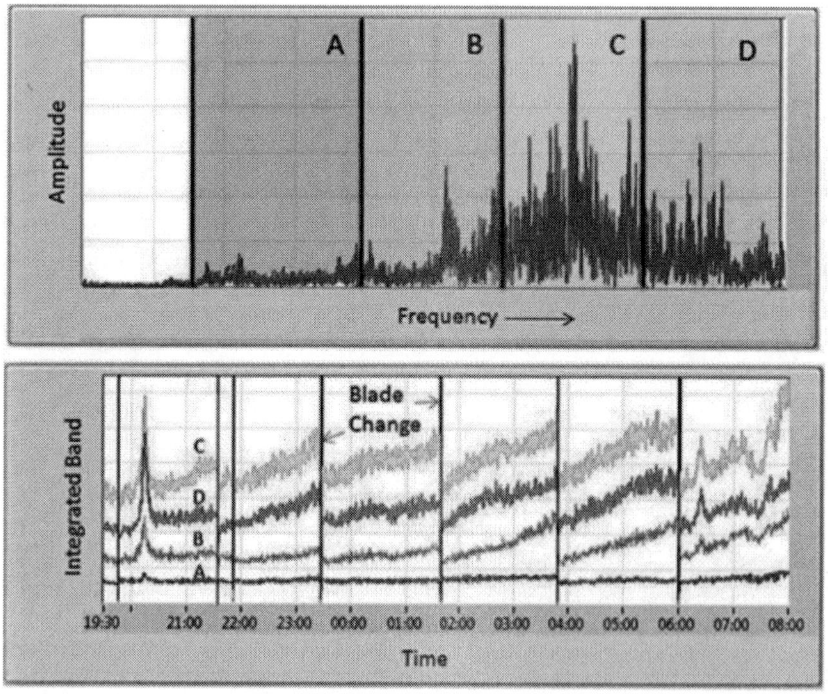

Figure 4 – *Top plot shows a frequency spectrum collected from the EWCD system with user selected markers indicating frequency bands A, B, C, and D used in recreating the trends shown in the lower plot. Blade changes are indicted on the lower plot by the vertical lines.*

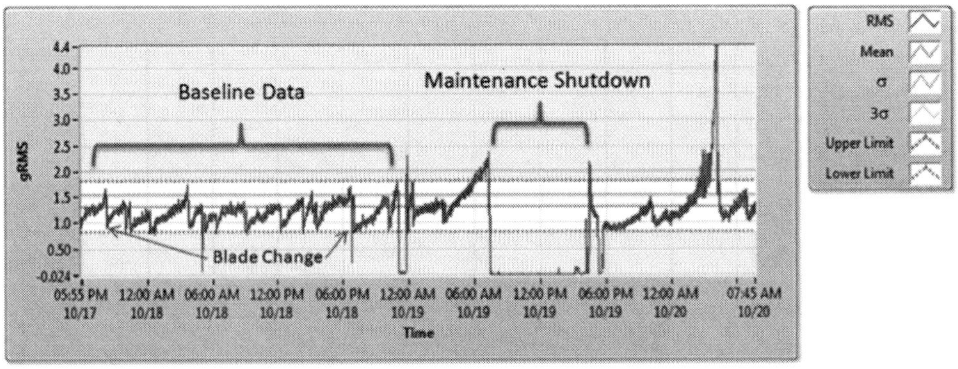

Figure 5 – *Trend showing outlier data in shaded region that is removed in the alarm level calculation. Baseline data spans a minimum and maximum range of 1.082 and shows the effect of blade changes that appear as a sharp drop in the gRMS value.*

For a tissue machine operating under normal conditions, i.e., no evidence of chatter, the natural variation in the gRMS trend defines a minimum and maximum range for the gRMS value. The alarming strategy developed in the EWCD HMI is based on first determining an nσ (ex. n=3) alarm level calculated from baseline data collected with outliers removed. Generally, outliers will appear in baseline data collected over several days and are easily filtered out using upper and lower limits in the Alarm Setup layer of the EWCD program. For example, outlier data occurs when the creping blade is disabled for maintenance procedures resulting in the gRMS trend value approaching zero. Alternatively, a transient jump in the gRMS can also occur driven by a process operating condition change, e.g., startup after maintenance shutdown (sheet moisture decreases). These different features are highlighted in Figure 5 with the outliers in the shaded area of the plot.

The second step in the EWCD alarm strategy is to define a threshold value to alert operators whether a Level-1 (Yellow) or Level-2 (Red) alarm event occurred. If the gRMS value is > nσ the following condition is triggered in the EWCD HMI:

$$\int (RMS - n\sigma)dt = \begin{cases} < n\sigma * threshold = Level - 1\,(Yellow) \\ > n\sigma * threshold = Level - 2\,(Red) \end{cases} \quad \text{Eq. 1}$$

where a Level-1 (Yellow) signals a cautionary state and a Level-2 (Red) alarm is a critical state. When a critical Level-2 (Red) alarm is active an inspection by the operator following a decision tree is the recommended action. The integrated value in Eq. 1 represents the excess vibration, i.e., the vibration energy above the nσ setting over an integrated period of time. This value is also tracked over time and displayed on the Alarm layer as the cumulative excess vibration.

Implementing this alarm level switching strategy reduces false positive alarms as well as indicating shifts in the mean gRMS trend. For example, a gRMS spike with short time duration will not necessarily trigger a critical Level-2 alarm unless the switching condition is met. Similarly, a gRMS value slightly > nσ will trigger a Level-2 alarm if the gRMS value stays above nσ for an extended period of time. In this case, the Level-2 alarm may not be critical, but rather indicating an overall shift in vibration energy occurred. Process changes such as grade, furnish, or production, can cause shifts in the overall gRMS that require the alarm parameters to be readjusted.

The visual display on the EWCD HMI alarm layer provides operators with a clear indication of the current vibration state of the machine. An example screen shot of the EWCD Alarm layer is shown in Figure 6 for an installation using two accelerometers mounted near the front and back side of the creping doctor. The EWCD system can accommodate up to four accelerometers if needed, e.g., wide Yankee dryers > 200 inches can benefit with the addition of a third or fourth sensor to enhance the sensitivity by reducing the zone coverage required for each sensor. For each sensor an alarm level indicator is displayed to show the vibration state of the sensor based on the alarming strategy. A green display indicates the vibration energy is below the alarm level and operating in its normal vibration range. Vibration energy exceeding the alarm level will trigger the alarming strategy to indicate either a yellow cautionary state or a critical red condition. The current gRMS value is reported on a color coded bar plot linked to the alarm level along with a digital display. Plot display options include the gRMS trend with the alarm level or current vibration frequency spectra for each sensor. In addition,

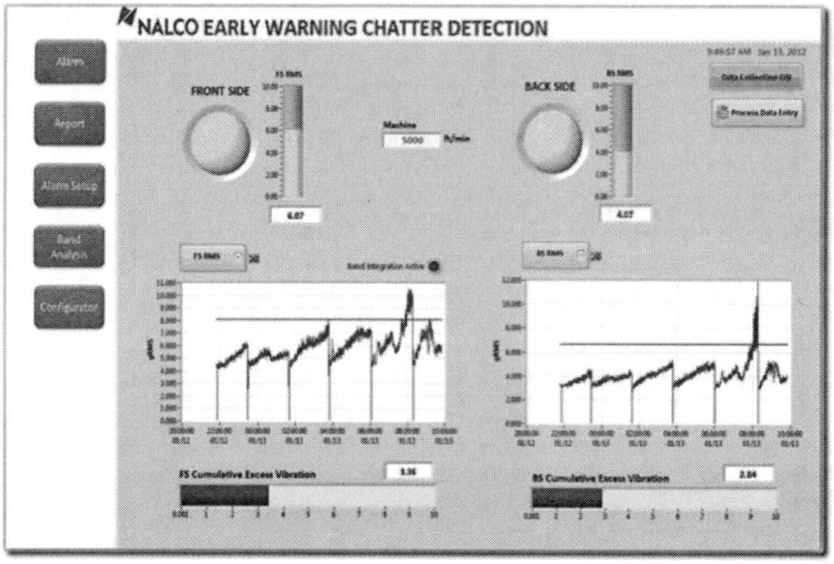

Figure 6 – *Sample Alarm layer HMI displays the current vibration alarm condition and historical gRMS trend or current frequency spectrum.*

the cumulative excess vibration determined from Eq. 1 is displayed to provide an overall indication of the historical level of vibration above the set alarm level. Lastly, a process data entry option is available to track changes in operating conditions, mechanical setup changes, comment on atypical events, and indicate whether chatter is visually observed in the coating. Data entered in this form can be later extracted with the report generation and used to identify trends or conditions leading to chatter, establish a better understanding of operational procedures that stress the creping operation, or alert operators that alarm level adjustment is needed due to a change in operating conditions.

The EWCD HMI also includes a report generation feature to provide a summary of the vibration data and alarm events collected over a user selected time period. Results reported in this layer are reduced to a simple format allowing the user to quickly assess the machine operating performance with regards to the severity and frequency of alarm events. This streamlined approach provides a first level screening to evaluate vibration energy changes related to either operating conditions or procedures. Conclusions drawn from this screening process may indicate required machine maintenance, reevaluation of operating procedures, or may warrant conducting a detailed analysis of the vibration data, e.g., identification of characteristic vibration frequencies, to identify the problematic source.

To highlight the different features in the report generation layer a sample screen-shot is shown in Figure 7. The screen is divided into sections depending on the number of measurement points. In this case, reporting is made for sensors mounted near the front and back side of the creping doctor for data collected over 9 hours where at 08:00 an alarm event occurred. The pie chart on the left hand side shows the analysis results reported as the percentage of time the machine operated within an acceptable vibration level (Green) and time spent in the cautionary Level-1 (Yellow) and critical Level-2 (Red) states for the selected time period. Machines operating with a high percentage of time in a critical state are at risk of irreversible chatter occurring. Additional information on the alarming frequency, i.e., number of times an alarm event occurred, and a breakdown of the average gRMS values with standard deviations for the alarm events are displayed on the right side. This breakdown provides a higher level of analysis useful in evaluating the severity of the alarm events, based on the magnitude of the mean and standard deviation. The horizontal bar plots for each side report the cumulative excess vibration for the selected time period. This gives a quick indication on the side-to-side differences and severity of the vibration. An additional visual check is also provided by displaying the gRMS trend plot and alarm level settings. Optional markers displaying where a Level-1 or Level-2 alarms occurred are also available on the trend plots.

Another optional reporting feature allows comparing the categorized reduced alarm event data collected at different time periods. Alarm event comparison is useful to evaluate how changes in operating conditions or procedures may have negatively or positively affected the vibration on the creping doctor. For example, comparisons can be made for different grades, changes in coating chemistry, furnish changes, or even to evaluate operating procedures from

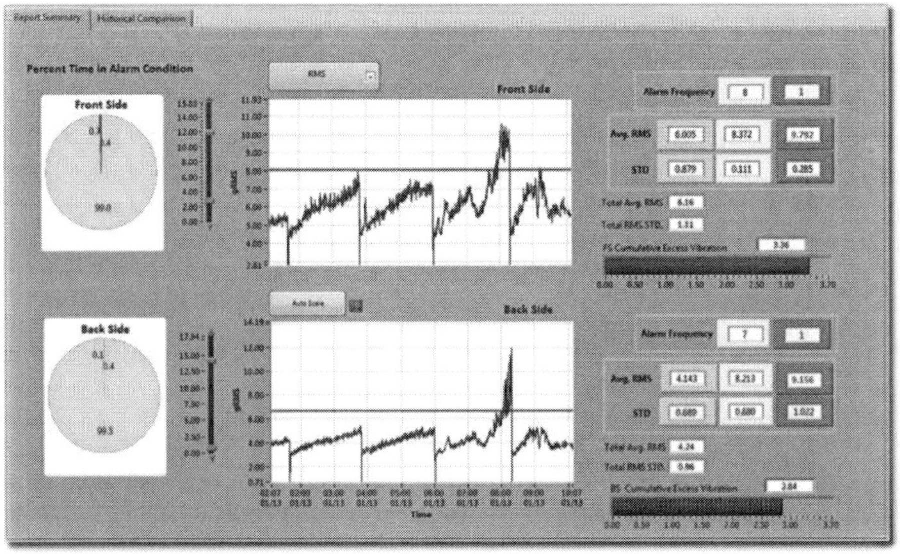

Figure 7 – *Report generation layer to summarize the gRMS vibration measurement quantities collected over a user selected time period.*

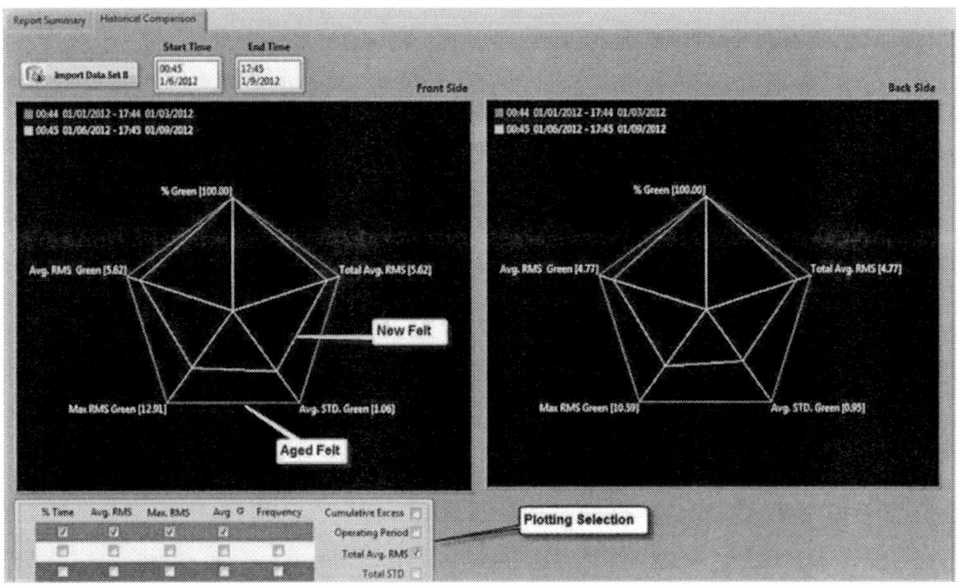

Figure 8 – *Radar plot illustrating changes in selected variables for data collected 3-days before and after a felt change.*

different shifts. The historical comparison is displayed using a radar chart to identify striking differences and commonalities. An example is shown in Figure 8 for data collected 3 days before and after a felt change. The user has the option to select which quantities to plot for comparison. In this example no alarm events occurred, but the average gRMS, standard deviation, and maximum gRMS values for data collected with an aged felt are all higher compared to the new felt data for both the front side and back side sensors.

All data displayed in the report generation layer can be exported to a formatted EXCEL file for printing and distribution or used for further analysis. An option is available to include a higher level of detail in the EXCEL report that includes all captured alarm events and process data entries. Captured alarm event data includes the timestamp when the event occurred, duration of the alarm event, average gRMS for the alarm event, standard deviation, maximum gRMS, and the cumulative excess vibration.

APPLICATION NOTES

This section presents a number of examples illustrating how the EWCD system was used to identify the onset of chatter. The examples demonstrate the sensitivity and response of the system to various mechanical, operational, and chemical changes that stressed the creping operation. Detection of the crepe blade vibration increasing and/or becoming unstable in these examples allowed operators to intervene early and perform the necessary corrective actions to minimize the severity of these chatter events.

Detecting Mechanical Runnability Problems

In the first example, a customer was experiencing a high frequency of sheet breaks along with trim plugging problems. The customer also had a EWCD system installed and observed unusual gRMS trends that showed a high level of instability on the back side of their machine. Time periods when the back side sensor gRMS stability became poor are shown in Figure 9. Here the gRMS trend shows the characteristic decrease when a new creping blade is installed followed by a gradual increase. However, the peak to peak value of the gRMS trend varied significantly at different periods, as indicated on Figure 9. The runnability problems experienced and the instability in the gRMS trend motivated operators to inspect the back side of the machine. This led to discovering a failed tube in the doctor blade holder resulting in poor profiling of the creping doctor and inefficient vibration dampening. After adjusting edge profiling screws and installing a shim behind the crepe blade, the gRMS signal peak-to-peak variation decreased by a factor of 5. In addition, the rapid gRMS increase observed after a blade change was reduced and stabilized Furthermore, the sheet break and trim plugging problems disappeared after conducting the temporary repair. With the improved runnability blade life also increased and ran >5 hours before a scheduled maintenance shutdown, at which time the tube was replaced. Prior to the repair blade changes averaged < 2 hours.

Moisture Variation

Dynamic changes in the tissue machine operating conditions during a startup are common and require careful attention to stabilize the process to minimize the potential for chatter. In this example, a sudden drop in sheet moisture occurred during the startup period after a felt change. The reduction in moisture hardened the coating resulting in more tangential resistance at the doctor blade. The gRMS trend shown in Figure 10 demonstrates the vibration effect when the sheet moisture content dropped. In the startup phase, a balance between the hood temperature and stock flow to the machine is critical. For example, if the hood temperature is ramping up faster than the desired stock flow, the sheet moisture content will become too low causing the coating to harden. Furthermore, sheet moisture content is often not measured during the startup period because the scanner is not yet operational. In this case, the EWCD can provide additional insight on the dynamic changes taking place. Here the EWCD alerted operators the gRMS level had reached a critical state. By taking the appropriate corrective action, which was to increase the stock flow, the sheet moisture level increased, and the overall gRMS level was reduced, thus averting a potential chatter event.

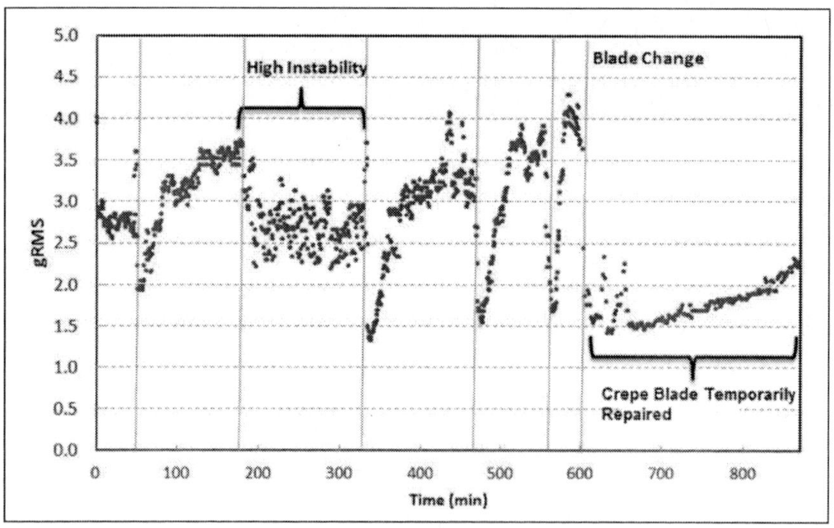

Figure 9 – *Trend exhibiting instability related to the crepe blade holder malfunctioning.*

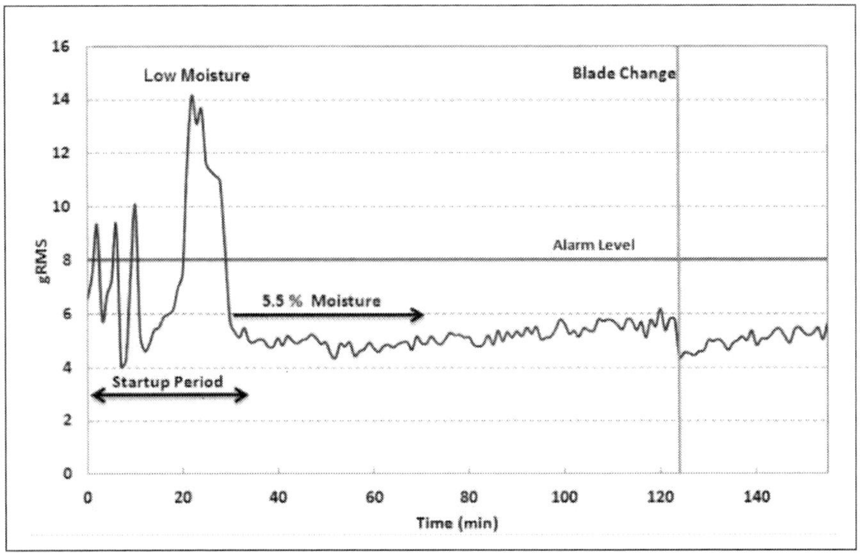

Figure 10 – *Effect of sheet moisture on crepe blade vibration during a startup operation.*

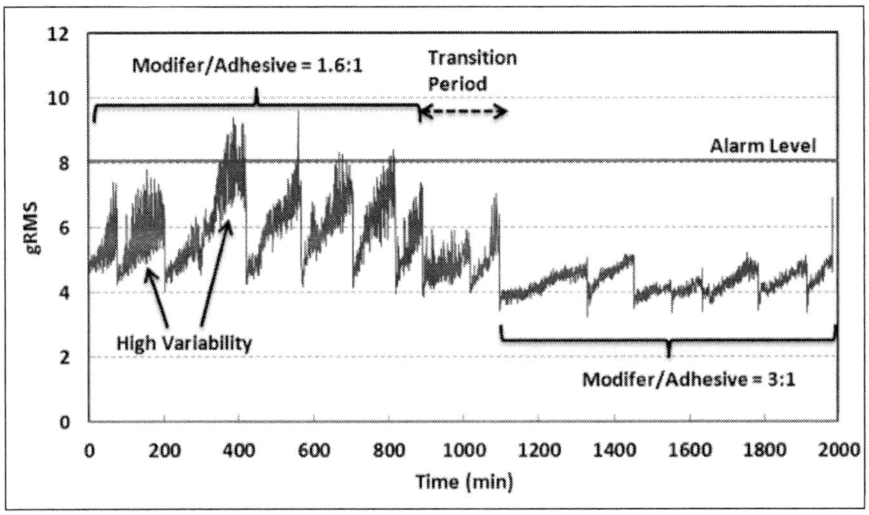

Figure 11 – *Effect of coating chemistry modifier/adhesive ratio on vibration stability and amplitude.*

Coating Chemistry Impact on Vibration

Coating hardness is also sensitive to the modifier or release/adhesive ratio, and if not adjusted properly can lead to chatter conditions. In this example, the modifier/adhesive ratio was adjusted off the optimal centerline value of 3:1 by the operator in an attempt to solve an on-going quality control problem. The operator adjusted the modifier/adhesive ratio down to 1.6:1 and improved the quality, but the change also hardened the coating. Figure 11 shows the gRMS trend operating at the different modifier/adhesive ratios. At the lower modifier/adhesive ratio the gRMS amplitude and variability significantly increased as a result of operating with a harder coating. At this condition micro chatter was visually observed in the coating near the front edge of the sheet. The EWCD system alerted operators of the problem leading to an adjustment of the modifier/adhesive ratio back to the centerline value. As shown in Figure 11, the overall gRMS level and variability decreased along with eliminating the sheet edge chatter. Additionally, the rise in gRMS observed significantly decreased after adjusting the modifier/adhesive ratio

CONCLUSIONS

1. The EWCD HMI provides an intuitive user-friendly environment that engages operators to improve management of the creping operation. Navigation through the different HMI layers is streamlined to develop alarm levels, select optimal vibration frequency bands for alarming, and summarize the results to track how the vibration on the creping operation changes over time.
2. The EWCD is a sensitive and fast response monitoring system to capture changes in operating conditions that can stress the creping operation. This provides additional insight to evaluate operational practices and their effect on the creping operation.
3. Early detection of events stressing the creping operation allows operators to intervene quickly and perform the necessary corrective actions to maintain runnability. Additionally, the EWCD system provides asset protection by minimizing the severity and duration that the creping operation is exposed to critical vibration levels.
4. Implementation of the EWCD vibration system is not restricted to creping process design constraints. The technology can be retrofitted to any creping system with flexibility in the number and location of the measurement points. Installation is straightforward typically done in less than one day during a maintenance shutdown.

REFERENCES

1. Archer, S., Grigoriev, V., Furman, G., Bonday, L., and Su, W., "Chatter and Soft Tissue Production: Process Driven Mechanisms," Tissue World Americas Conference, March 11-14, 2008, Miami Beach Florida.
2. Archer, S., Furman, G., and Von Drasek, W., "Yankee Surface Chatter," Tissue World 2011, Nice, France, March 2011.
3. Alessandrini, A. and Pagani, P., "Chatter Marks: Origin, Evolution and Influence of the Creping Doctors," Tissue World 2003, Nice, France March 2003.
4. Corboy, W., "Vibration Induced Yankee Surface Wear- An Overview of Chatter," Tissue World 2003, Nice, France March 2003.
5. Goldman, S., Vibration Spectrum Analysis: a practical approach, New York, Industrial Press Inc., 1999.

ACKNOWLEDGEMENTS

The authors would like to thank Tom McCllymont, Alberto de Martino, and Dominic Luneau from Nalco, and Ed Bondoc and Paul Miele from SKF for their contributions to this work.

ICT Iberica case study

Gain sustainable savings by uncovering the Mysteries of Chatter

ict Ibérica

BTG RAISING YOUR PRODUCTIVITY

Presented by:
Antonio JORDAN – ICT Iberica
Florent BOUGEROLLE – BTG Eclépens

Co-author:
Xavier Escaler – CDIF-UPC

Tissue World Miami, March 20th/21st 2012

BTG RAISING YOUR PRODUCTIVITY

Chattermarks: Evidence of a misaligned process

ict Ibérica

Blade strategy
Blade holder settings
Yankee edge control
Yankee coating
Yankee drying control
SPR: Yankee crown fit
Felt dewatering
Sheet formation
Wet-end management
...

= *Unstable process*
Poor performance
High chatter risks

Ceramic artwork by Alice Abrams

2

ICT Group

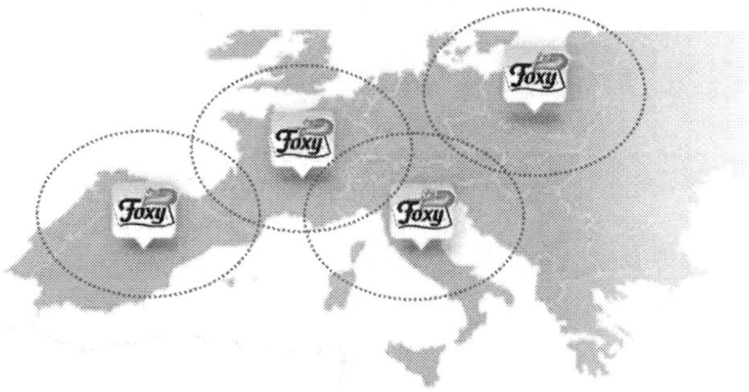

- European group
- 7% EU market share
- 9 machines 470,000t/y
- 5th EU tissue producer

3

ICT Iberica – Spain

- PM21 Andritz (2005)
- 70,000 t/y
- 1,800 m/min.
- Converting facility
- Toilet & towel paper

4

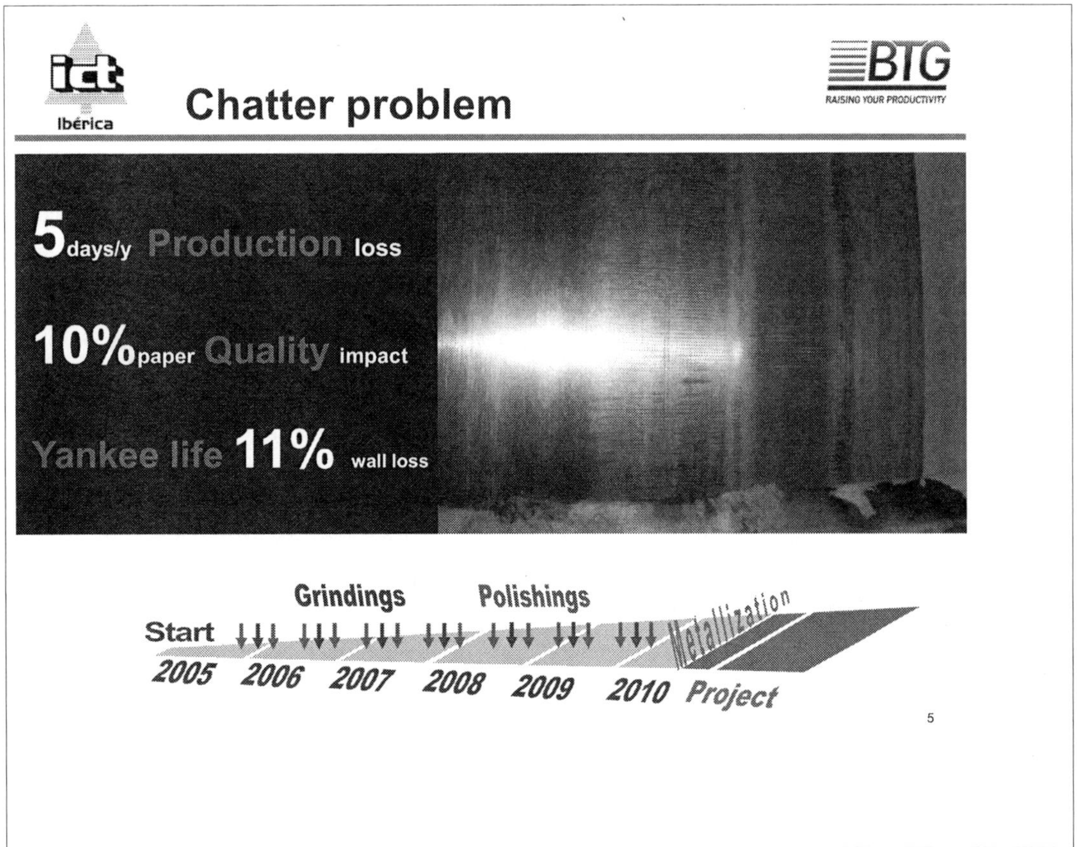
ict
Ibérica
Chatter problem
BTG
RAISING YOUR PRODUCTIVITY
5 days/y Production loss
10% paper Quality impact
Yankee life 11% wall loss
Grindings
Polishings
Start
Metallization
2005
2006
2007
2008
2009
2010
Project
5

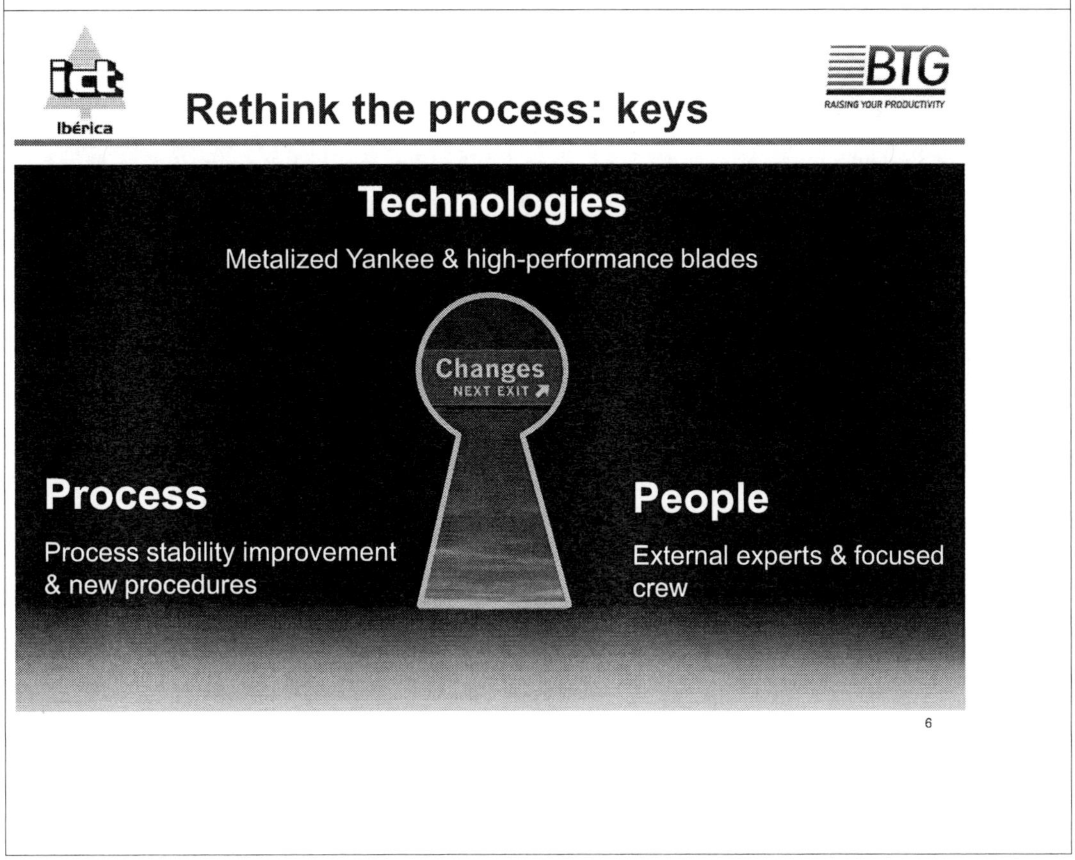
ict
Ibérica
Rethink the process: keys
BTG
RAISING YOUR PRODUCTIVITY
Technologies
Metalized Yankee & high-performance blades
Changes
NEXT EXIT
Process
Process stability improvement & new procedures
People
External experts & focused crew
6

BTG RAISING YOUR PRODUCTIVITY

Ceramic blade is the safest option for chatter

ict Ibérica

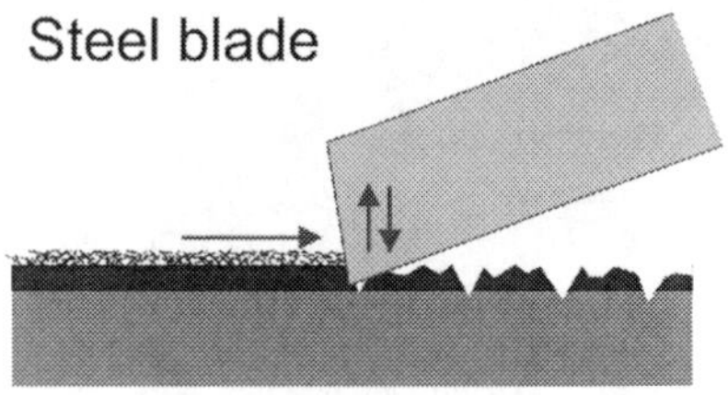

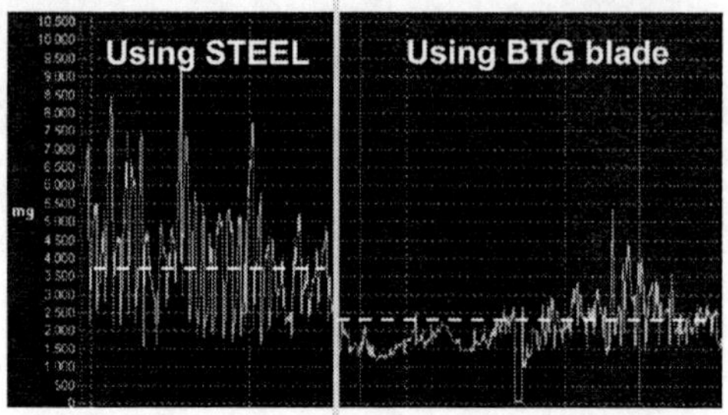

- **Fewer blade changes**
- **Coating stabilization**
- **Cleaner edges**
- **Yankee safety**

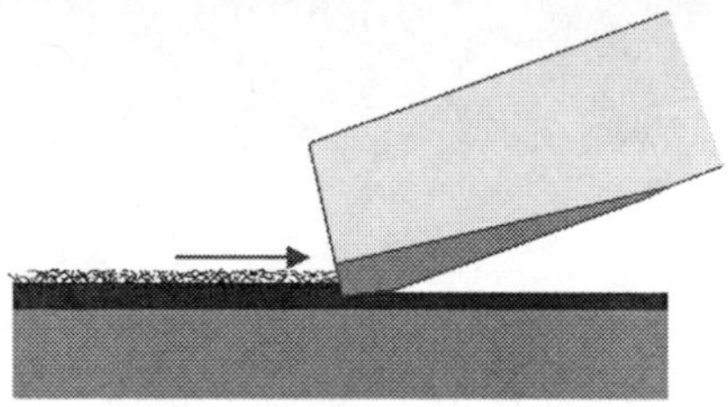

BTG ceramic blade

7

BTG RAISING YOUR PRODUCTIVITY

Vibration: too fast wear development

ict Ibérica

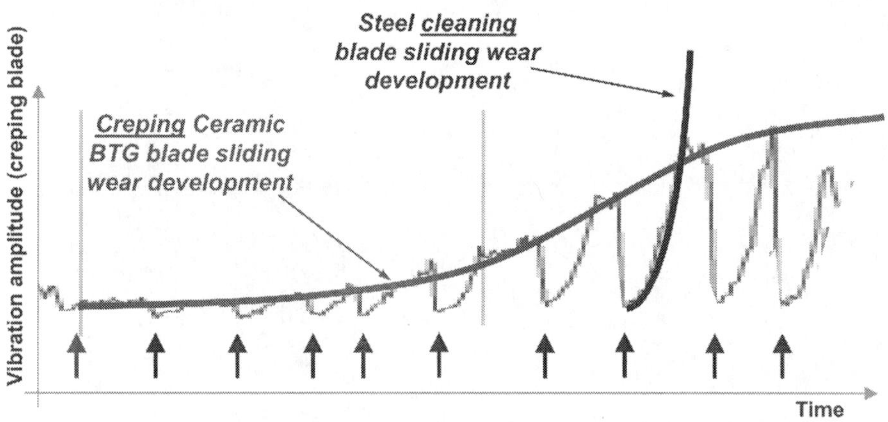

- **High frequency vibrations in coating**
- **Low frequency vibrations in contamination**

8

Reduce chatter potential thanks to fines management

Avoid fines going into the coating *here*... **… by fixing them onto the fiber *here***

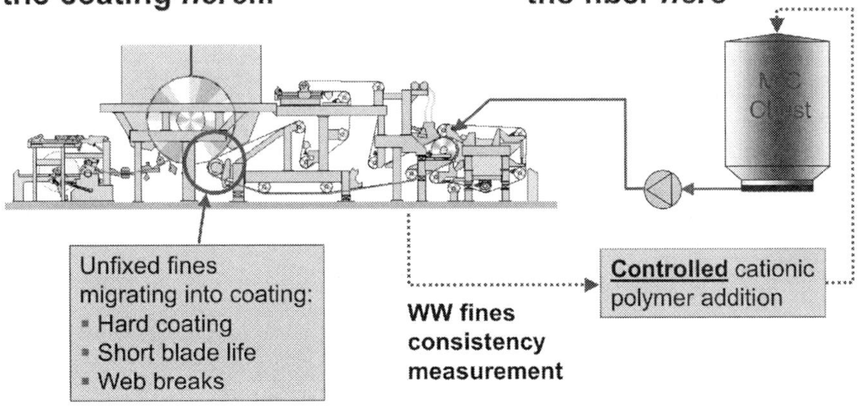

9

Rethink Process Procedures: Examples

- **Blade change procedure**

  when, why, how…

Vibration KPI

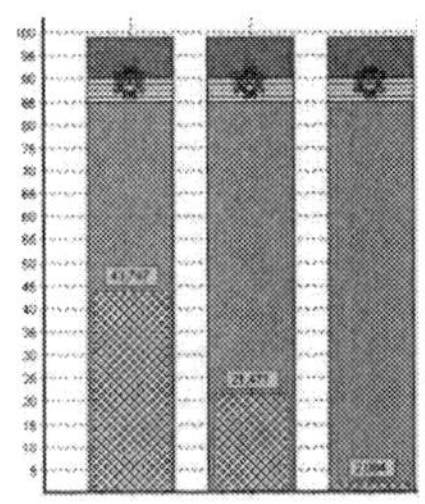

- **Machine cleaning procedure**

  when, where, how…

Time KPI

10

Project results

1,5y NO CHATTER

+5days/y AVAILABILITY

-30% lost time/day

PRODUCTIVITY

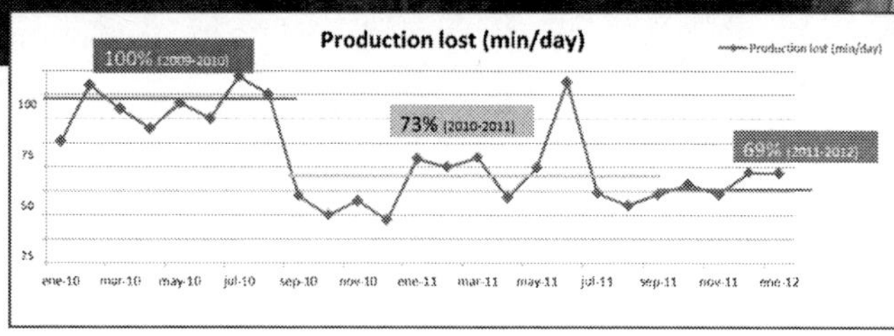

11

WE SOLVED THE CHATTER PROBLEM!

ICT Iberica – www.foxy.it.en
a.jordan@ictes.eu

BTG – www.btg.com
florent.bougerolle@btg.com

CDIF – www.cdif.es
escaler@mf.upc.edu

12

Steps Towards the Sustainable Value Chain of Tomorrow

Session 1: Sustainability in the Tissue Business
Suhas Apte
VP Sustainability
Kimberly-Clark Corporation
March 21, 2012

AMERICAS TISSUE WORLD

Kimberly-Clark

Abstract:

* The historical sustainability focus of manufacturing companies, optimizing operational footprints and minimizing environmental impacts, is evolving to one of ensuring sustainability across their entire supply chain (from raw materials to shelf). Going forward, our scope needs to be expanded to consider our business's value chain in a way that we become part of the solutions addressing the greater global challenges affecting us and our stakeholders. This will need to be demonstrated under the heightened expectation of transparency expected by our consumers and customers. We should expect that these trends will only be exacerbated in a 2050 World, inhabited by 9 billion consumers, where sustainable production and consumption is no longer a nice to have but rather a requirement. Kimberly-Clark's head of global sustainability will outline steps your company can undertake to proactively address existing and emerging sustainability related global value chain challenges and opportunities.

AMERICAS TISSUE WORLD

Kimberly-Clark

Agenda

* Sustainability @ K-C
* Global Value Chain Challenges
* Steps to a Sustainable Value Chain

Kimberly-Clark

K-C's Strong Sustainability History

* Time-tested founding values
* Multinational – global expansion
* Created five of the eight "essential" categories
* Employer of choice
* "Sustainability pioneer" promoting the strategy that *what's good for business, must also be good for our people and planet*

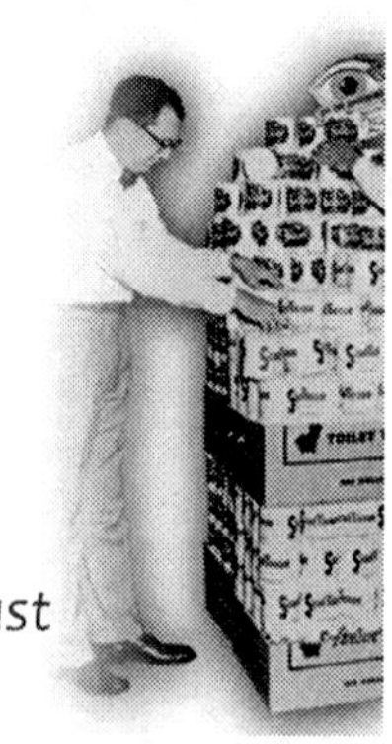

Kimberly-Clark

Our Structure

Consumer

$16 Billion

Professional

$3 Billion

Health Care

$2 Billion

AMERICAS TISSUE WORLD

Kimberly-Clark

A Global Health & Hygiene Leader

- 53,000-plus employees worldwide
- $21 Billion in Net Sales in 2011
- Well-known global brands – HUGGIES®, KLEENEX®, SCOTT®, KOTEX®, PULL-UPS®, DEPEND®
- #1 or #2 position in more than 80 countries
- Operations in 36 countries; sold in more than 175 countries
- Every day, nearly a quarter of the world's population trust K-C brands and the solutions they provide

AMERICAS TISSUE WORLD

Kimberly-Clark

Public Recognition

- Governance Metrics International – Best-in Class corporate governance standards
- FTSE4Good Index Series for the ninth consecutive year
- Perfect rating in the Human Rights Campaign 2011 Corporate Equality index
- #8 on the 2011 Corporate Responsibility Officer 100 Best Corporate Citizens List (3rd consecutive year)
- #9 on list of Fortune 500 companies using green power
- American Business Ethics Award winner for large companies

Kimberly-Clark

Global Value Chain Challenges

- Workplace Rights – Extended Supply Chain
- Climate Change – RM Supply Disruption
- Tissue Fiber Availability and Responsible Sourcing
- Energy Cost Increases
- Increasing Regulation
- Competition for Water
- Alternative Products – Innovation
- Global Sourcing

Kimberly-Clark

Workplace Rights – Extended Supply Chain

- Reputation and brand
- Customer loyalty
- Recruitment/retention
- Investor Attraction
- Legal risk
- Security
- Shutdowns/Disturbances
- 'Social license to operate'

AMERICAS TISSUE WORLD

Kimberly-Clark

Climate Change – RM Supply Disruption

- Over half of global supply chains were disrupted by bad weather in 2011 (41 per cent by telecommunication problems and 21 per cent by transport network disruption).
- 49 per cent of businesses reported a loss of productivity from such disruption, while the cost of business was raised for 38 per cent, and 32 per cent lost revenue.
- 21 per cent of businesses were disrupted by the natural disasters in Japan and New Zealand.

AMERICAS TISSUE WORLD

*Business Continuity Institute
http://www.supplychainstandard.com/Articles/3823/51pc+of+supply+chains+disrupted+by+weather.html

Kimberly-Clark

Tissue Fiber Availability

Utilization Rates for Eucalyptus & NBSK High

	Eucalyptus	NBSK
Total Global Capacity (end 2010)	16,285,000	12,945,000
Total Global Demand	15,505,000	11,925,000
Global Utilization	95.21%	92.12%

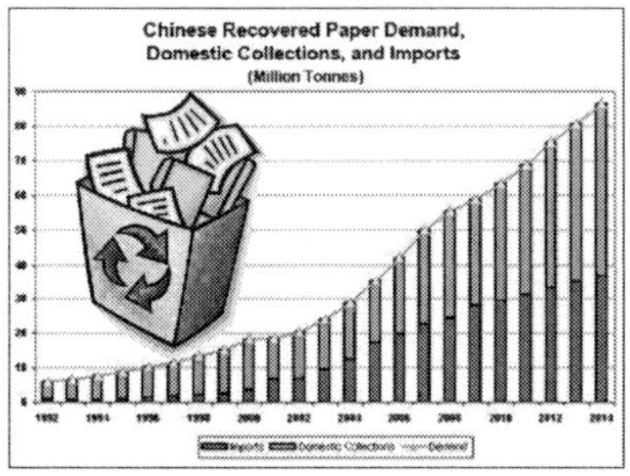

Kimberly-Clark

Responsible Fiber Sourcing

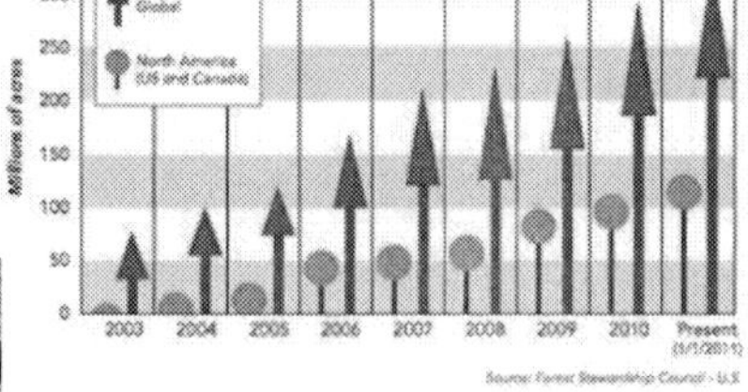

Kimberly-Clark

Increasing Energy Demand

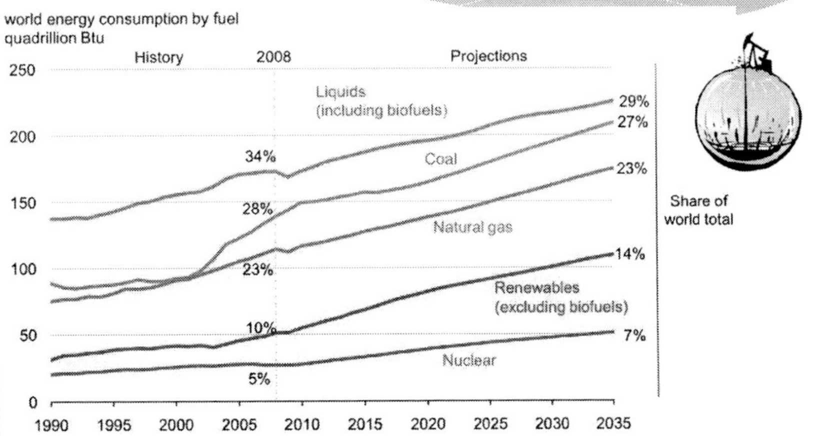

Source: EIA, International Energy Outlook 2011

Kimberly-Clark

Oil Price Forecast

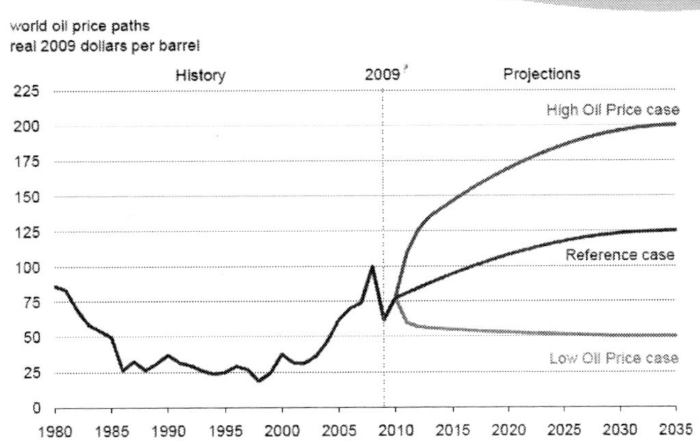

Source: EIA, International Energy Outlook 2011

Kimberly-Clark

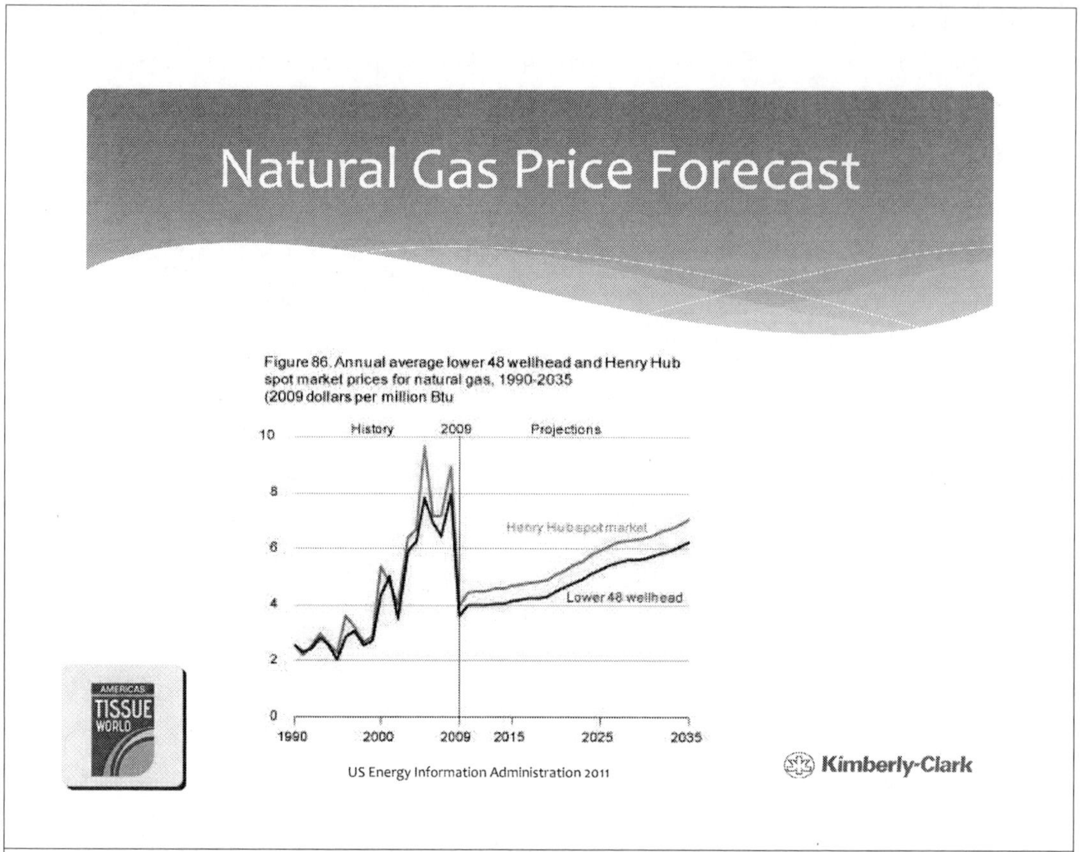

Natural Gas Price Forecast
Figure 86. Annual average lower 48 wellhead and Henry Hub spot market prices for natural gas, 1990-2035
(2009 dollars per million Btu
History
2009
Projections
Henry Hub spot market
Lower 48 wellhead
10
8
6
4
2
0
1990
2000
2009
2015
2025
2035
US Energy Information Administration 2011
AMERICAS
TISSUE
WORLD
Kimberly-Clark

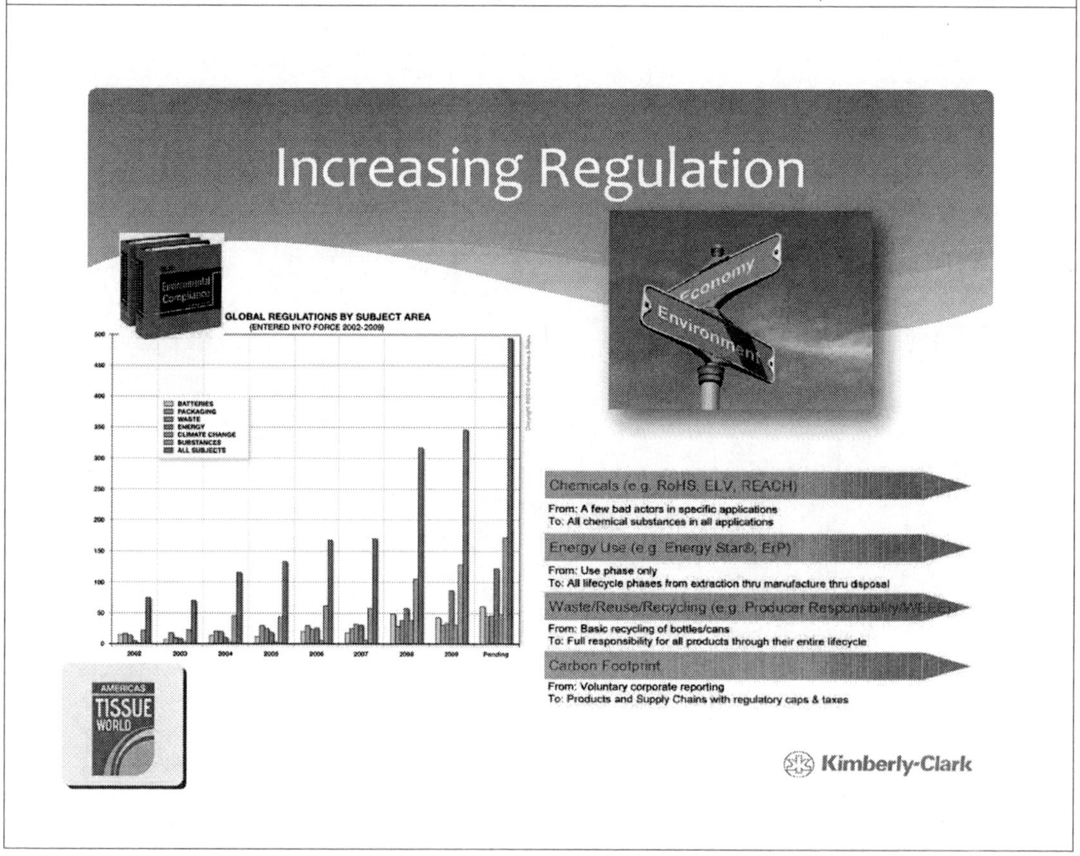

Increasing Regulation
Environmental Compliance
GLOBAL REGULATIONS BY SUBJECT AREA
(ENTERED INTO FORCE 2002-2009)
BATTERIES
PACKAGING
WASTE
ENERGY
CLIMATE CHANGE
SUBSTANCES
ALL SUBJECTS
Economy
Environment
Chemicals (e.g. RoHS, ELV, REACH)
From: A few bad actors in specific applications
To: All chemical substances in all applications
Energy Use (e.g. Energy Star®, ErP)
From: Use phase only
To: All lifecycle phases from extraction thru manufacture thru disposal
Waste/Reuse/Recycling (e.g. Producer Responsibility/WEEE)
From: Basic recycling of bottles/cans
To: Full responsibility for all products through their entire lifecycle
Carbon Footprint
From: Voluntary corporate reporting
To: Products and Supply Chains with regulatory caps & taxes
AMERICAS
TISSUE
WORLD
Kimberly-Clark

Competition for Water

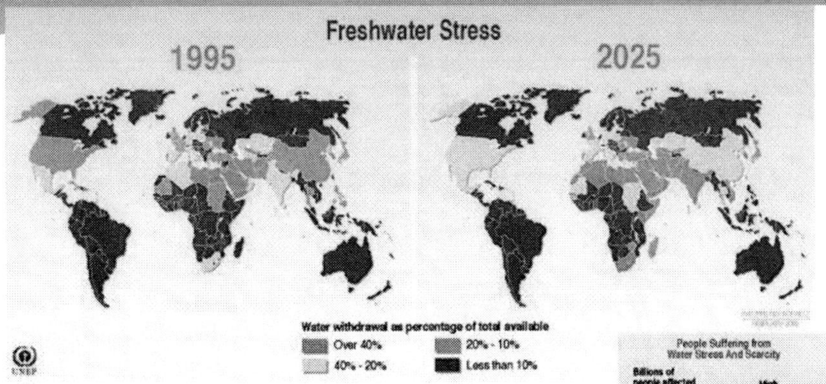

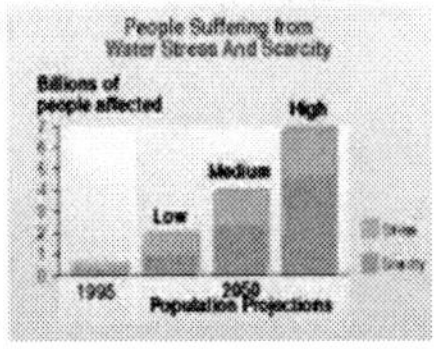

AMERICAS TISSUE WORLD

Kimberly-Clark

Global Sourcing

Global Recovered
Paper Market

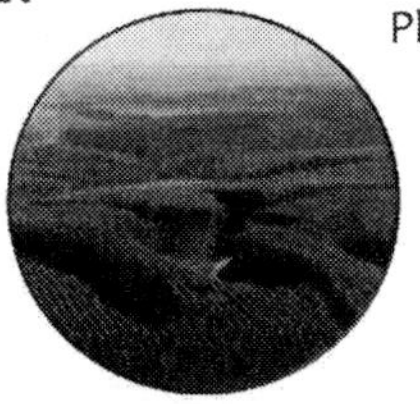

Eucalyptus
Plantations

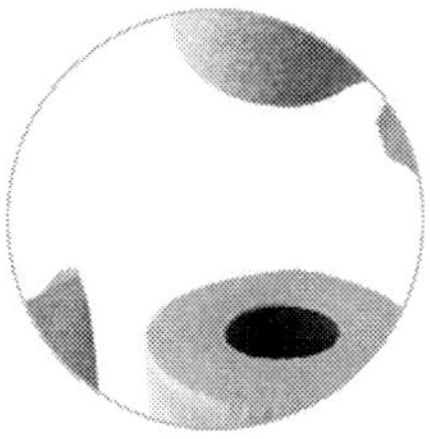

Hardrolls
Sourced from
Abroad

Finished Product
Exported/Imported

AMERICAS TISSUE WORLD

Kimberly-Clark

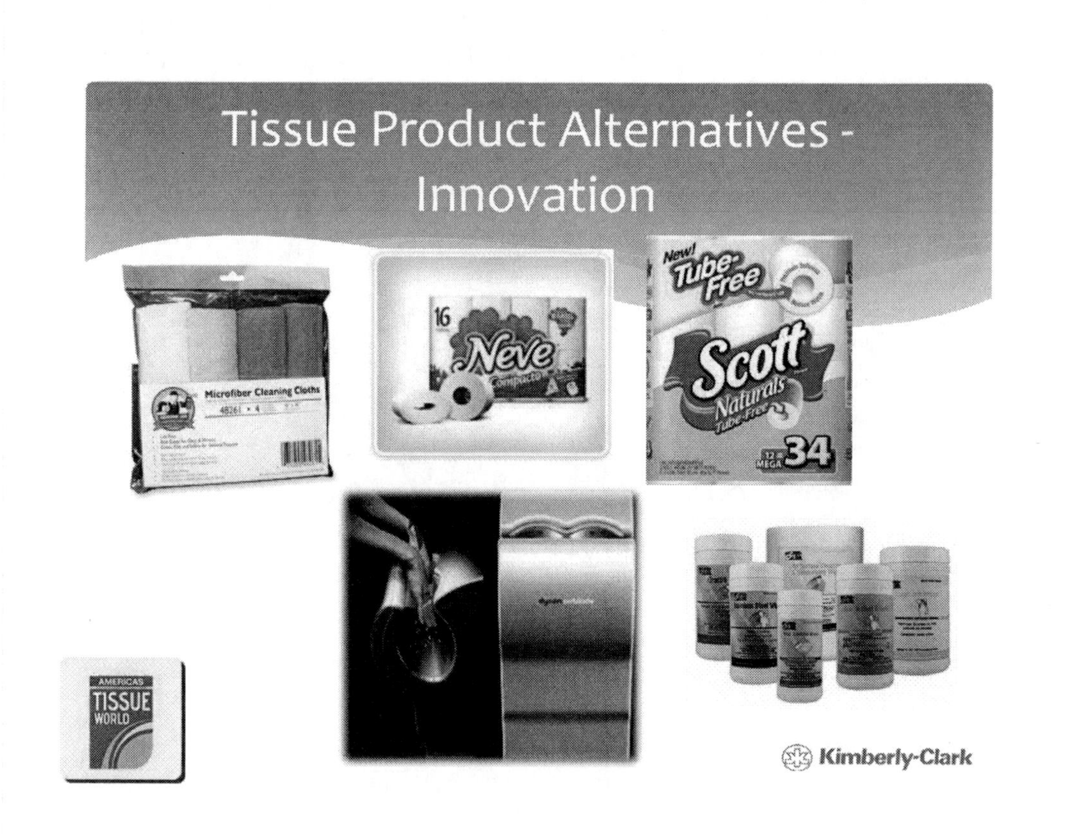
Tissue Product Alternatives - Innovation
Microfiber Cleaning Cloths
16
Neve
New! Tube-Free
Scott
Naturals
Tube-Free
MEGA 34
AMERICAS
TISSUE
WORLD
Kimberly-Clark

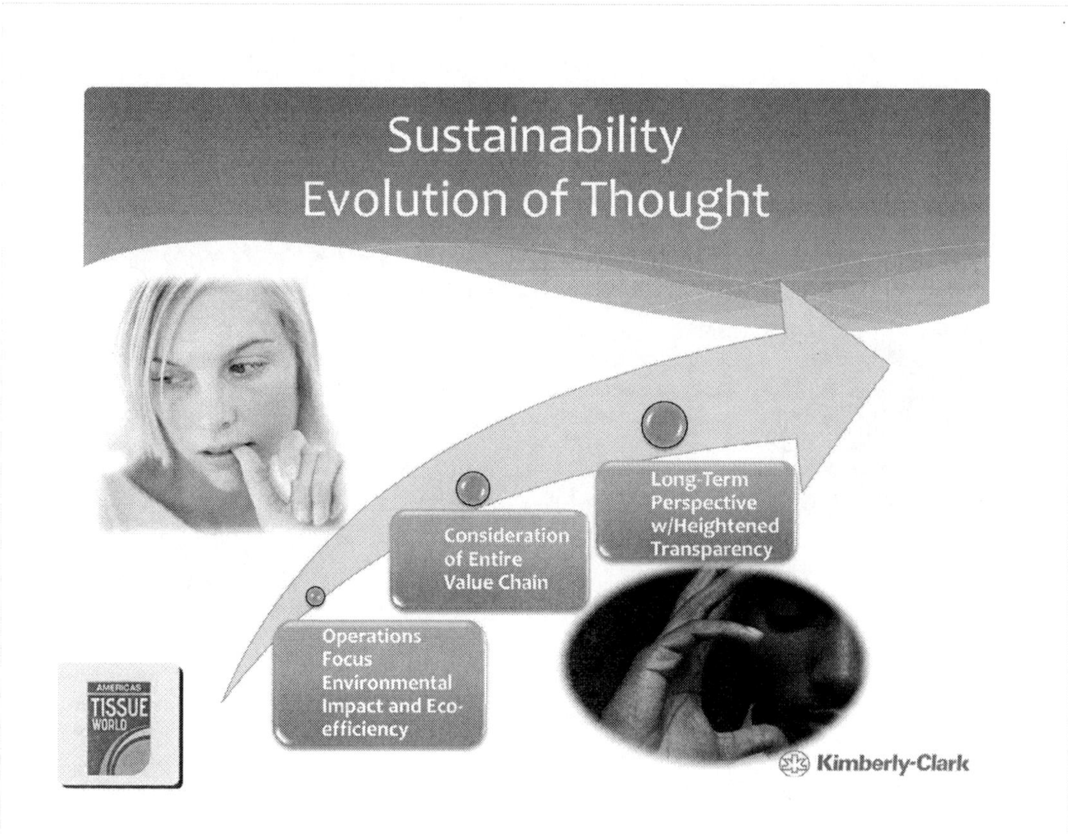
Sustainability
Evolution of Thought
Operations Focus Environmental Impact and Eco-efficiency
Consideration of Entire Value Chain
Long-Term Perspective w/Heightened Transparency
AMERICAS
TISSUE
WORLD
Kimberly-Clark

Sustainability Steps

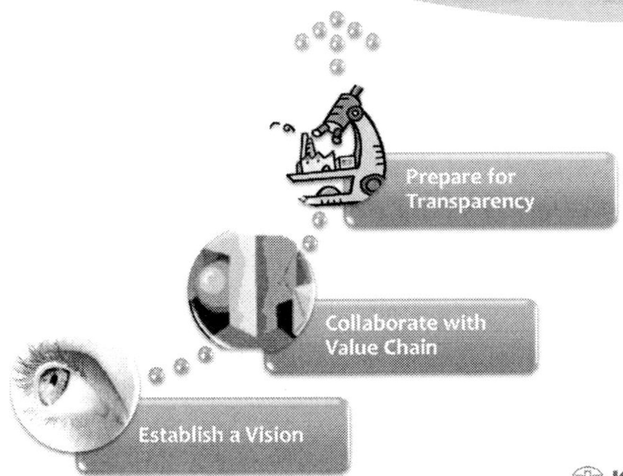

Kimberly-Clark

Establish a Vision

* Strategize Sustainability - 10+ Years Out
* Consider Scenario Planning – Futures Project
* Pressure Test Your Strategies
* Project Your Future Value Chain

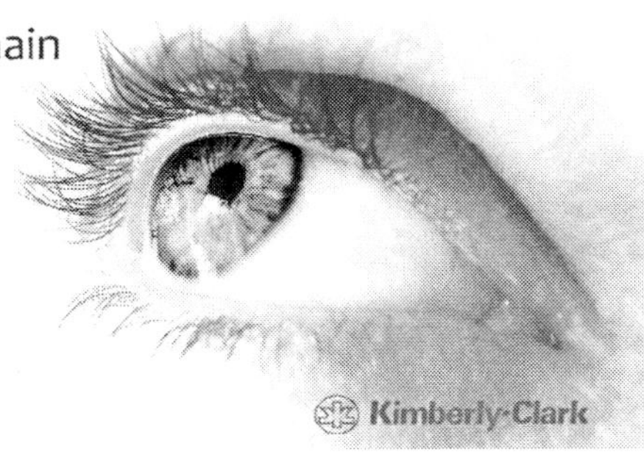

Kimberly-Clark

Collaborate w/ Value Chain

* Raw Material to Shelf
* LCA Mindset
* Collaborate to a Sustainable Future

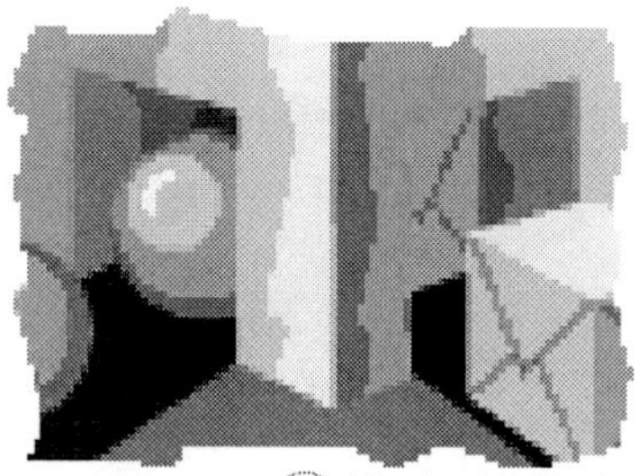

Prepare For Transparency

* Sustainability Reporting
* Consumer and Customers Empowered
* Global Economy – Highest Standards Prevail
* Social Media

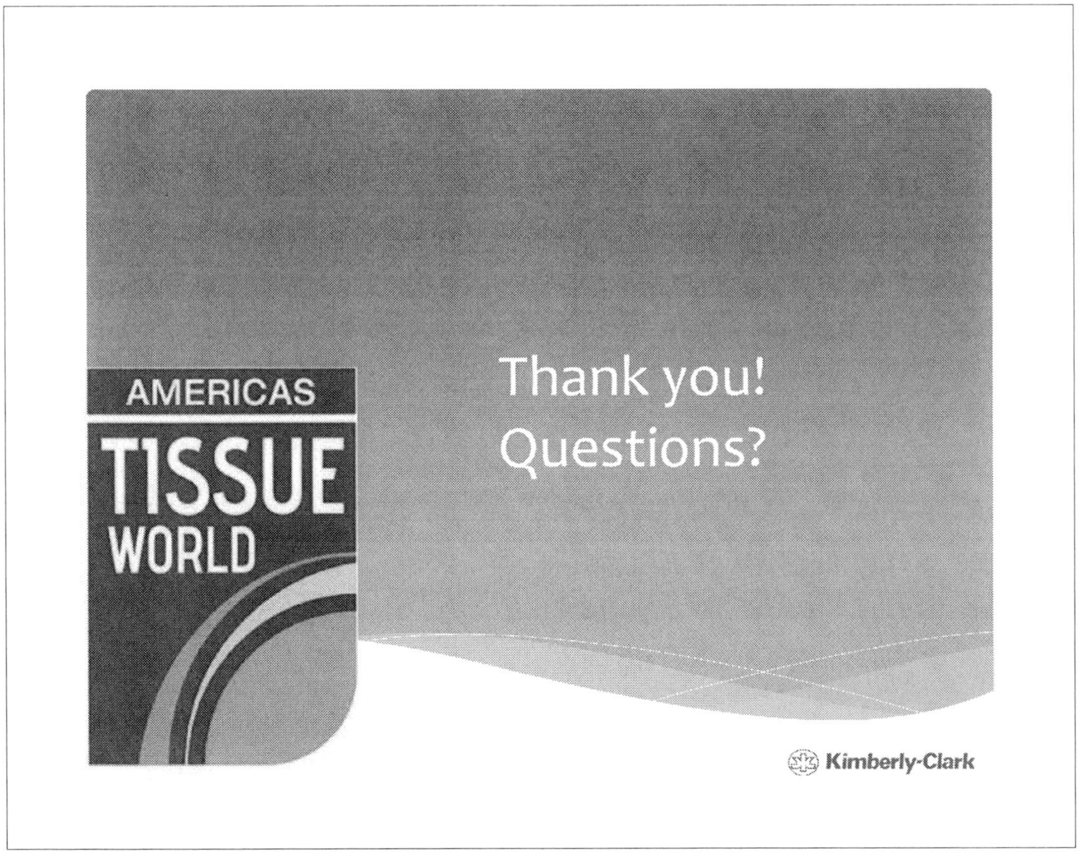
AMERICAS
TISSUE
WORLD
Thank you!
Questions?
Kimberly-Clark

"No Reason To Be Afraid, Necessarily"

Rolf Skar
Senior Forest Campaigner
Tissue World Americas -- March 21, 2012

GREENPEACE www.greenpeace.org

Greenpeace is:

- **The world's largest environmental campaigning organization**
- **Mission-driven**
- **Integrated internationally**
- **Independent: "Nulli Permanentes Amici, Nulli Permanentes Inimici"**

GREENPEACE www.greenpeace.org

Greenpeace is not:

- The pawn of Northern governments / interests
- The pawn of Southern governments / interests
- Out to get you
- Crazy...all of the time

GREENPEACE www.greenpeace.org

We Have Issues...

- Climate change
- Species extinction
- Toxics / human health
- Human rights / social justice

GREENPEACE www.greenpeace.org

We Tried to Tell You

- Sustainability is here to stay
- Sustainability can be good for business
- Change is inevitable
- Ignoring this is bad for business

GREENPEACE www.greenpeace.org

Getting It Right

- Be proactive
- Be transparent
- Don't "greenwash"
- Be ambitious
- Put the right people on the job

GREENPEACE www.greenpeace.org

What the Future Holds For:

- **Your business**
- **Conservation efforts**
- **Collaboration**
- **Laggards**

GREENPEACE www.greenpeace.org

Making Sense of Environmental Certifications Programs
Suzanne Blanchet
President and CEO of Cascades Tissue Group
Cascades
GREEN BY NATURE

1

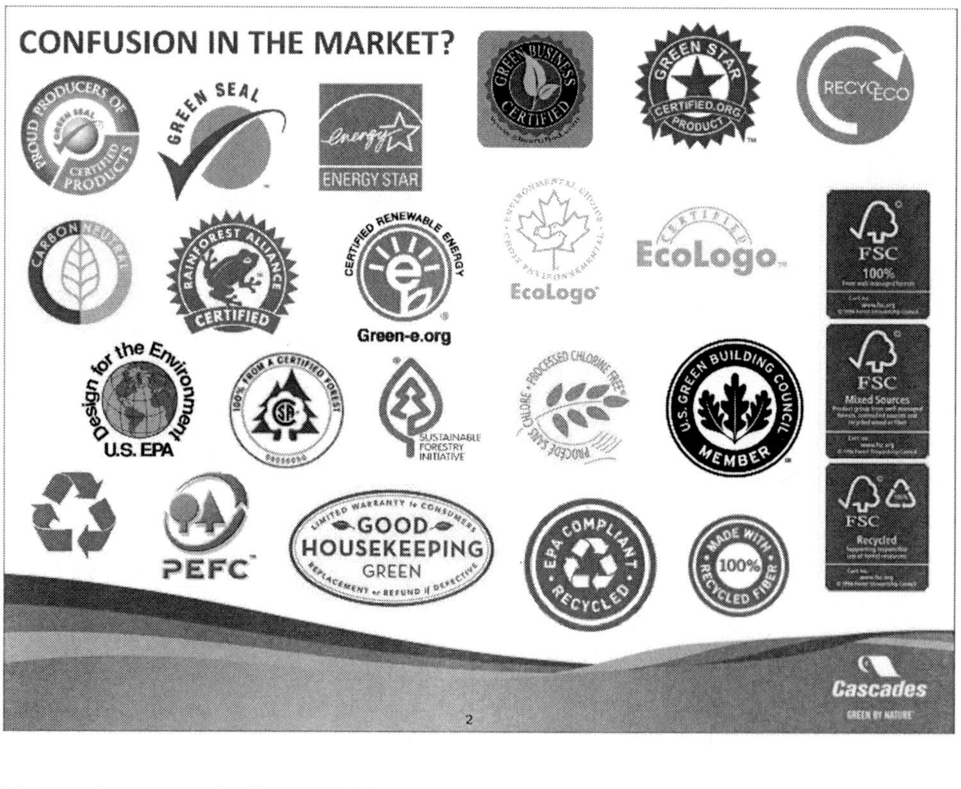
CONFUSION IN THE MARKET?
GREEN SEAL
ENERGY STAR
GREEN BUSINESS CERTIFIED
GREEN STAR CERTIFIED.ORG PRODUCT
RECYC ECO
RAINFOREST ALLIANCE CERTIFIED
CERTIFIED RENEWABLE ENERGY
Green-e.org
EcoLogo
FSC 100%
FSC Mixed Sources
FSC Recycled
Design for the Environment U.S. EPA
SUSTAINABLE FORESTRY INITIATIVE
U.S. GREEN BUILDING COUNCIL MEMBER
PEFC
GOOD HOUSEKEEPING GREEN
EPA COMPLIANT RECYCLED
MADE WITH 100% RECYCLED FIBER
Cascades
2

NGOs are making their own recommendations

There's even an app for that!!

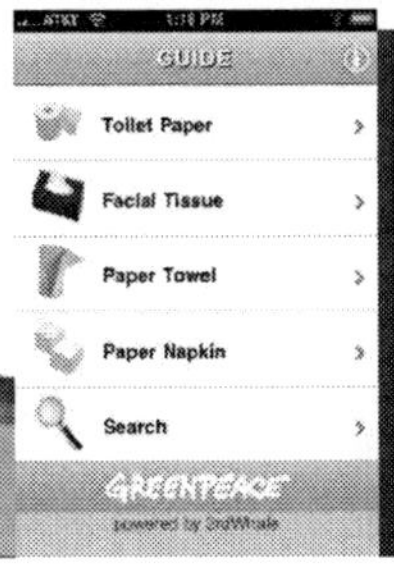

Cascades
GREEN BY NATURE

Ideal Third-Party Certification

- It protects consumers from myths, misconceptions, misleading information and corrupt manufacturers.
- It is credible, recognized and reputable, with scientific expertise while avoiding conflicts of interest
- It must raise the bar and stimulate manufacturers to always reduce their environmental footprint. Certified products should be less than 25% of products sold in a given market.
- Its standards have to be revised on a regular basis involving credible NGOs and industry experts.
- It audits manufacturers on site every year or every 2 years

NOT MANY OF THEM RESPECT THESE GUIDELINES

4

Cascades
GREEN BY NATURE

ENVIRONMENTAL CRITERIA

Environmental Criteria	US EPA's CPG	CFPA Processed Chlorine Free	Green Seal	Environmental Choice (EcoLogo) EcoLogo	FSC FSC	Green-e Green-e.org
On site audits		Every 2 years	Not mandatory	Once	Yearly	
Resource Use for Production						
Minimum Recycled Content	X	X	X		X	
Minimum Post-Consumer Content	X	X	X		X	
Chain of Custody for All Raw Materials		X			X	
Energy Use for Production		X	X	X		
Certified Renewable Energy Use/Offset for Production						X
Certified Sustainable Forestry (virgin fiber)		X			X	
Water Quality						
Prohibits Chlorine Compounds		X	X			
Effluent Content Restrictions		X		X		
Water Usage		X	X			
Water Permits/ Violations		X				
Waste Reduction						
Net Removal of Solid Waste				X		
Beneficial Use of Residual Waste		X				
Air Quality						
Regulatory Emissions Compliance				X		
Air Permits/ Violations		X				
Other Product Requirements						
Sq. ft. minimum requirements			X			
Packaging Requirements		X	X			
No Pigments, Dyes or Fragrances		X	X			
Product Performance Standards		X	X			

Cascades
GREEN BY NATURE

5

ENVIRONMENTAL CRITERIA – Forest Certifications

	American Tree Farm System	Canadian Standards Association	Forest Stewardship Council FSC	Programme for Endorsement of Forest Certification Schemes PEFC	Sustainable Forestry Initiative
Basics					
Scope	Family woodland owners in the US.	Focus on all forest types in Canada	Focus on all forest types throughout the world	Mutual recognition body that endorses national systems throughout the world	Primarily focused on medium and large-scale forests in the US and Canada
Total area	24 million acres	179.9 million acres	331.9 million acres	551 million acres	179.7 million acres
Accreditation					
Reviewer	IAF accredited auditors, ANSI-ASQ National Accreditation Board (ANAB) and Standards Council of Canada (SCC)	Standards Council of Canada (SCC)	Accreditation Services International (ASI)	The relevant national accreditation body which is a member of International Accreditation Forum	ANSI-ASQ National Accreditation Board (ANAB) or in Canada, the Satndards Council of Canada (SCC)
Evaluation Process	Auditors application process which includes a on-site and witness assesment process	Task group reviews application and forest management audit	ASI conducts office audits and the witnessing of one trial audit in the field prior to gaining FSC accreditation	A task group examines documentation, qualifications of reviewers and dispute resolution procedures	Auditors application process which includes a on-site and witness process
Monitoring	Annual surveillance audits	Annually assessed	Annual office and field audit	Annual office and field audit	Annual office and field audit
Renewal	Every 3 years	Every 3 years	Every 5 years	Every 3 years	Every 5 years
Product tracking and claims					
On-product label	Fiber from ATFS certified lands is included in SFI and PEFC chain-of-custody systems as certified content	Fiber from CSA certified lands is included in PEFC chain-ofcustody systems as certified content	Yes, 3 product labels: 1) FSC pure label for 100% certified product group; 2) FSC mixed label for product groups from well managed forests, controlled sources and recycled wood or fiber 3) FSC recycled label for product groups with 100% recycled content	Yes, 2 product labels: 1) PEFC Certified with a minimum 70% PEFC certified content; 2) PEFC Certified & Recycled with a minimum 70% comes from a certified source, and/or postconsumer recycled material	Yes, 4 product labels are available: 1) Certified Content Label for volume based credit; 2) Certified Content Label for average percentage calculations; 3) Recycled Content Claims for including percentage of recycled content, and; 4) Fiber Sourcing/Procurement Label for companies certified under the fiber sourcing requirements

Cascades
GREEN BY NATURE

6

WATER

- **North American industry average for virgin fiber***
  85.5 m^3 (20,520 USG)/ ton
- **Green Seal's maximum requirement**
  80.2 m^3 (19,250 USG)/ ton
- **North American industry average for recycled fiber***
  43.1 m^3 (10,352 USG)/ ton
- **Cascades' pulp & paper mill average**
  12.4 m^3 (3,276 USG) / ton

Cascades uses 85% less water than the Green Seal's requirement

* *EPN (Environmental Paper Network)*

Cascades
GREEN BY NATURE

7

EARTH

- **EcoLogo**: if pulp used is made from primary wood fiber, they require that the pulp be derived from "forests that may be demonstrated to be managed under a corporate code of sustainable forest practices". NO 3rd party "certified sustainable forestry" required.
- **Green Seal and PCF:** requires 100% recycled fiber with a minimum post-consumer material content
- **FSC MIX:** Products manufactured with a combination of FSC-certified virgin fiber (from FSC-certified forests), controlled sources and/or recycled wood or fiber. This label can include the Mobius loop.
- **FSC Recycled:** Products manufactured with 100 percent recycled fiber of which at least 85 percent is postconsumer (PCW) recycled material.

Cascades processed 2.6 million short tons of recycled fiber in 2010 – avoiding the cut of over 37.6 million trees

Cascades
GREEN BY NATURE

8

ENERGY

- **North American industry average for virgin fiber***
  38.7 millions BTUs (44.9 GJ)/ ton
- **North American industry average for recycled fiber***
  21.7 millions BTUs (25.2 GJ)/ ton
- **Green Seal's maximum requirement**
  17 millions BTUs (19.7 GJ) / ton
- **Cascades' pulp & paper mill average**
  9 millions BTUs (10.5 GJ) / ton

Cascades uses 47% less energy than the Green Seal's requirement. It purchases 20,00 MWh of Green-e certified renewable energy annually.

** EPN (Environmental Paper Network)*

9

Cascades
GREEN BY NATURE

BLEACHING

- **Elemental Chlorine -** Uses elemental chlorine to bleach pulp. In the U.S. elemental chlorine was phased out as of 1998 per EPA's Cluster Rule.
- **Elemental Chlorine Free ("traditional ECF")** - Replaces elemental chlorine with chlorine dioxide.
- **ECF with extended or oxygen delignification ("enhanced ECF')** - Removes more of the lignin before bleaching, thus reducing energy and chemical use during bleaching process. (Final stage uses chlorine dioxide.)
- **Enhanced ECF with ozone or hydrogen peroxide** - Uses ozone or hydrogen peroxide as brightening agent in initial stages of bleaching processes. (Final or near final stage uses chlorine dioxide.)
- **Processed Chlorine Free (PCF)** - Completely substitutes oxygen-based compounds for chlorine compounds **/ No bleaching**

Cascades Tissue Group operates its mills with a PCF process. 7 of its paper mills are also PCF certified by the CFPA

Source: EPN (Environmental Paper Network)

10

Cascades
GREEN BY NATURE

LEED (Leadership in Energy and Environmental Design)

- A product CANNOT be LEED certified. However, some manufacturers add their USGBC membership logo on product packaging.
- Towel and tissue products can contribute to attaining credits under the LEED®-EB: O&M (operation & maintenance) rating system only.
  - 2 credits and 2 prerequisites for towel and tissue dispenser systems
  - 4 credits and 3 prerequisites for recycled content
  - 1 credit for renewable energy/climate neutral properties
  - 40 points minimum required for certification (and 9 prerequisites)

Cascades received the LEED-NC Gold certification for its Lachute, QC facility and LEED-CI for its Waterford, NY offices

Cascades
GREEN BY NATURE

11

NEW ENVIRONMENTAL CALCULATOR

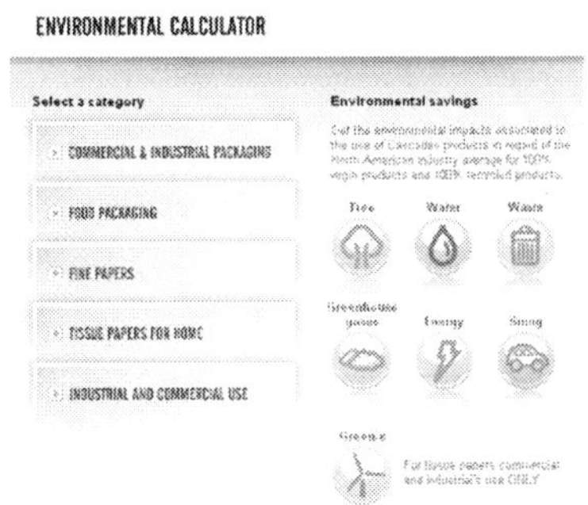

www.cascades.com/calculator

Cascades
GREEN BY NATURE

12

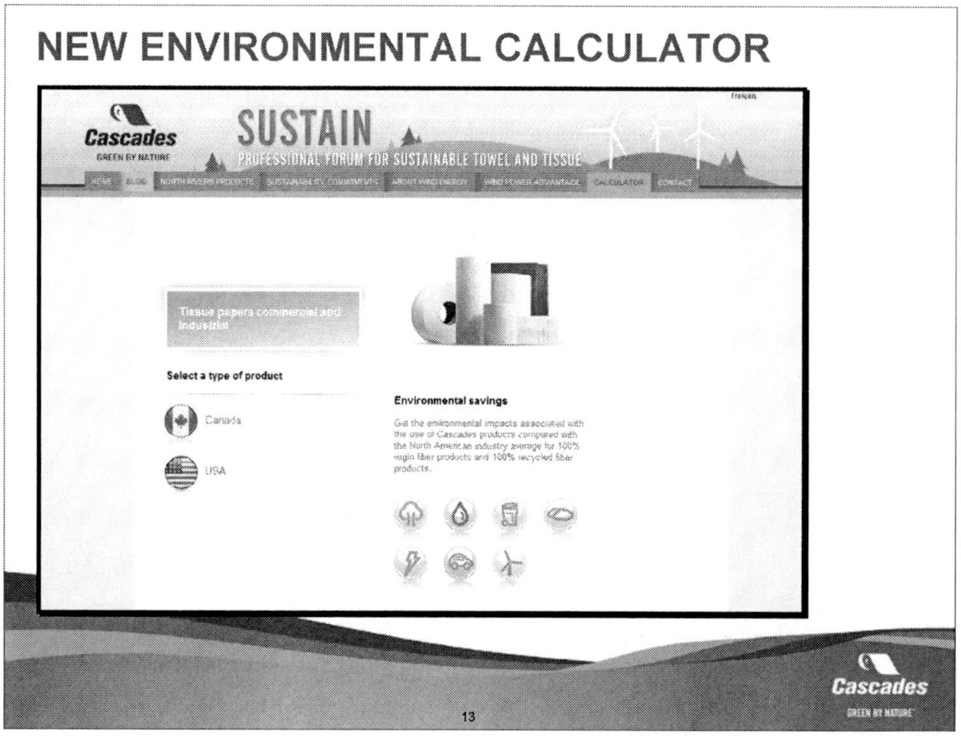
NEW ENVIRONMENTAL CALCULATOR
Cascades
GREEN BY NATURE
SUSTAIN
PROFESSIONAL FORUM FOR SUSTAINABLE TOWEL AND TISSUE
CALCULATOR
CONTACT
Tissue papers commercial and industrial
Select a type of product
Canada
USA
Environmental savings
Get the environmental impacts associated with the use of Cascades products compared with the North American industry average for 100% virgin fiber products and 100% recycled fiber products.
13
Cascades
GREEN BY NATURE

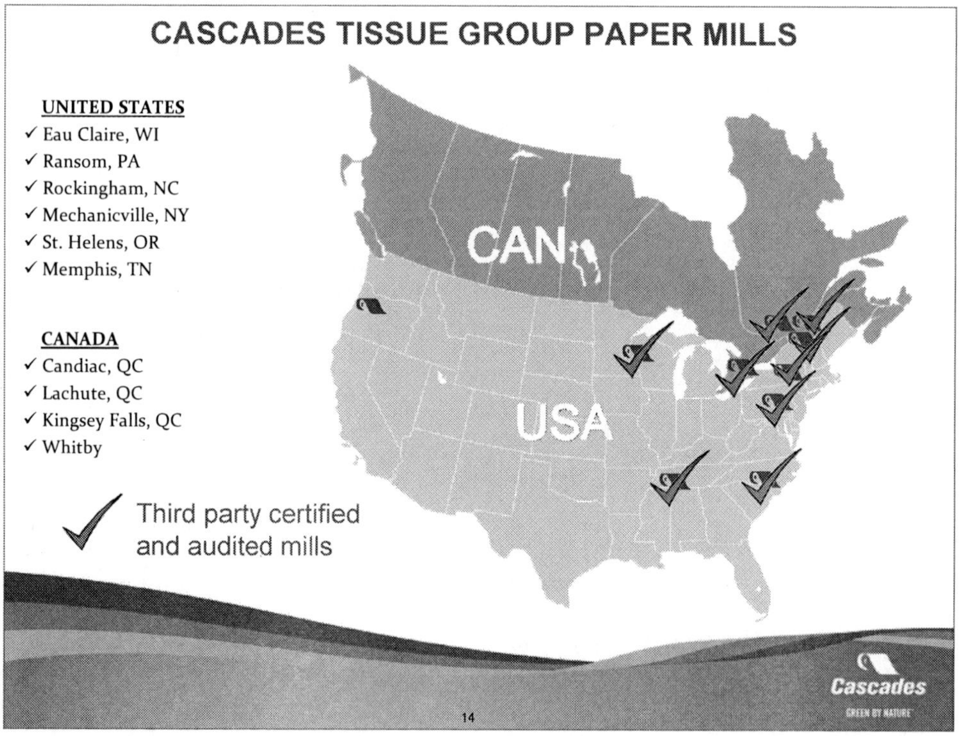
CASCADES TISSUE GROUP PAPER MILLS
UNITED STATES
✓ Eau Claire, WI
✓ Ransom, PA
✓ Rockingham, NC
✓ Mechanicville, NY
✓ St. Helens, OR
✓ Memphis, TN
CANADA
✓ Candiac, QC
✓ Lachute, QC
✓ Kingsey Falls, QC
✓ Whitby
CAN
USA
Third party certified and audited mills
14
Cascades
GREEN BY NATURE

Sustain.cascades.com

Professional Forum for Sustainable Towel & Tissue

Cascades

GREEN BY NATURE

15

Sustainable Development Objectives 2012

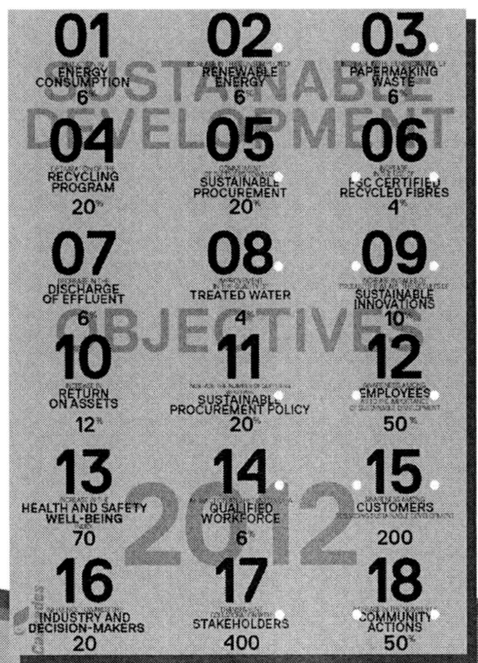

Cascades

GREEN BY NATURE

16

Forest Stewardship Council

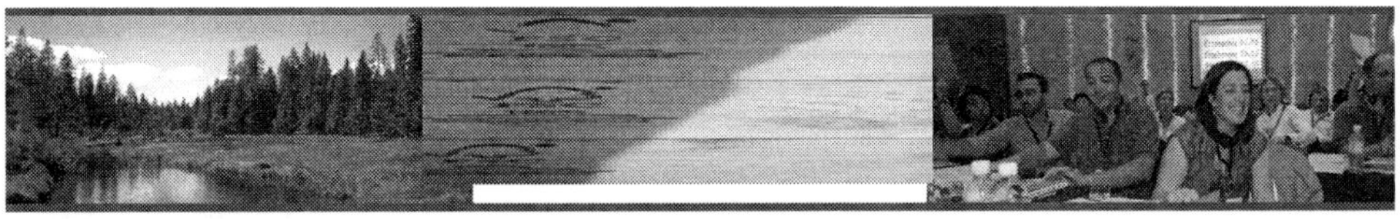

Forest Stewardship Council

Where North American Brands and NGO's Join Forces

Tissue World Americas - March 21, 2012

Corey Brinkema

c.brinkema@us.fsc.org

 September 2011

Forest Stewardship Council

The Gold Standard in forest certification

- Global certification and eco-labeling system initiated in 1993
- 365 million acres certified for forest management; 140 million in North America
- 23,000 companies certified for Chain of Custody; 5,000 in North America
- Roughly 5% US market share of forest products traded

 2

FSC

Forest Stewardship Council

Diverse voices make FSC stronger

- Governs via open, transparent democracy
- Broad and deep support of major conservation and social advocacy groups
- Balanced, comprehensive, globally consistent, and locally relevant standards that:
  - Protect rare habitats
  - Prevent deforestation
  - Protect rights of indigenous people and local communities
  - Prohibit highly hazardous chemicals
  - Maintain and enhance ecological function

 3

FSC

Forest Stewardship Council

National Wildlife Federation · WWF · Sierra Club (Founded 1892) · The Nature Conservancy — Protecting nature. Preserving life.

The Wilderness Society · Greenpeace · Defenders of Wildlife · NRDC — The Earth's Best Defense

FSC is the only forest certification system recognized and supported by North America's leading conservation organizations.

 4

Forest Stewardship Council

FSC

US system growth

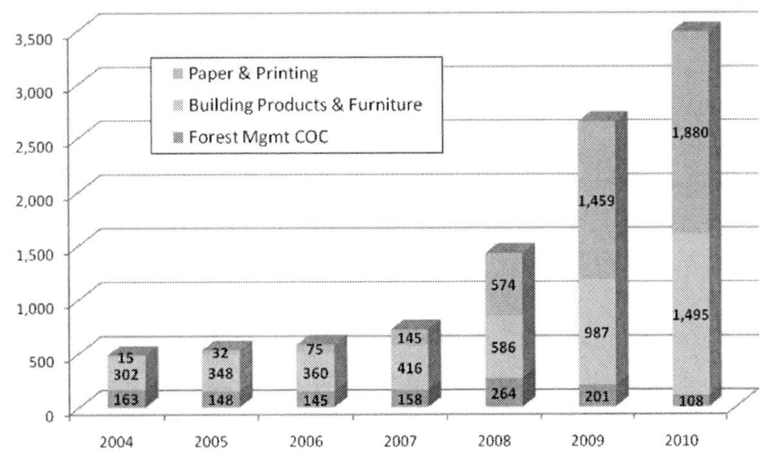

 6

Forest Stewardship Council

FSC

The land base is responding

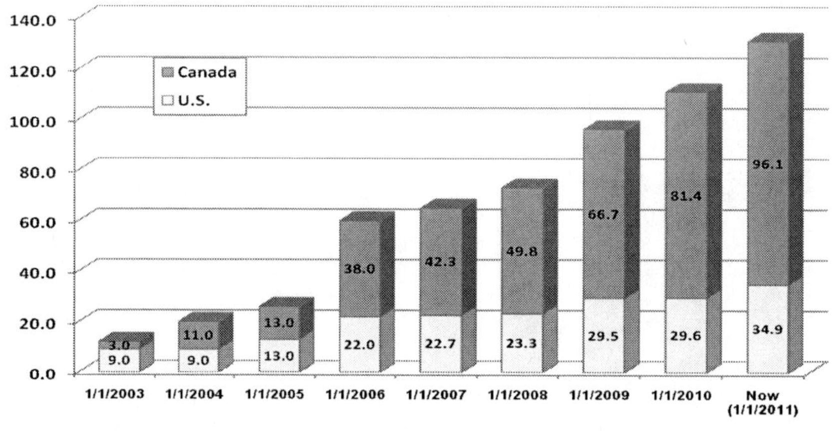

 7

Forest Stewardship Council

FSC

FSC inroads in the North American tissue sector

- Early leadership of Kimberly-Clark (2009)
  - Sourcing commitment
  - K-C Professional
  - Family Care – consumer Kleenex nearly 100% FSC-certified in the US
- Clearwater Paper (2010)
  - Private label bath and facial tissue
- Kruger (2011)
  - Cashmere, Purex

 8

Forest Stewardship Council

FSC

Kruger's rationale for choosing FSC

- "Recognition as the leading global authority on responsible forest management"
- "We believe the FSC standard to be the most stringent"
- "Measure of accountability with FSC certification that does not exist with other certifications.
- "Support by a wide variety of non-government organizations such as Greenpeace, WWF, Sierra Club, David Suzuki Foundation, etc. was important"

* From Tissue World Magazine Jun 2011 interview with Steven Sage

 9

Forest Stewardship Council

FSC

Will the US consumer care?

- Our qualitative and quantitative research says emphatically "YES"
- 58% of dark and middle greens in the US find FSC certification very or extremely influential in tissue product purchasing decisions
- Trusted brands + leading environmental NGO's = compelling and powerful combination

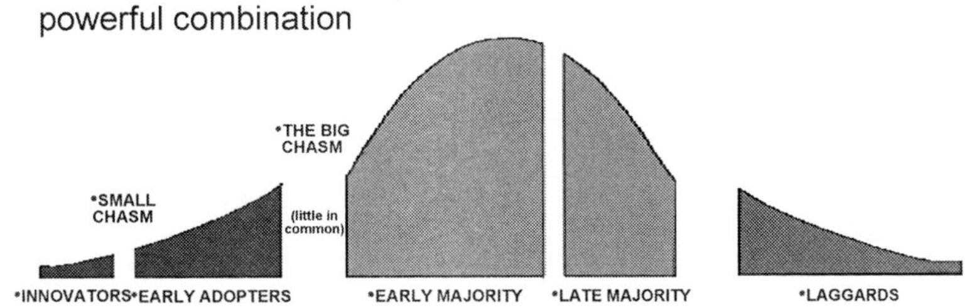

 10

Forest Stewardship Council

FSC tissue fiber supply present & future

- Canadian and Brazilian forest managers provide the bulk of FSC certified tissue fiber today
- Recent opportunities open up supply chains in US Southeast
  - Private industrials
  - TIMO's
  - Family forests
- Multiple drivers converging to make business case for forest owners (LEED, DIY retailers, office paper, packaging, and tissue)

Forest Stewardship Council

Thank you

Corey Brinkema, President
FSC-US
email: c.brinkema@us.fsc.org
phone: 612-353-4511
www.us.fsc.org

EcoLogo™

PART OF THE UL GLOBAL NETWORK

Environmental Leadership in the Tissue Paper Sector – Beyond Forest Certification

Tissue World Americas

21st March 2012

Susan Rutherford, Sustainability Analyst

EcoLogo Program

CERTIFIED.
SUSTAINABLE.
LEADERSHIP.

EcoLogo™

PART OF THE UL GLOBAL NETWORK

Outline

Performance

Paper life cycle

Leadership practices

Business and environmental performance

Ecolabeling

Summary

2

EcoLogo™
PART OF THE UL GLOBAL NETWORK

Performance – of the tissue paper production

Conference theme is "Boosting performance for better quality, sustainability and profitability".

Focus **has** to be the performance of the *entire* process that leads to the production of tissue paper.

3

EcoLogo™
PART OF THE UL GLOBAL NETWORK

Paper life cycle – stages

Raw material extraction
Raw material processing
Manufacturing – paper production, converting
Packaging
Distribution and sale
Use
Disposal

4

EcoLogo™
PART OF THE UL GLOBAL NETWORK

Life cycle

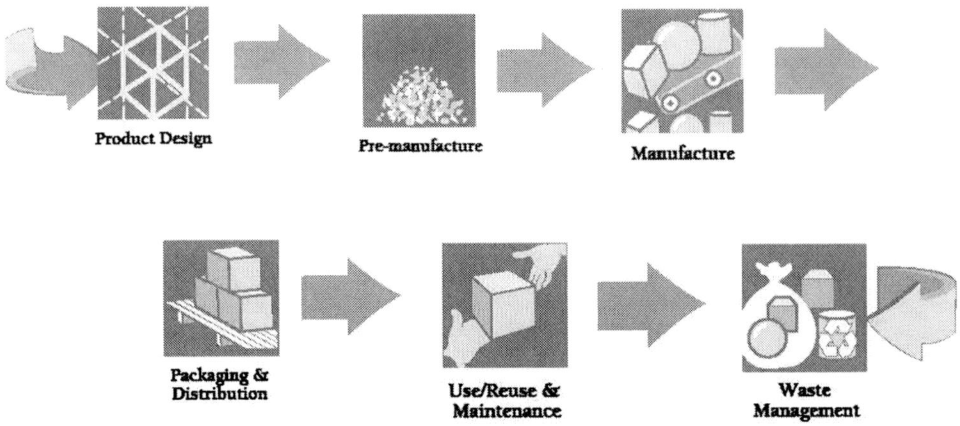

EcoLogo™
PART OF THE UL GLOBAL NETWORK

Paper life cycle – potential impacts

- Physical effects on land base – forests, soil, aquatic systems - species habitat
- Natural resource & energy consumption
- Toxicants – environmental & human health effects from use of chemicals, materials processing
- Waste – input/output considerations – recycling, diversion, disposal

6

EcoLogo™
PART OF THE UL GLOBAL NETWORK

Leadership practices

Sourcing fibre responsibly

Maximizing efficiencies and re-use possibilities through design improvements and smarter sourcing

Reducing overall waste streams and value – recycling, energy and heat recovery / cogeneration, better packaging

7

EcoLogo™
PART OF THE UL GLOBAL NETWORK

Leadership practices

Managing / reducing energy and water use

Maximizing air and water quality

Maximizing toxicological safety – avoiding adverse effects on the environment and on humans, from chemicals, materials processing

8

EcoLogo™
PART OF THE UL GLOBAL NETWORK

Business and environmental performance

Conference theme is "Boosting performance for better quality, sustainability and profitability".

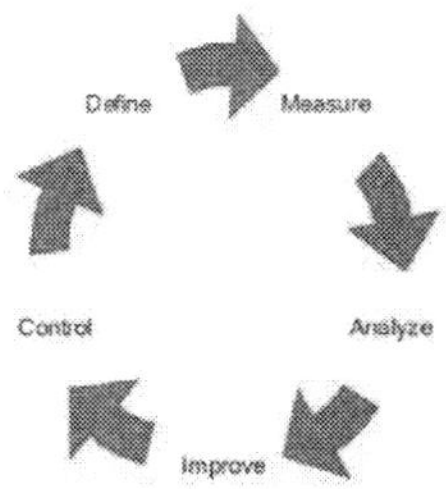

9

EcoLogo™
PART OF THE UL GLOBAL NETWORK

The landscape: considerations

- Triple bottom line
- Full cost accounting
- Corporate responsibility and accountability / transparency
- Green specifiers in purchasing
- Environmental liability

10

EcoLogo™
PART OF THE UL GLOBAL NETWORK

Connections

Decisions in one business / operational unit impact the other

Means → integrated approach and decision-making is key

Becoming more transparent to buyers that increasingly demand best practices

Ongoing improvement

11

EcoLogo™
PART OF THE UL GLOBAL NETWORK

Where do ecolabels fit in?

Market transformation tool – leaders lead, the majority follows

ISO 14024 Type 1 ecolabeling

Life cycle based, multi-attribute criteria

Credible, transparent, performance 3rd party verified

12

EcoLogo™
PART OF THE UL GLOBAL NETWORK

Updating a well-known standard

- EcoLogo is revising its standard for sanitary papers category
- New consolidated standard for all sub-categories
- Please participate: more information at www.ecologo.org

13

EcoLogo™
PART OF THE UL GLOBAL NETWORK

Communicating your successes

- Remember nothing is truly "green" except untouched nature
- Best message - environmental preferability
- Stay out of trouble – substantiate, avoid greenwash
- Multi-attribute green + traditional product claims
- Type 1 ecolabel – a trusted mark that signals 3rd party verified performance on multiple attributes

14

EcoLogo™
PART OF THE UL GLOBAL NETWORK

Environmental leadership in the tissue paper sector – summary

Performance - across the entire process

Leadership means all the important impacts not just one or two

Leadership opportunities at every life cycle stage

Business success is connected

A credible eco-label signals leading performance

15

EcoLogo™
PART OF THE UL GLOBAL NETWORK

THANK YOU. QUESTIONS?

Contact:

Susan Rutherford

srutherford@ecologo.org

Tel. 800-478-0399 ext. 229

www.ecologo.org

CERTIFIED.
SUSTAINABLE.
LEADERSHIP.

SCA
c/o Life

How customers can leverage your Sustainability

Don Lewis
President, SCA Americas

Tissue World Americas - March 21, 2012 - Don Lewis

3rd largest hygiene products company in the world

Sales in 100 countries

SCA
c/o Life

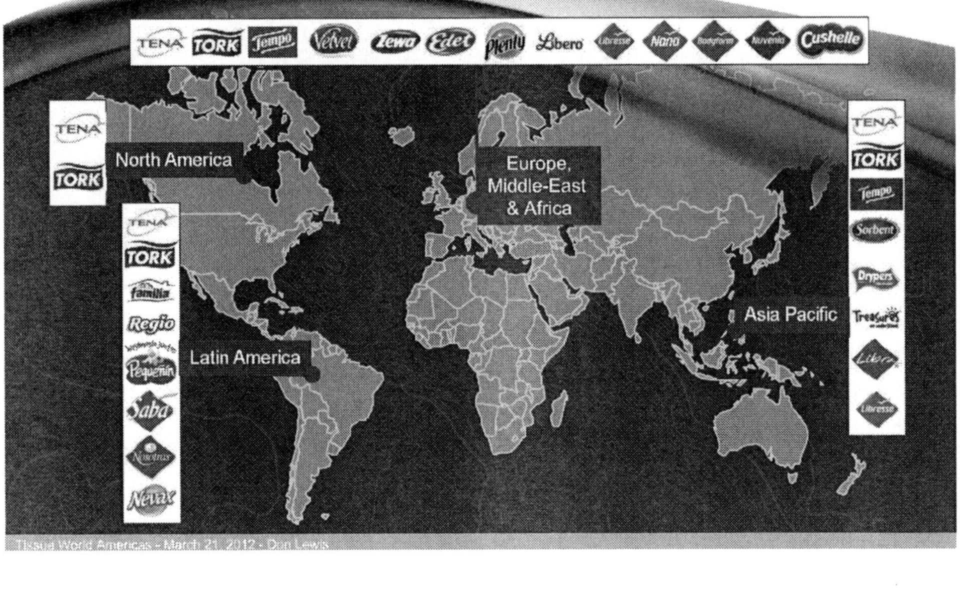

Making a difference – The role of SCA in society

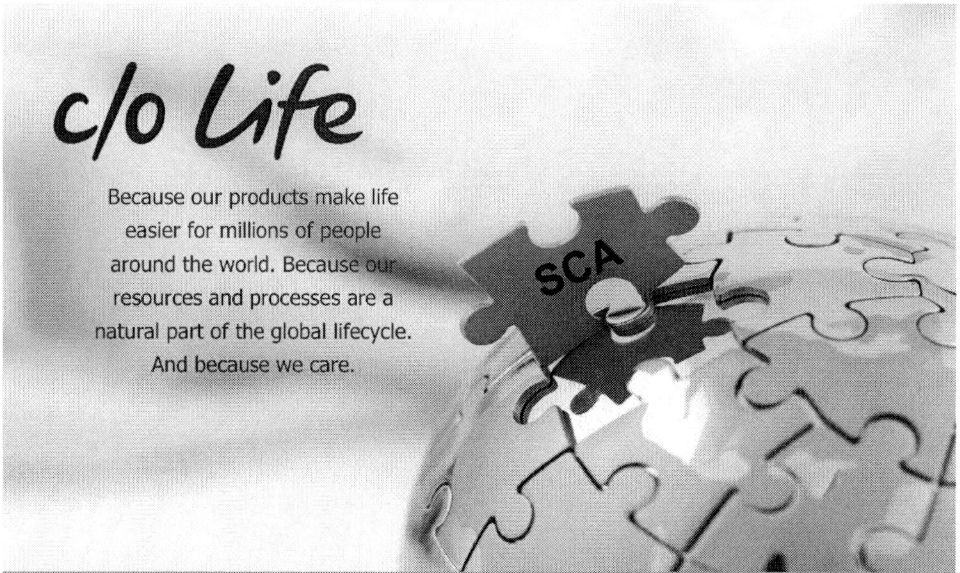

Tissue World Americas - March 21, 2012 - Don Lewis

Our unique DNA
Embedded in our Corporate Brand

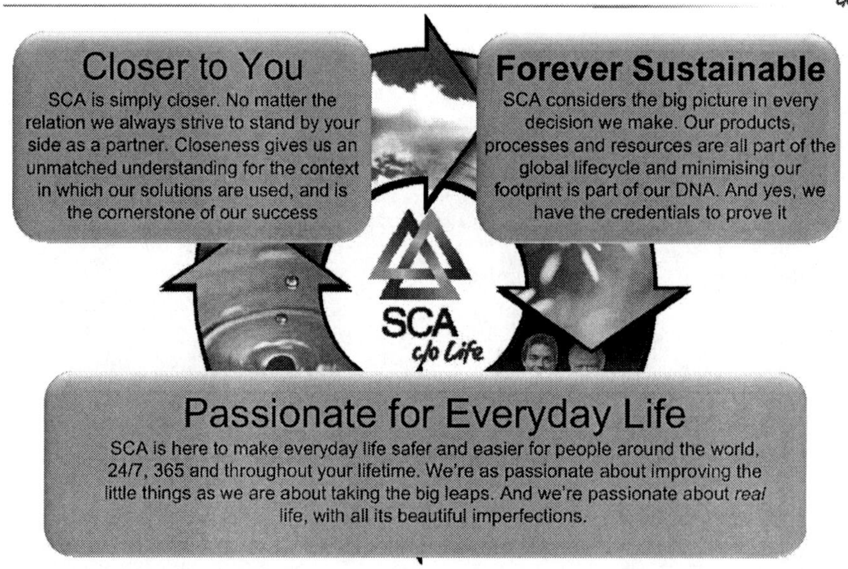

Tissue World Americas - March 21, 2012 - Don Lewis 4

Sustainability Focus

With an 80 year history, it's in our DNA

Tissue World Americas - March 21, 2012 - Don Lewis

Sustainability Focus

With an 80 year history, it's in our DNA

Tissue World Americas - March 21, 2012 - Don Lewis

Sustainability Focus

With an 80 year history, it's in our DNA

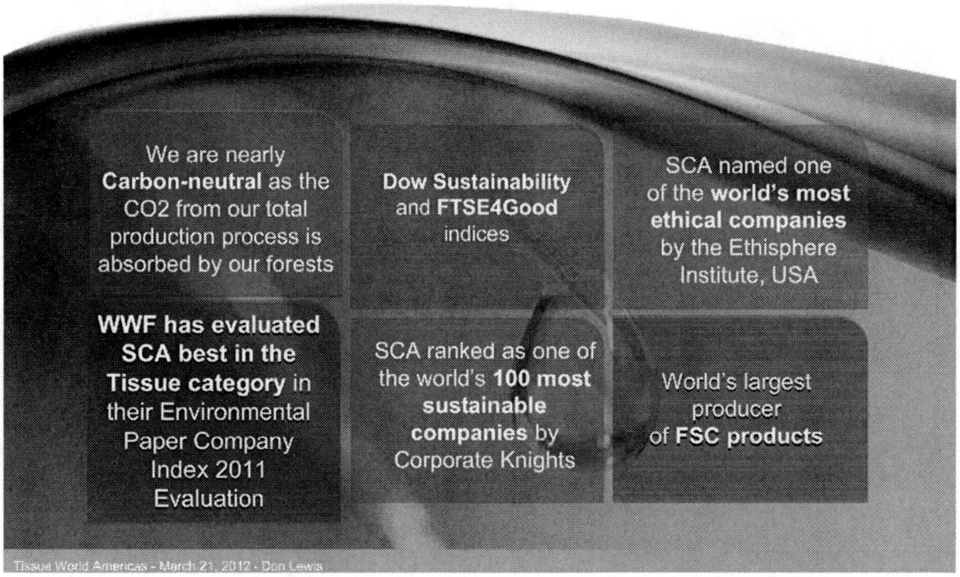

SCA's Tork® professional hygiene business in North America

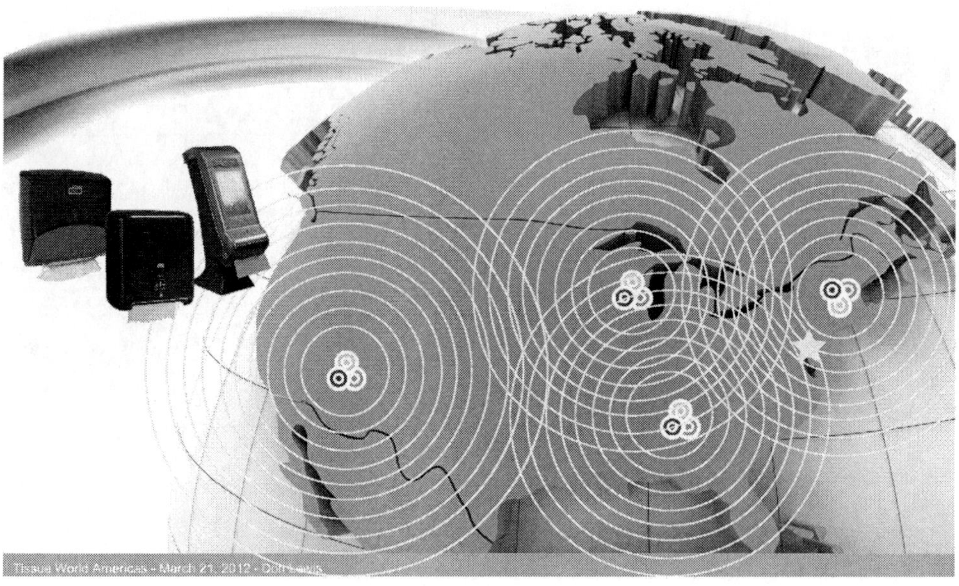

TORK Half the napkins in U.S. Restaurants

SCA c/o Life

In North America, 100% Recycled

SCA c/o Life

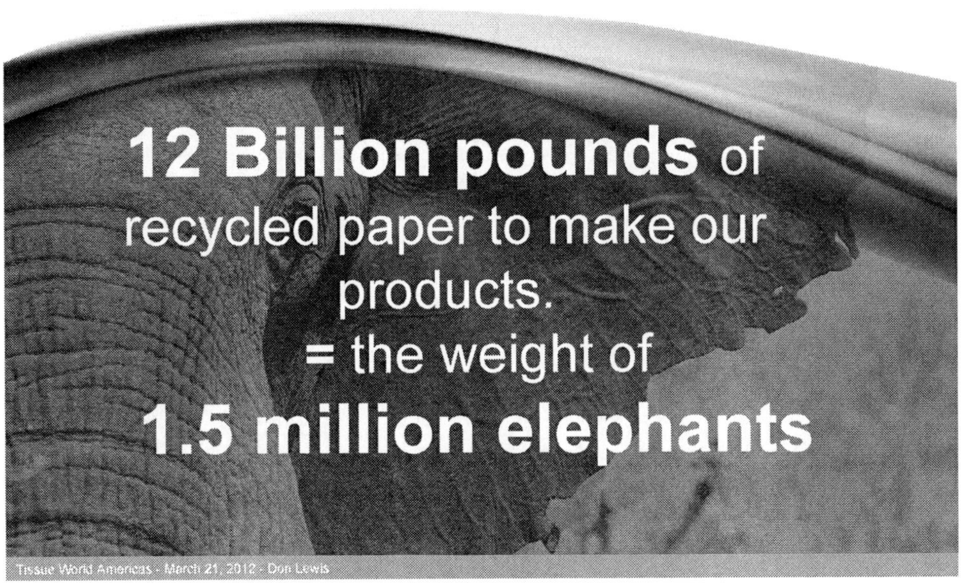

What today's consumers say about sustainability in restaurants

When asked how the economy impacted green spending, **75% say they are on the same level**

Primary reason for green purchases is to **help the environment** – that's the motivator

Half will choose restaurants because they are green if all other things are equal

About half expect to **pay a little more** in this type of restaurant – up to $3 on a bill of $20

Best way to communicate green is
1) on the menu
2) on internal signs
3) and on the web

Results from Harris Interactive survey, April 2011

Tissue World Americas - March 21, 2012 - Don Lewis

Green Facilities, SCA's and Yours

Tissue World Americas - March 21, 2012 - Don Lewis

SCA
c/o Life

c/o Life Because our products make life easier for you and for millions of people around the world. Because our resources and the way we work are natural parts of the global lifecycle. And because we care.

Market Developments and Trends in North and South America

Esko Uutela, Principal, Tissue, RISI

March 21-23, 2012, The Miami Beach Convention Center, Miami, Florida, USA

Structure of Presentation

- Global Tissue Market Trends
- World Demand Forecast and Supply Outlook
- North American Trends
- North American Market Outlook
- Latin American Trends
- Latin American Market Outlook
- Conclusions

GLOBAL OUTLOOK

RISI

World Tissue Consumption 2011

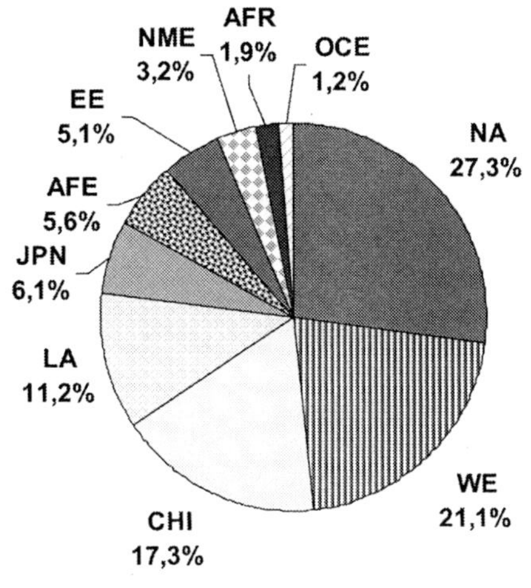

Global consumption 30.4 million tonnes (preliminary figure)

RISI

World Tissue Consumption 1991-2011

Growth increasingly concentrating on emerging markets

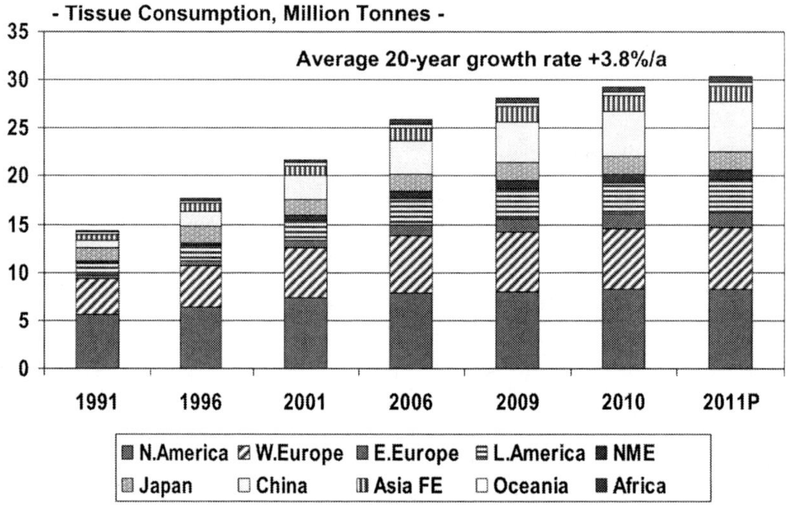

Past Long-Term Growth Rate of World Tissue Consumption

Back on normal growth track after small break due to 2009 recession

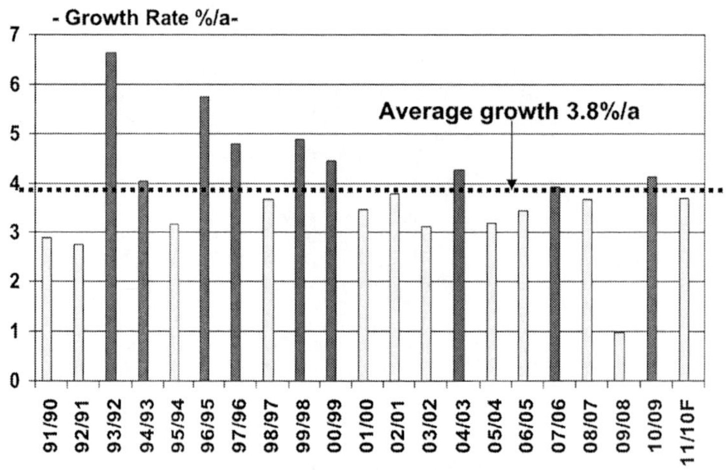

Per Capita Consumption of Tissue by World Region in 2010

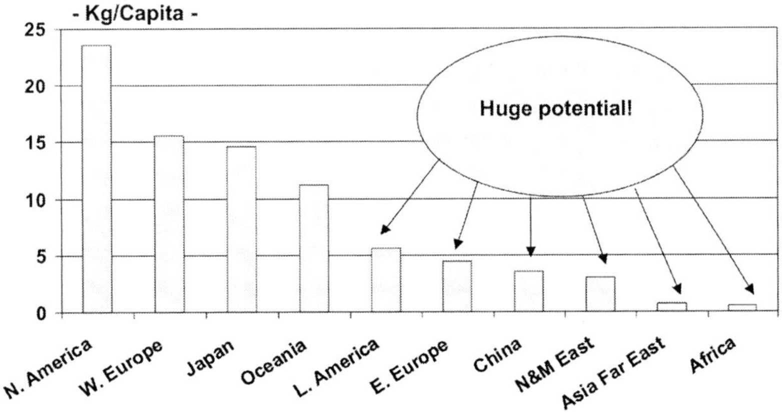

RISI

Expected Growth Rate of World Tissue Consumption, 2006-2021

Strong global demand growth expected to continue

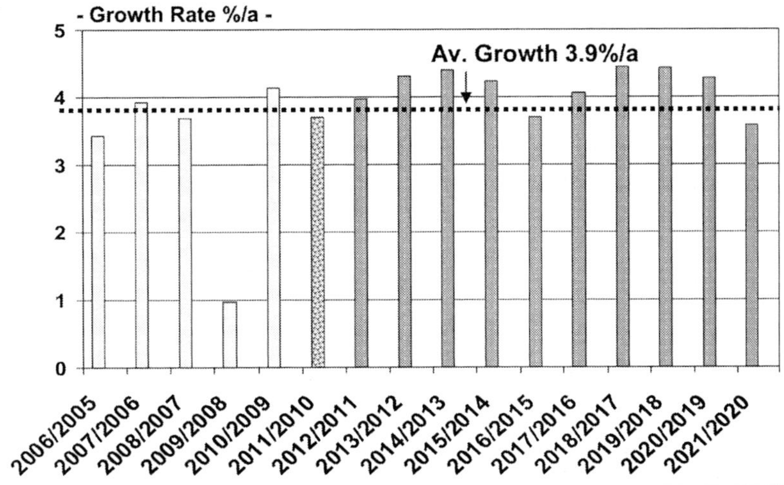

RISI

Expected Regional Volume Growth of Tissue Consumption 2011-2021

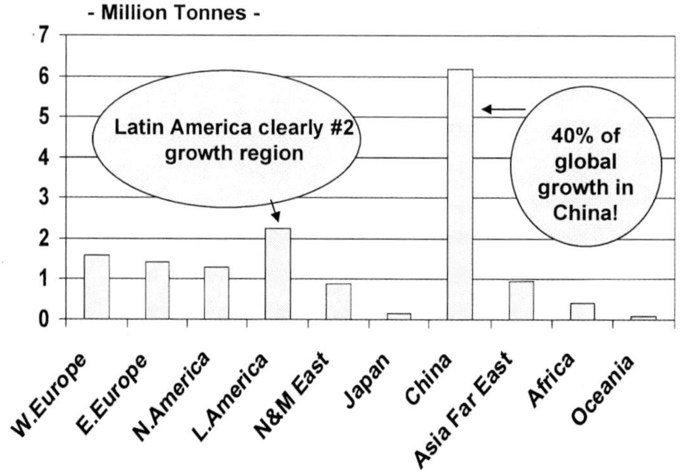

Total expected global growth 15.2 million tonnes

World Tissue Capacity Change*,1996-2016

Will overcapacity threaten with too many projects coming on stream?

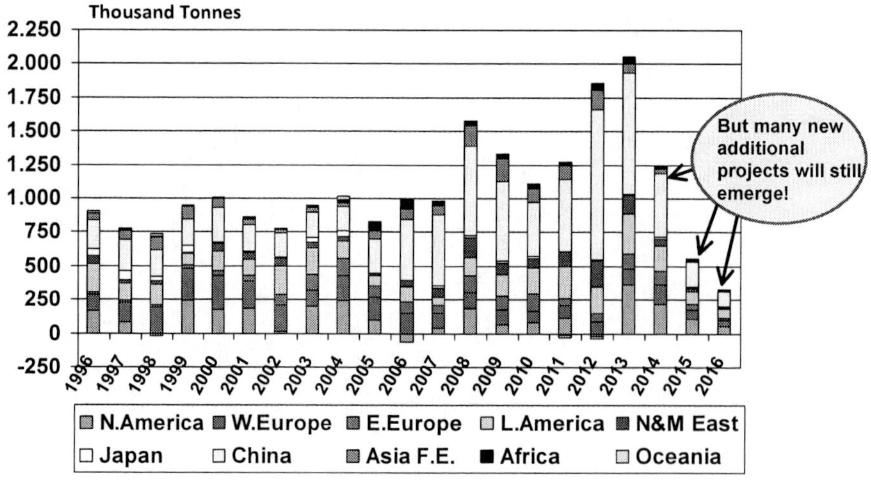

** Based on committed projects only*

NORTH AMERICAN TRENDS

North American Tissue Market Growth Has Slowed Down in the Past Five to Six Years

Focus on product value and lightweighting cutting volume growth

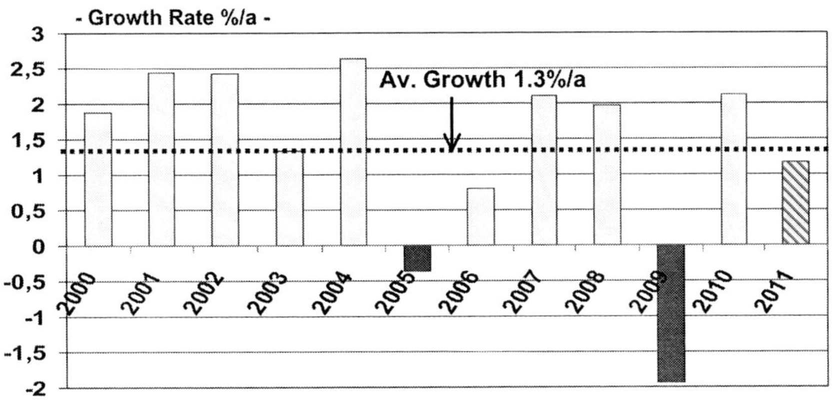

Quality Upgrading the Dominating Feature

- Major trend toward ultra and premium quality products in the consumer tissue sector continues -- retailers led by Wal-Mart increasingly require high quality products similar to the leading brands -- household towels and toilet tissue in key position.
- Three-ply toilet tissue, introduced first in Canada and later on in the US market by Georgia-Pacific, has had some success -- it can offer good quality but now with continuing high raw material costs it is not attractive for producers as the product simply contains too much fiber to keep margins.
- New structured/textured tissue making technologies have made their breakthrough with one Voith ATMOS system already working in Canada and at least three more projects sold in North America.
- Quality upgrading rapidly also hitting the AfH tissue sector with the first new technology AfH PM projects announced -- allowing recovered paper use so that environmental certification not a major problem (unlike TAD).
- Some price increases achieved, but principally they do not cover any more than just raw material price escalation.

RISI

Retailer Labels Showing Reverse Trends in the USA and Canada

Gradually growing share in the USA, brands regaining through promotional activities and increasing US imports in Canada

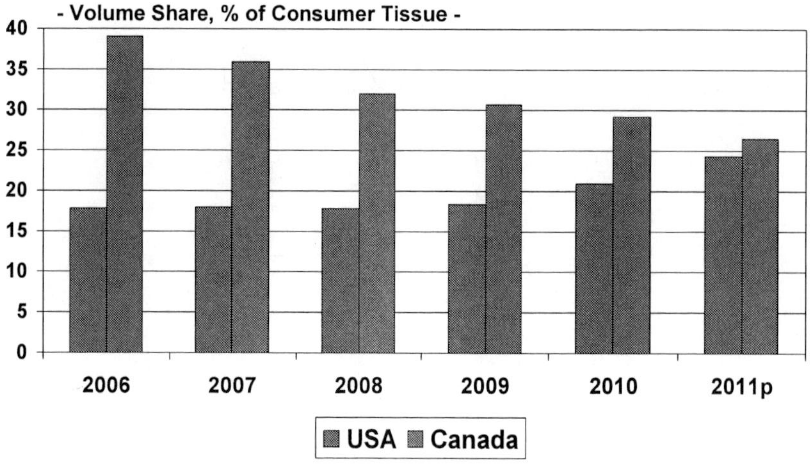

RISI

Sustainability Issues Gaining Importance

- **Environmental certification getting increasingly important in the AfH business -- several large buyers do not purchase products without environmental certificates from renowned organizations.**
- **Sustainability has been a prime concern by NGOs such as WWF and Greenpeace also in the consumer tissue sector -- the recent campaigning against Oasis Brands/Mercury Paper using base paper from Indonesia because of APP's claimed rain forest and tiger habitat destruction there -- has resulted in stopping sales of these brands by a number (but not all) of retailers.**
- **Kimberly-Clark had long-lasting debate with Greenpeace concerning the use of pulp made from wood grown in ancient forests in Canada but finally managed to convince the NGO and has thereafter been promoting a lot with its brands that meet the strictest environmental certification criteria set by FSC (Forest Stewardship Council).**
- **Another example is the recent announcement of Cascades Tissue for the launch of "Cascades Moka" AfH toilet tissue later this year -- after having the "Moka" successful napkin line in the market since 2004. With "beige is the new green" Cascades is trying something long considered the unthinkable: brown (beige) toilet paper.**

RISI

Trends in Recovered Paper Utilization Rates in Tissue Manufacture in Selected Countries

Enhanced sustainabilty does not necessarily mean more RCF use

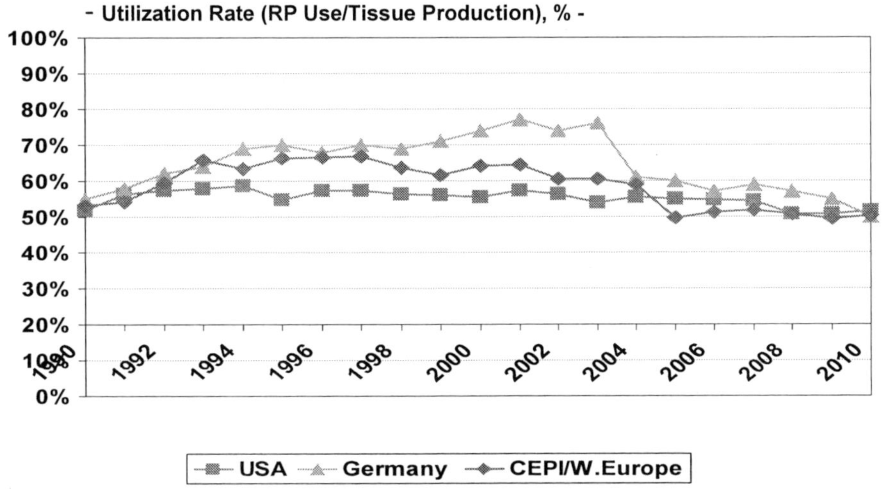

RISI

NORTH AMERICAN MARKET OUTLOOK

RISI

No Return to High Growth Rates Expected in the North American Tissue Market

Population growth contributing two-thirds to market growth

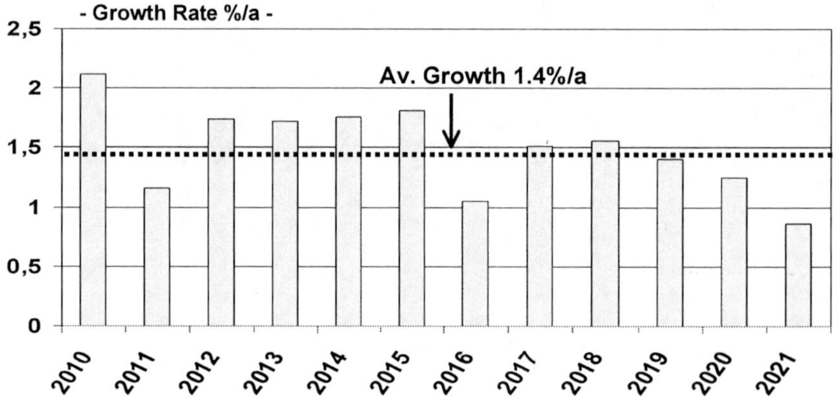

RISI

Quality Upgrading Driving New Investments

- New investments are needed in the consumer tissue sector as buyers target the highest quality category. Now also in the retailer label segment, successful suppliers need the capability of offering ultra quality products -- this means the need to use TAD or alternative new, advanced tissue making technologies.
- At least four new TAD PMs will come on stream in 2012-2013, plus about the same number of projects involving alternative technologies, led by Voith's ATMOS concept but not ignoring G-P's self-developed "f-TAD" technology that is still largely wrapped in secrecy for many issues, including the quality level to be reached.
- But what will happen with the mills relying on the conventional tissue technology? It is too early to draw a final conclusion, but most likely requirements for higher quality will make it difficult to compete -- capacity closures are likely to follow.
- The North American tissue industry is in an interesting phase, moving toward technology that decreasingly resembles that of traditional papermaking. We are curious to see what the next moves by the main players will be -- will they be someting totally new?

RISI

Major Tissue Capacity Changes (>10,000 short tons/year) in North America, 2010-2011

2010:	
• Marcal Manufacturing, Elmwood Park, NJ	-15,000 t/a
• Irving Tissue, Fort Edward, NY (TAD PM)	35,000 t/a
• Procter & Gamble, Box Elder, UT (TAD PM)	80,000 t/a
	100,000 t/a
2011:	
• Blue Heron, Oregon City, OR 32,000 t/a	-
• First Quality Tissue, Anderson, SC (TAD PM)	70,000 t/a
• Kimberly-Clark Corp., Everett, WA	-208,000 t/a
	-170,000 t/a

Note: temporarily idled PMs not included

RISI

Major Tissue Capacity Changes (>10,000 short tons/year) in North America, 2012-2013

2012:

• Georgia-Pacific, Crossett, AR (f-TAD)	25,000 t/a*
• First Quality Tissue, Anderson, SC (TAD PM)	70,000 t/a
• Gorham Paper and Tissue, Gorham, NH	35,000 t/a
• Florelle Tissue, Brownville, NY	15,000 t/a*
• Tak Investments, Franklin, VA (PM modification)	70,000 t/a
• Clearwater Paper, Shelby, NC (TAD PM)	70,000 t/a
• Georgia-Pacific, Port Hudson, LA (f-TAD)	35,000 t/a*
	320,000 t/a

2013:

• K.T.G. (USA)Kruger, Memphis, TN (TAD PM)	66,000 t/a
• Wausau Paper, Harrodsburg, KY (ATMOS)	75,000 t/a
• Confidential (new ATMOS PM)	75,000 t/a*
• South Carolina Tissue, Snelling, Barnwell, SC	33,000 t/a
• Procter & Gamble, Box Elder, UT (TAD PM)	80,000 t/a
*Note: ATMOS for SCA Menasha PM3 in 2013, no capacity change * = estimated*	329,000 t/a

Growth of Ultra Quality Tissue Capacity in North America 2008-2013

Capacity share around 37% in 2013 based on committed projects

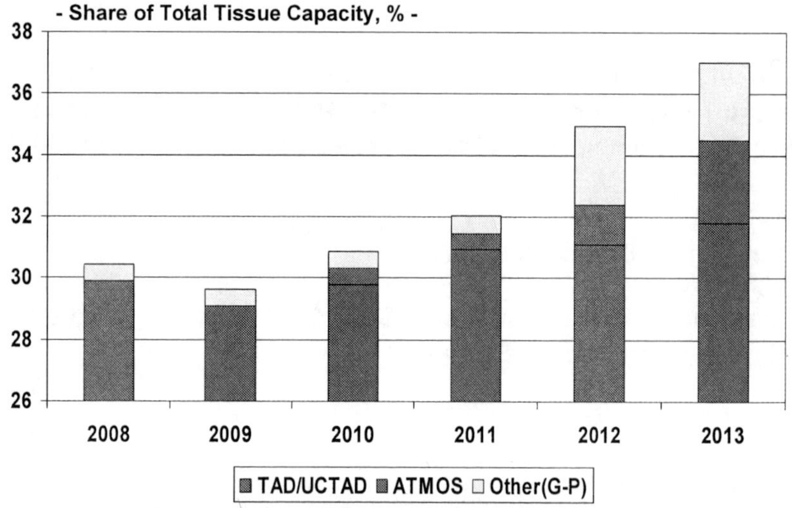

Monthly Utilization Rate in the US Tissue Industry, Parent Rolls Production/Capacity

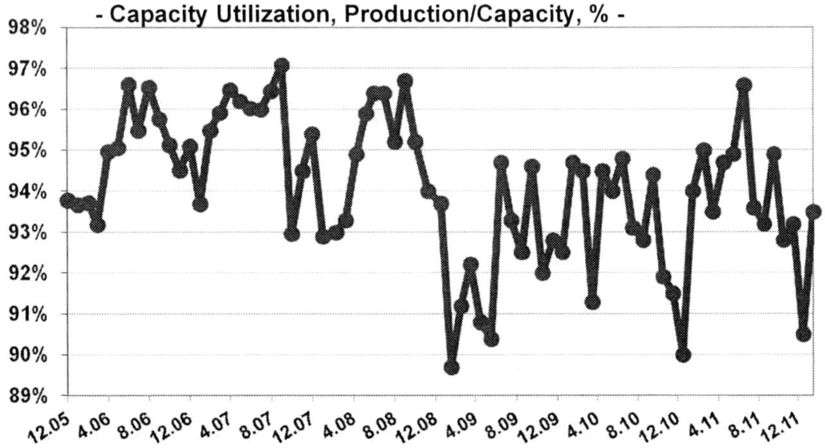

Net Capacity Change* and Average Capacity Utilization in the North American Tissue Industry Until 2016

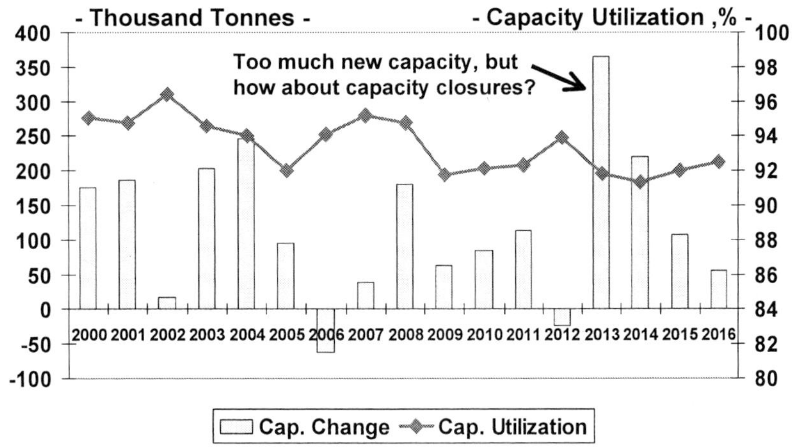

* *based on committed projects only*

RISI

LATIN AMERICAN TRENDS

Main Latin American Tissue Markets, 2011

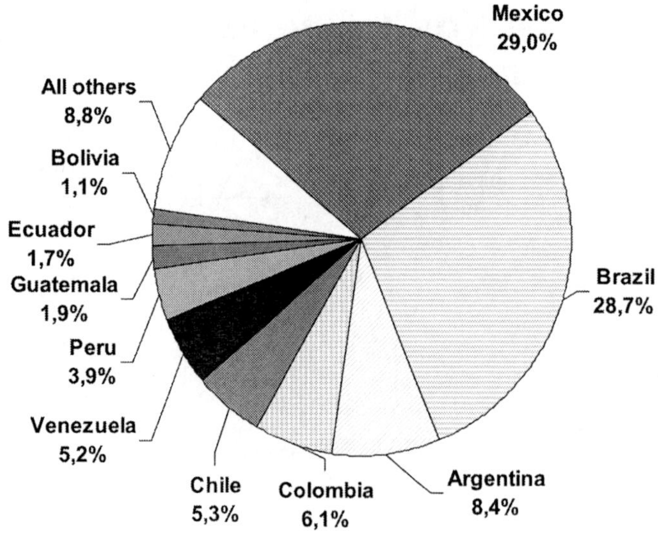

Total market size: 3.4 million tonnes

Annual Growth Rates of Latin American Tissue Market since 2000

Market growth shows major variation from year to year

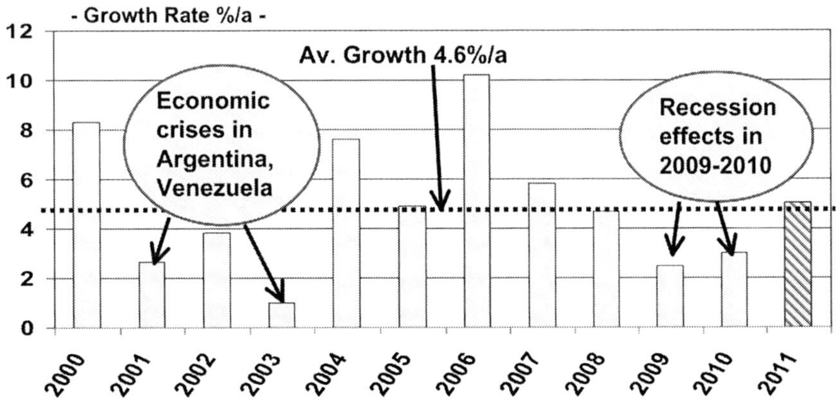

Per Capita Tissue Consumption in the Main Latin American Markets, 2010

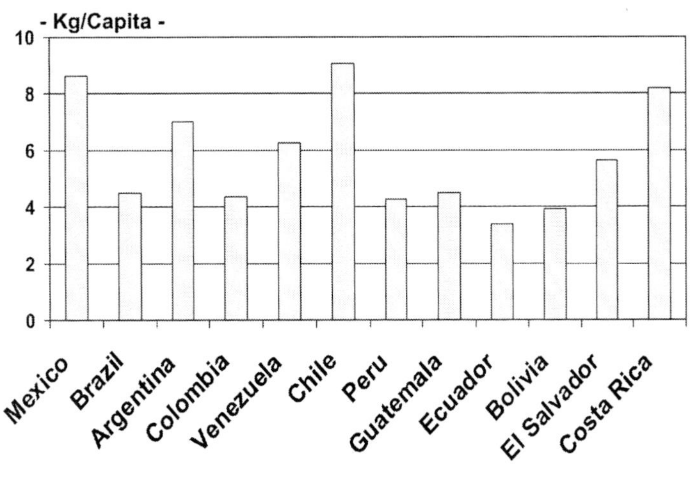

Growth Rate of Latin American Tissue Consumption by Main Country, 2001-2011

Very rapid growth in some smaller countries

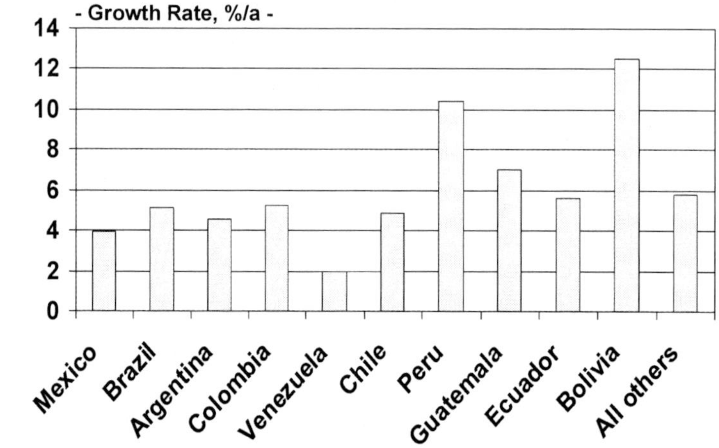

Development of Latin American Foreign Trade Balance in Tissue, 2000-2011

Exports suffering for competitive reasons, imports from Asia a threat but trade balance showing improvement in 2009-2011

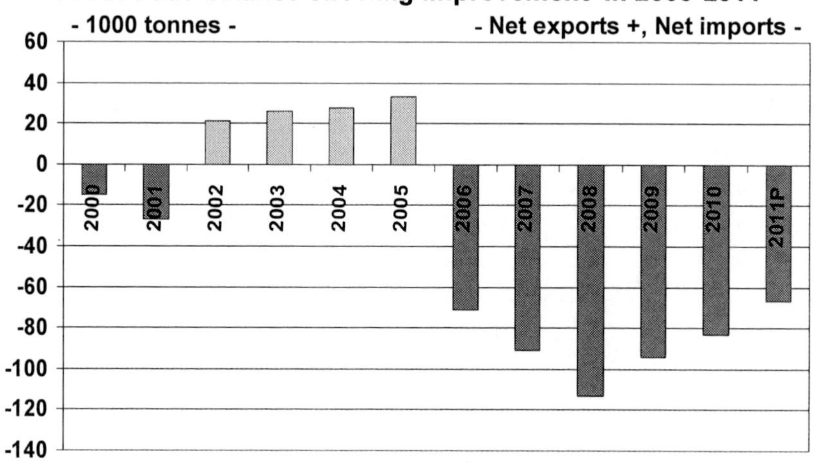

LATIN AMERICAN MARKET OUTLOOK

RISI

Latin American Tissue Market in a Very Positive Expansion Phase

Growth slightly acceleratring due to expected strong growth in Brazil

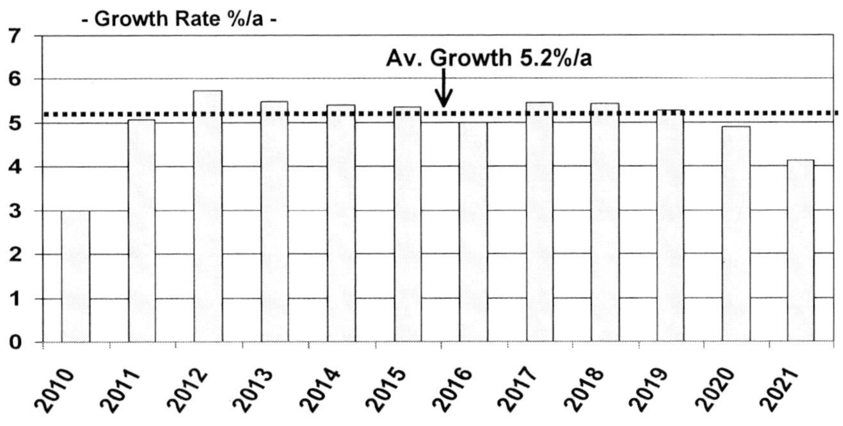

RISI

Major Tissue Capacity Changes (>10,000 metric tons/year) in Latin America, 2010

2010:

- K-C de Centroamerica, Sitio del Nino, El Salvador — 33,000 t/a
- Papelera Samseng, Pilar, Buenos Aires, Argentina — 35,000 t/a
- Canoinhas de Papel, Canoinhas, SC, Brazil — 22,000 t/a
- Absormex (CMPC), Altamira, Mexico — 60,000 t/a
- Protisa Colombia (CMPC), Bogota, Colombia — 30,000 t/a
- Cesar Iglesias, San Pedro, Dominican Republic — 17,000 t/a
- SCA, Ciudad Sahagun, Mexico — 60,000 t/a
- Mili, Tres Barras, SC, Brazil — 33,000 t/a

290,000 t/a

RISI

Major Tissue Capacity Changes (>10,000 metric tons/year) in Latin America, 2011

2011:

- Papel Tangara, Pinheiro Preto, SC, Brazil — 10,000 t/a
- SCA, Ecatepec, Est. do Mexico, Mexico — -70,000 t/a
- K-C Argentina, Bernal, Buenos Aires, Argentina — 10,000 t/a
- Melhoramentos (CMPC), Caieiras, SP, Brazil — 54,000 t/a
- Protisa Peru (CMPC), Santa Anita, Lima, Peru — 15,000 t/a
- INCASA, Quito, Ecuador (???) — 11,000 t/a

30,000 t/a

RISI

Major Tissue Capacity Changes (>10,000 metric tons/year) in Latin America, 2012

2012:

- **SEPAC, Mallet, PR, Brazil** — **33,000 t/a**
- **IPEC, Itaocara City, RJ, Brazil** — **12,000 t/a**
- **Mili, Tres Barras, SC, Brazil** — **10,000 t/a**
- **Dama-Pel, Guarulhos, SP, Brazil** — **25,000 t/a**
- **Productos Familia (SCA), Cajica, Bogota, Colombia** — **35,000 t/a**
- **IPEL, Indaial, SC, Brazil** — **27,000 t/a**
- **Papelera Samseng, Sao Paulo state, Brazil** — **35,000 t/a**
- **Carta Goias, Anapolis, GO, Brazil** — **30,000 t/a**
- **Inpaecsa, Guayaquil, Ecuador** — **15,000 t/a**

222,000 t/a

RISI

Major Tissue Capacity Changes (>10,000 metric tons/year) in Latin America, 2013

2013:

- **Celulosa Campana, Lima, Buenos Aires, Argentina** — **30,000 t/a**
- **BIPACEL, Manaus, AM, Brazil** — **17,000 t/a**
- **Papelera Samseng, Sao Paulo state, Brazil** — **35,000 t/a**
- **CMPC Tissue, Talagante, Santiago, Chile** **60,000 t/a**
- **Mili, Tres Barras, SC, Brazil** — **70,000 t/a**
- **FAPSA, Mexicali, Baja California Norte, Mexico** — **30,000 t/a**

242,000 t/a

Note: K-C Brasil is also likely to build a new tissue PM around 2013, probably at its Mogi das Cruzes mill in Sao Paulo, but details are still missing and so the plan is not yet included on the list of committed projects.

RISI

Net Capacity Change* and Average Tissue Capacity Utilization in Latin America

Capacity closures and project delays could improve the outlook

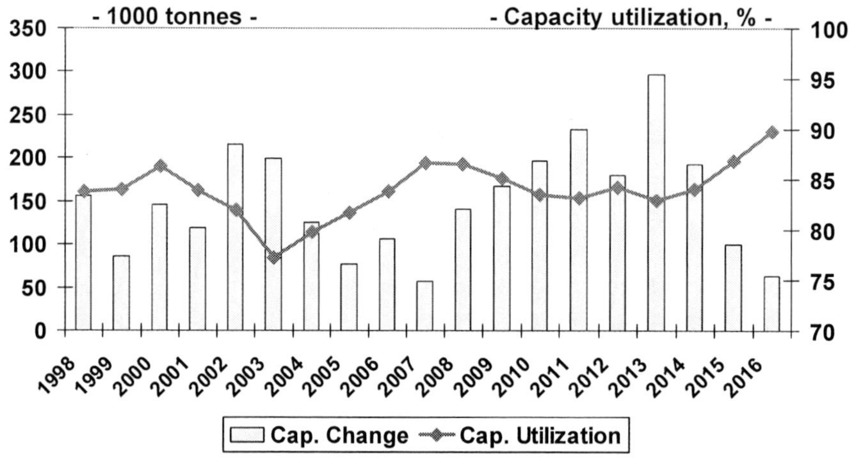

* based on committed projects only

RISI

Conclusions

- Globally, the tissue business is focusing its expansion in emerging markets -- China in particular, followed by Latin America. Growth in these regions will outstrip the growth in North America or Western Europe. The focus of the global tissue business expansion is clearly moving to the emerging countries.
- In North America, the new capacity already in the pipeline and the additionally planned largely concentrates on the highest quality segments: the focus is more on selling performance (or product area) than product weight. Older capacity closures are expected. Sustainability is developing into a major issue.
- There are good long-term growth prospects for tissue in Latin America, with Brazil showing the largest expansion potential, but also Mexico, Argentina, Colombia, Peru, Chile and many smaller markets offer new opportunities for tissue suppliers.
- Brazil's positive tissue outlook is due to the large population base and strong invasion of economic welfare into less developed regions in Brazil northeast and inland, along with continuing progress in product penetration and quality in the traditional consumption centre south.

RISI

Thank you for your attention!

For more information:

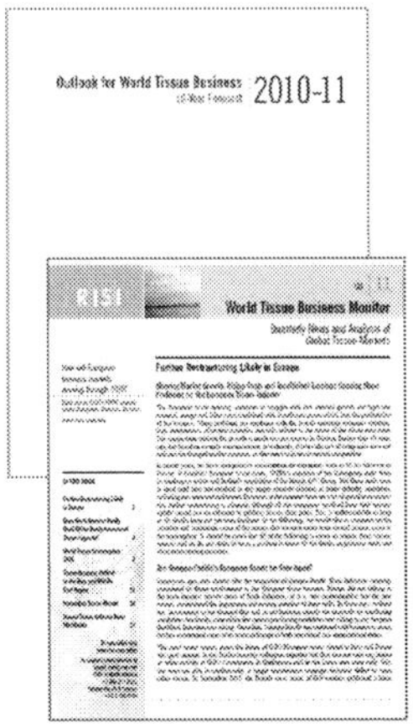

Outlook for World Tissue Business
www.risi.com/owtb

World Tissue Business Monitor
www.risi.com/wtbm

U.S. Tissue Monthly Data
www.risi.com/ustm

RISI

Fiber supply – where will fiber for tissue making come from?

PÖYRY

Tissue World Americas
Miami Beach, FL
March 21, 2012
By Soile Kilpi

Storyline

- Fiber for tissue
  - Sizing the usage
  - Fiber portfolio; regions and producers
- Tissue making and:
  - Integration to pulp
  - Recovered fiber
  - Market pulp
- Future perspective
  - Tissue production outlook
  - Fiber for tissue 2020
- Wrap up

FIBER FOR TISSUE; TYPES, REGIONS, PRODUCERS

PÖYRY

Global tissue market is 28 million tons

Product categories

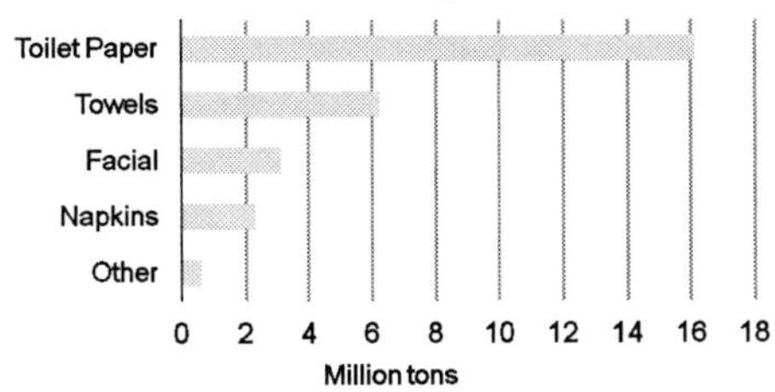

Market segments

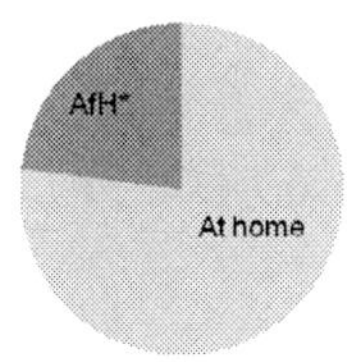

Regional consumption

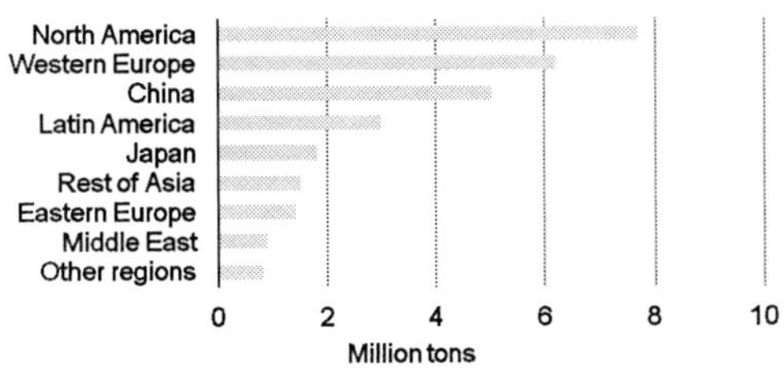

*AfH = away from home

PÖYRY

Tissue industry is a major user of different fibers

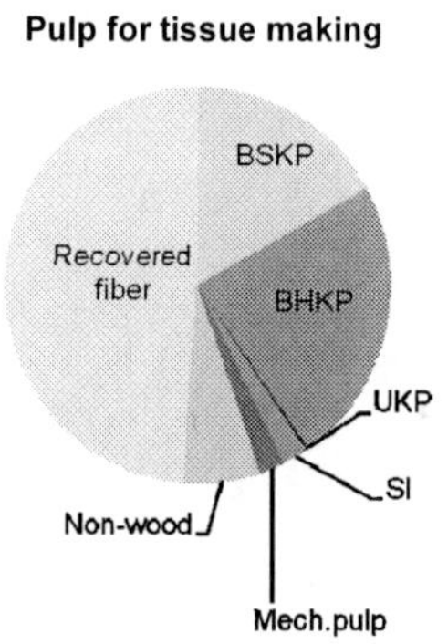

- Recovered fiber (RCF) and bleached chemical sulphate pulp dominate the fiber supply
- Recovered fiber consists of different grades; Sorted office paper (SOP), mixed paper, old news & mags, board, old corrugated containers (OCC), ...
- Pulp is bleached hardwood (euca, maple, mixed, etc.) and/or softwood (pine, spruce, etc.). Each has its own characteristics in terms of strength, softness, ...
- Non-wood pulp (based on bagasse, bamboo, reed/straw) is a key raw material in China
- Mechanical pulp is mainly high-yield BCTMP.

North America and Europe have similar fiber bases for tissue making. China has non-wood fiber.

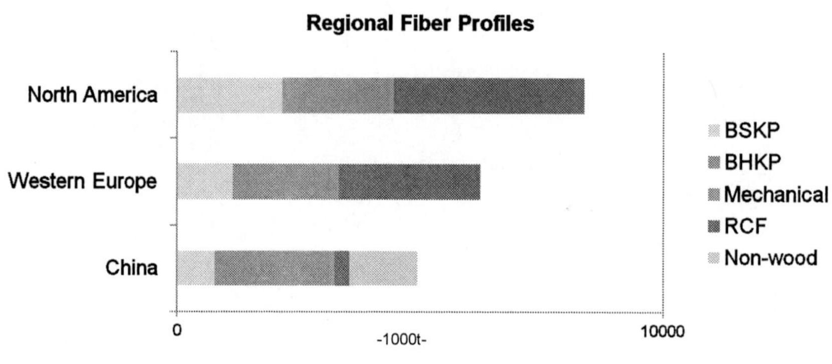

Fiber choices vary by producer and production platform

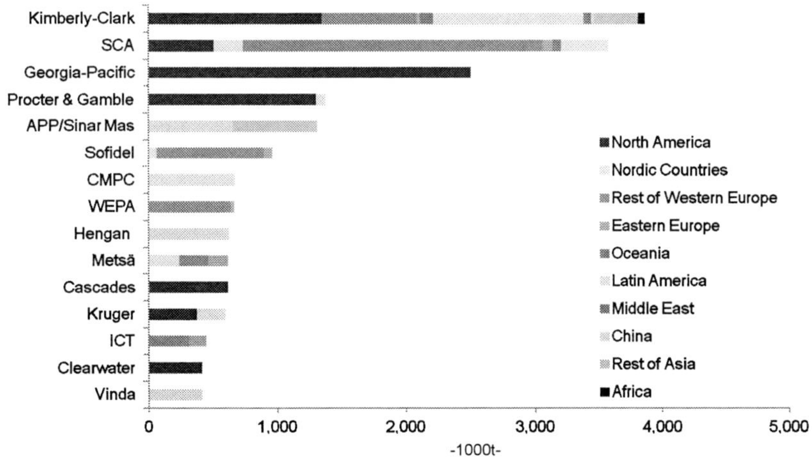

Top producers have diverse fiber strategies

Driven by market segment/product quality focus (At-home and Away-from-Home), scale and technology

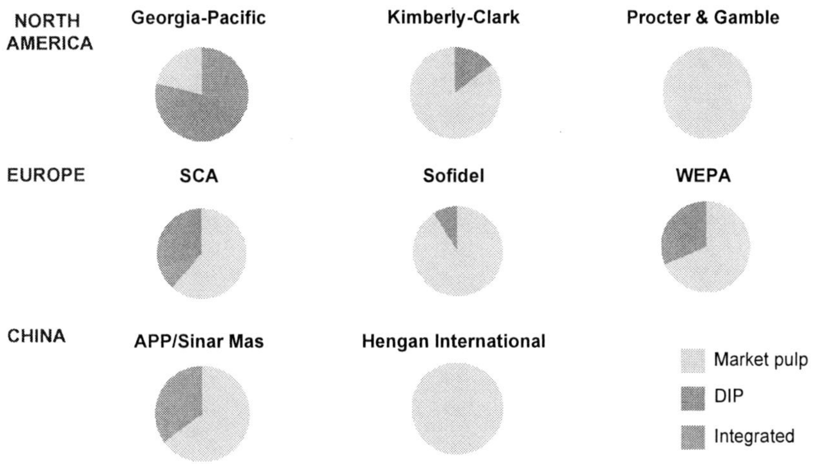

Different fiber suited for different segments – US example

- At-home products are mainly based on virgin fiber
- Away-from-home products based largely on recycled fiber
- Most large producers serve both segments; and have deinking lines, use market pulp and / or integrated pulp
- Certifications and sustainability are important to consumers and retailers

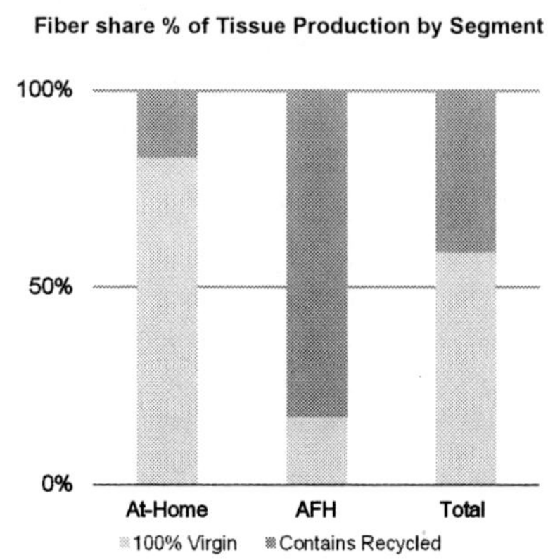

INTEGRATED PULP

Tissue making based on integrated pulp supply accounts for 10-15 % globally

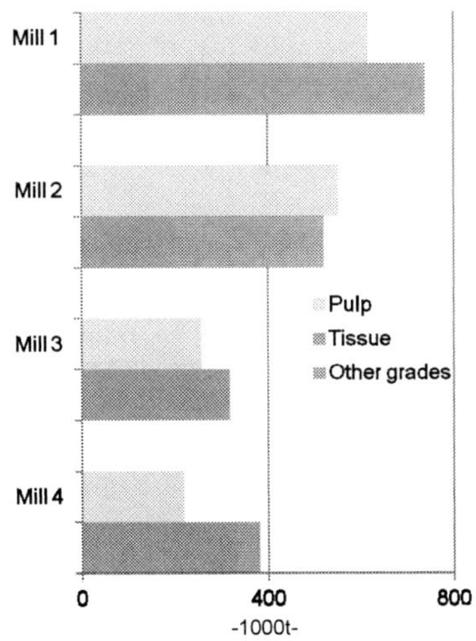

Mill concepts vary – as examples show:

- Larger scale >500, 000 t/a pulp mill and a number of PMs; producing tissue and other grades e.g.
  - UFS
  - Unbleached kraftliner
  - Sack paper
  - SBS
  - MG paper
  - Coated paper
- Pulp production for tissue manufacture mainly; smaller scale of <300,000 t/a pulp
- Potential for increased tissue supply at the large scale mills? To manage declining segments (UFS, coated paper) in NA and Europe? Or location too remote?

PÖYRY

New tissue machine investments at mills leveraging access to own pulp

Small portion of new tissue investments will be at sites with opportunities to utilize own pulp and reduce market pulp output.

Example Situation	Pulp capacity -1000t/a-	Region	Comments
Market DIP mill	65	North America	Focus on tissue growth, add a second PM, integrate
Market BHKP mill	150	Africa	Diversify, forward integrate to tissue, scale uncompetitive in market BHKP
Market non-wood pulp mill	50	China	Additional tissue PM, reduce market pulp
Large UFS integrate with market pulp	700	China	Diversify into tissue, utilize pulp more internally
Large market pulp mill, one board PM	1800	China	Diversify into tissue, utilize pulp more internally
Tissue mill with market pulp	200	Russia	Grow still in tissue, use more of own pulp

PÖYRY

Non-wood in Chinese tissue manufacture

Non-wood pulp is a key feature of the Chinese tissue industry
Good enough quality! Tree free! Readily available residue

- Non-wood fiber – straw, reed, bamboo, bagasse account for 20-25% of tissue making fiber
- This fiber base is domestic and readily available – but type of species vary by region
- Non-wood pulp has maintained its level of output although its share has clearly declined
- Closures taking place at small straw pulp based mills due to environmental reasons – but development of other non-wood pulp has gained momentum
- Planned new mix is 1/3 straw, 1/3 bamboo, 1/3 reed/bagasse

- Non-wood will maintain volume, but not offer enough growth – what are the implications to Chinese tissue industry?
  - Rely more on imported pulp?
  - More recovered paper?
  - More BCTMP?

RECOVERED FIBER

US case - Tissue production growth has outpaced recovered paper utilization

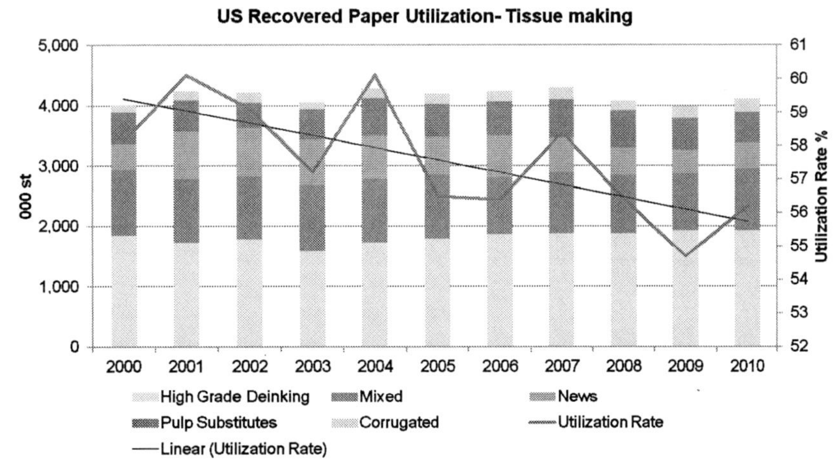

Source: AFPA

North America - Deinking lines mainly at AfH tissue mills

Consolidated set of producers lead the RCP based tissue making. Access to own collection or long term relationships with SOP, etc. suppliers. Market DIP sold to different segments – also tissue.

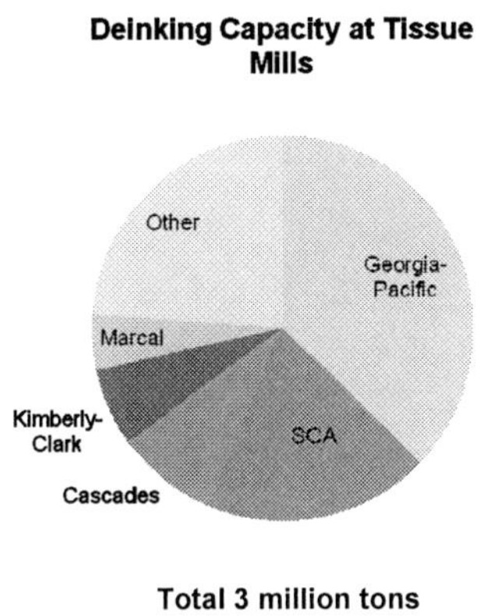

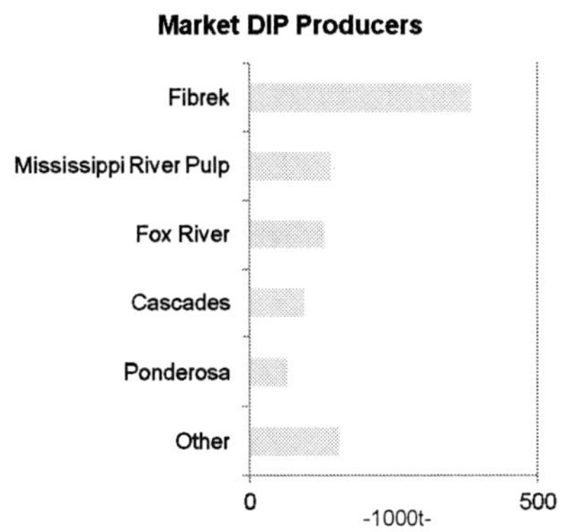

Europe - Few large tissue producers using recovered fiber and large number of smaller users

Deinking Capacity at European Tissue Mills

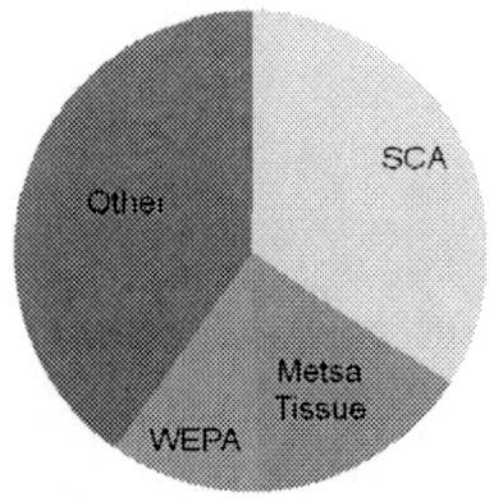

Total: 2.5 million tons

Market DIP producers

- A limited number of companies selling DIP market pulp (some tissue mills, and printing and writing mills)
- Not a growing supply base

MARKET PULP

Tissue industry accounts for 25% of BHKP and 19% of BSKP usage – and select market pulp suppliers are focused on tissue

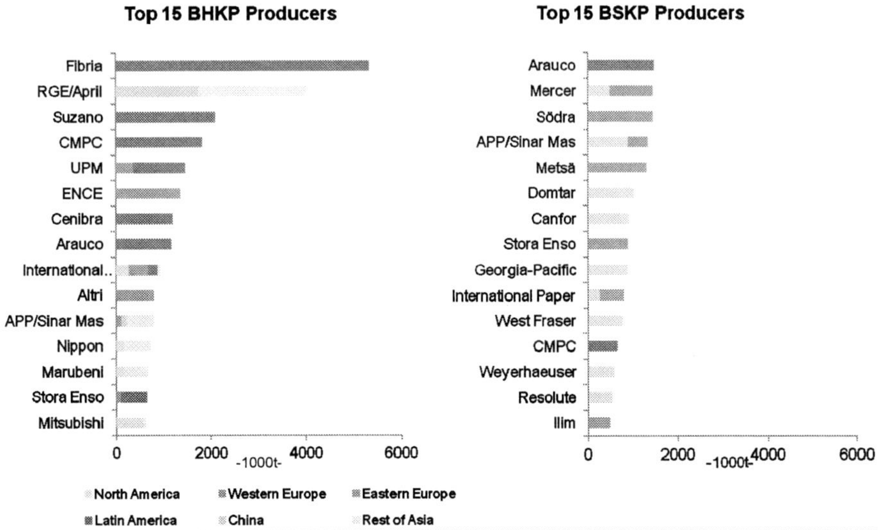

Tissue makers and market pulp

Region	Market pulp consumption for tissue making	Buyer structure	Outlook
North America	Nearly 3 millions tons	Consolidate set of big pulp buyers	New tissue machine investments based on market pulp, (deinked pulp) Certified pulp, sustainability Potential to replace recycled fiber
Europe	Over 3 millions tons	Fragmented set of buyers and smaller mills	New tissue machine investments based on market pulp, (deinked pulp) Certified pulp, sustainability Quality creep
Asia	Nearly 4 million tons	Diversified set of buyers	Rapidly growing tissue industry Most investments based on market pulp Potential to replace non-wood pulp

Market BCTMP supply is for diverse end uses, tissue being one of them

Top 10 BCTMP Market Pulp Producers

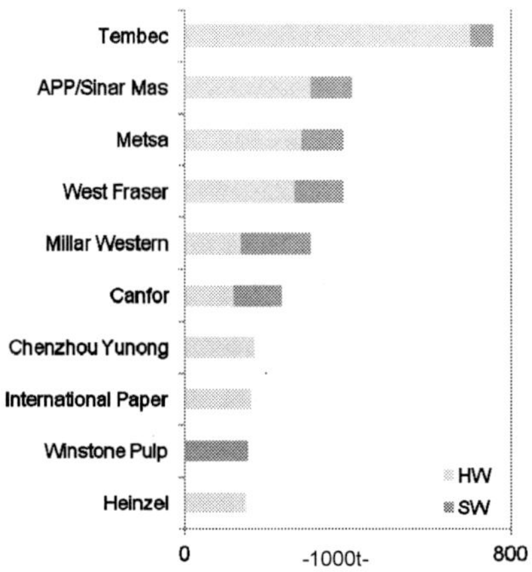

End Uses of BCTMP

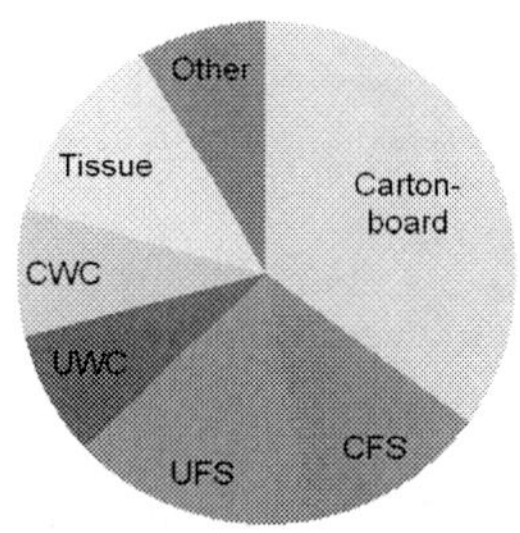

Total 3.9 million tons

PERSPECTIVE

Tissue production will continue to grow – and tissue makers will become even bigger users of fiber

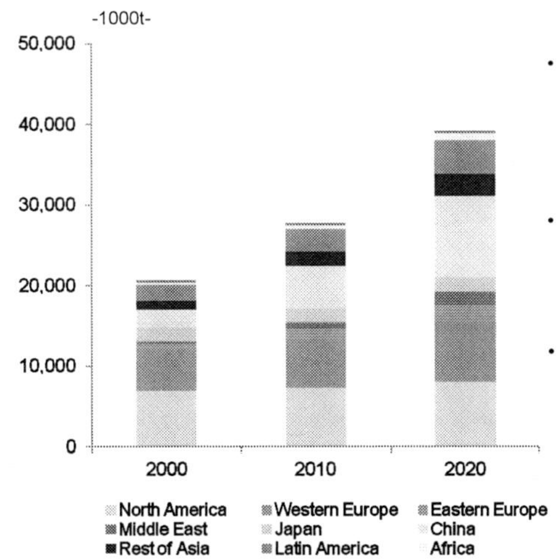

- North America: steady rate of new investments, shut down of uncompetitive capacity, TAD investments, US South, sustainable production
- Western Europe: steady rate of new investments, shut down of uncompetitive capacity, need for certified products, sustainability
- China: existing producers expanding, new entrants, larger scale mills, shut down of small scale operations

Fiber for tissue will increasingly be virgin fiber and come from market pulp mills – driven by production growth in China

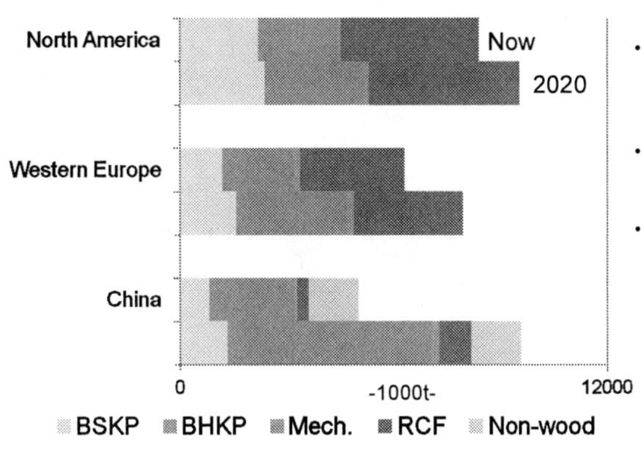

- North America: increased usage of market pulp, some integrated pulp, expanding on type of RCP used
- Western Europe: increased usage of market pulp
- China: increasing usage of market pulp, increasing integrated pulp usage, ramping up RCP

Wrap up

- Recovered fiber and bleached virgin fiber/chemical pulp are the options for tissue makers
- Top producers differ in their fiber strategies somewhat – driven by segment and product focus
- Fiber supply is based on three main platforms and their future outlook varies:
  - Integrated pulp; characterized by large scale mill operations. Mills could be focused on tissue production or have a number of other grades and market pulp at same sites. Investments to continue in this area and integration to pulp will be a viable concept for many tissue producers.
  - Recovered fiber; not keeping up with the expansion in tissue production in North America and Europe. However, some growth in China.
  - Market pulp; tissue an increasingly lucrative segment for market pulp producers as other end use segments in decline especially in North America and Europe. Future market pulp supply, especially BHKP/Euca will grow via large scale investments in Latin America.
- In summary
  - Virgin fiber will gain share and recovered fiber will lose share in the global tissue fiber supply

PÖYRY COPYRIGHT@PÖYRY Fiber supply - where will fiber for tissue making come from? 25
March 21, 2012

THANK YOU!

PÖYRY

Soile Kilpi, Director
soile.kilpi@poyry.com
646-651-1547

PÖYRY COPYRIGHT@PÖYRY

ICT Iberica case study

Gain sustainable savings by uncovering the Mysteries of Chatter

Presented by:
Antonio JORDAN – ICT Iberica
Florent BOUGEROLLE – BTG Eclépens

Co-author:
Xavier Escaler – CDIF-UPC

Tissue World Miami, March 20th/21st 2012

Chattermarks: Evidence of a misaligned process

Blade strategy
Blade holder settings
Yankee edge control
Yankee coating
Yankee drying control
SPR: Yankee crown fit
Felt dewatering
Sheet formation
Wet-end management
...

=

Unstable process
Poor performance
High chatter risks

Ceramic artwork by Alice Abrams

2

Ibérica

BTG
RAISING YOUR PRODUCTIVITY

ICT Group

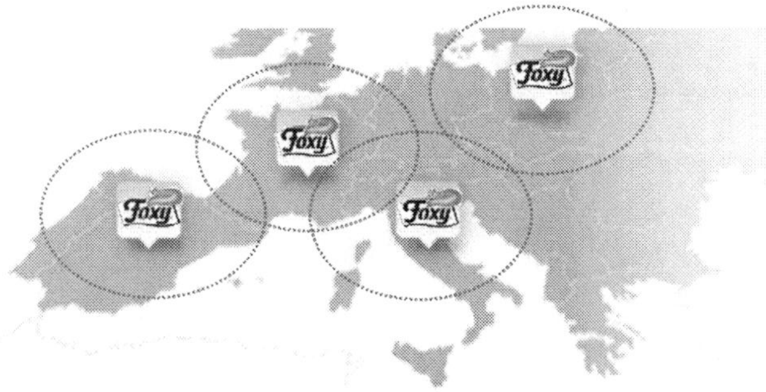

Foxy®

- European group
- 7% EU market share
- 9 machines 470,000t/y
- 5th EU tissue producer

3

Ibérica

BTG
RAISING YOUR PRODUCTIVITY

ICT Iberica – Spain

- PM21 Andritz (2005)
- 70,000 t/y
- 1,800 m/min.
- Converting facility
- Toilet & towel paper

Foxy®

4

Chatter problem

ict Ibérica — BTG RAISING YOUR PRODUCTIVITY

5 days/y **Production** loss

10% paper **Quality** impact

Yankee life **11%** wall loss

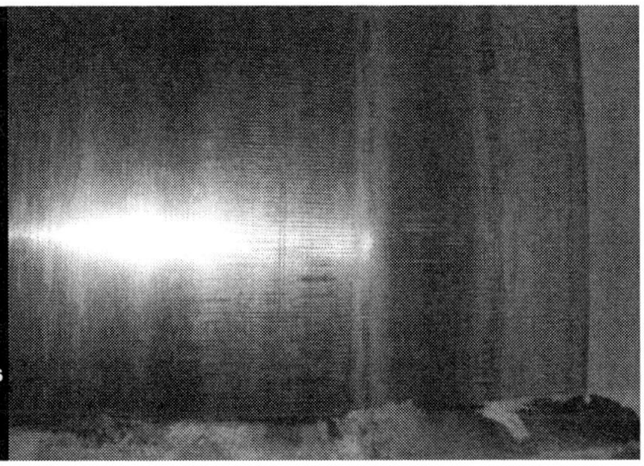

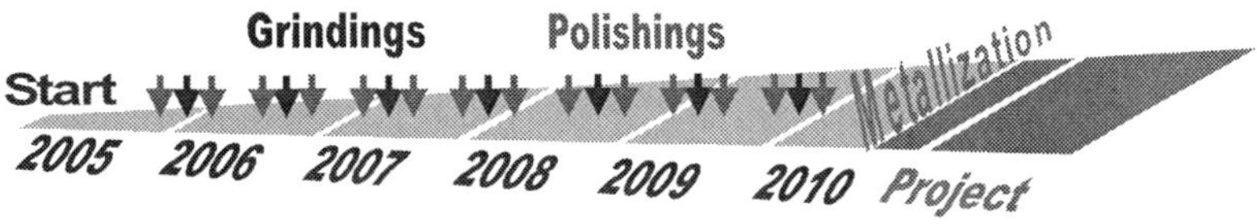

Rethink the process: keys

ict Ibérica — BTG RAISING YOUR PRODUCTIVITY

Technologies

Metalized Yankee & high-performance blades

Process

Process stability improvement & new procedures

People

External experts & focused crew

Ceramic blade is the safest option for chatter

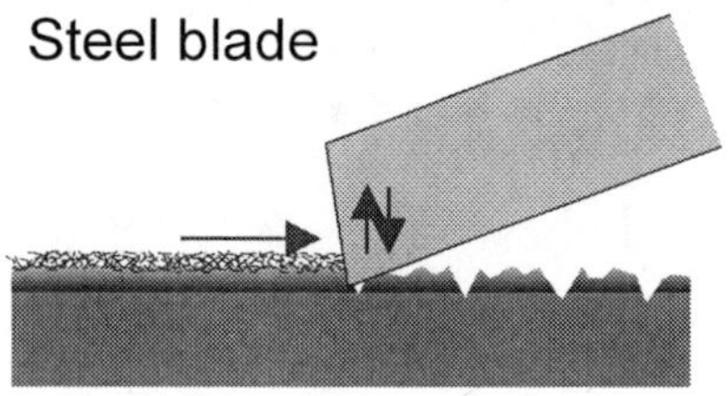

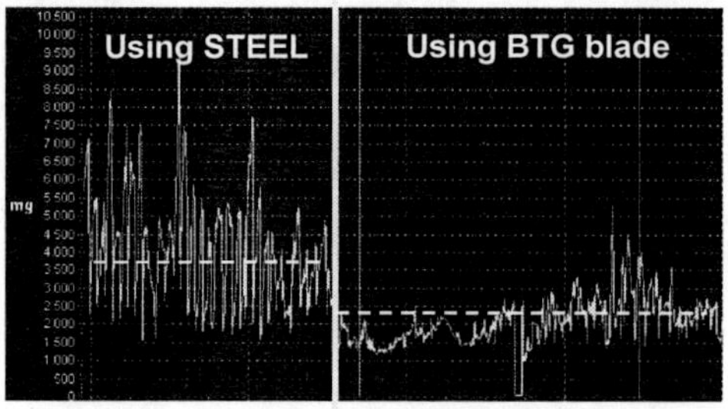

- **Fewer blade changes**
- **Coating stabilization**
- **Cleaner edges**
- **Yankee safety**

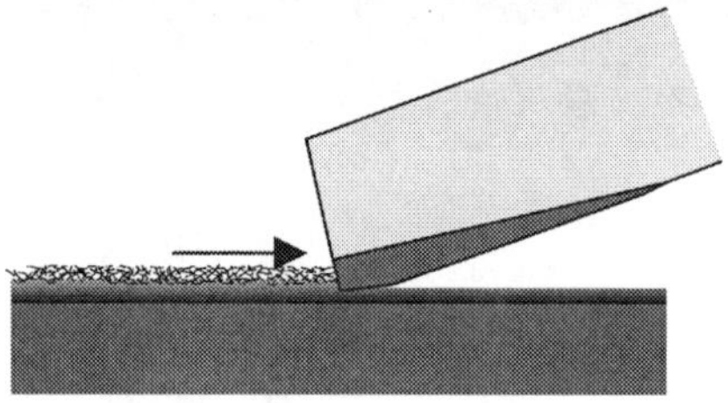

Vibration: too fast wear development

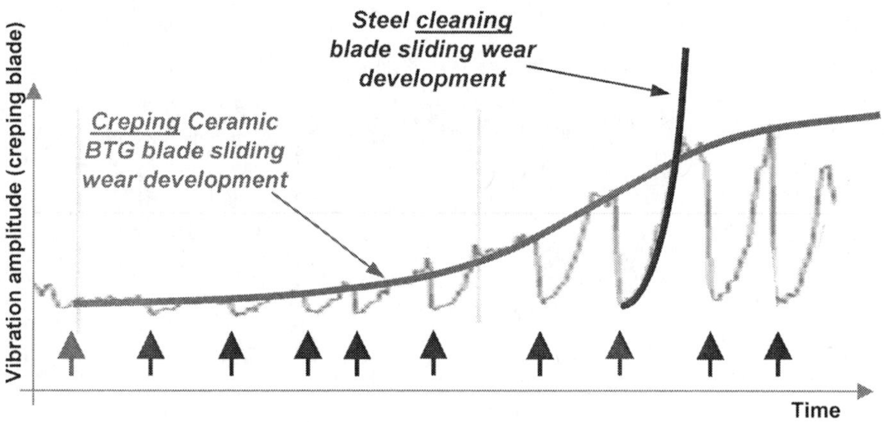

- **High frequency vibrations in coating**
- **Low frequency vibrations in contamination**

Reduce chatter potential thanks to fines management

BTG RAISING YOUR PRODUCTIVITY — ict Ibérica

Avoid fines going into the coating *here*...

... by fixing them onto the fiber *here*

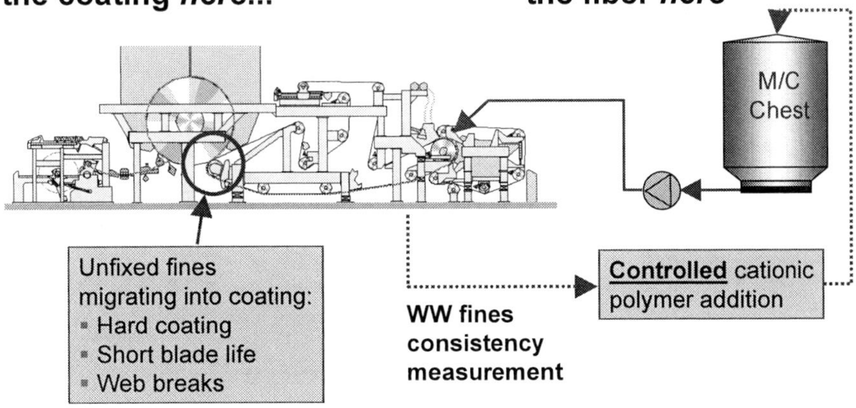

9

Rethink Process Procedures: Examples

ict Ibérica — BTG RAISING YOUR PRODUCTIVITY

- **Blade change procedure**

  when, why, how...

Vibration KPI

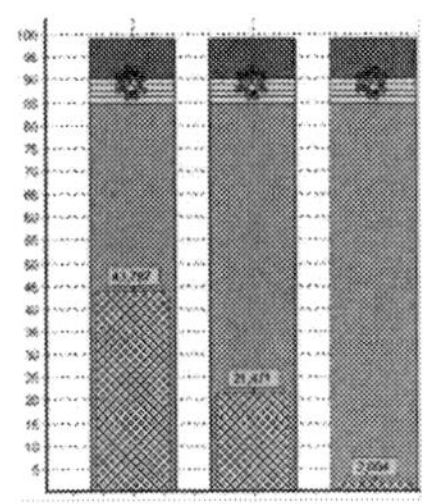

- **Machine cleaning procedure**

  when, where, how...

Time KPI

10

ict Ibérica

Project results

BTG RAISING YOUR PRODUCTIVITY

1,5y NO CHATTER

+5days/y AVAILABILITY

-30% lost time/day

PRODUCTIVITY

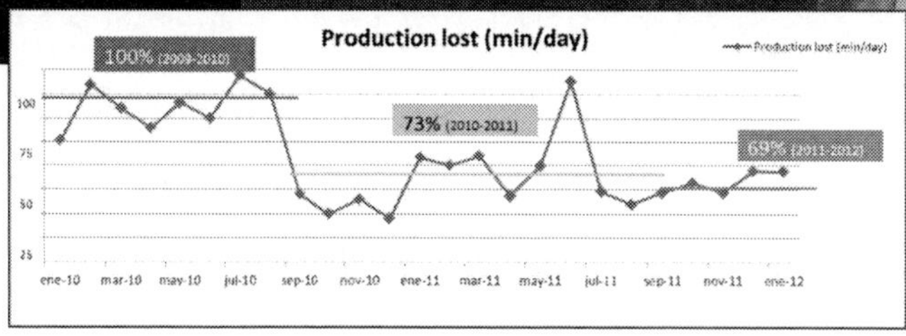

11

WE SOLVED THE CHATTER PROBLEM!

ICT Iberica – www.foxy.it.en
a.jordan@ictes.eu

BTG – www.btg.com
florent.bougerolle@btg.com

CDIF – www.cdif.es
escaler@mf.upc.edu

ict Ibérica

BTG RAISING YOUR PRODUCTIVITY

CDIF UPC

12

World´s first tissue machine with a compact wet end with active degassing

Topi Helle and Juha Iso-Herttua
Aikawa Fiber Technologies (AFT) Oy, Finland

Joan Vila and Carles Jord
LC Paper, Spain

AFT
AIKAWA GROUP

POM Compact Wet End Deliveries – over 70

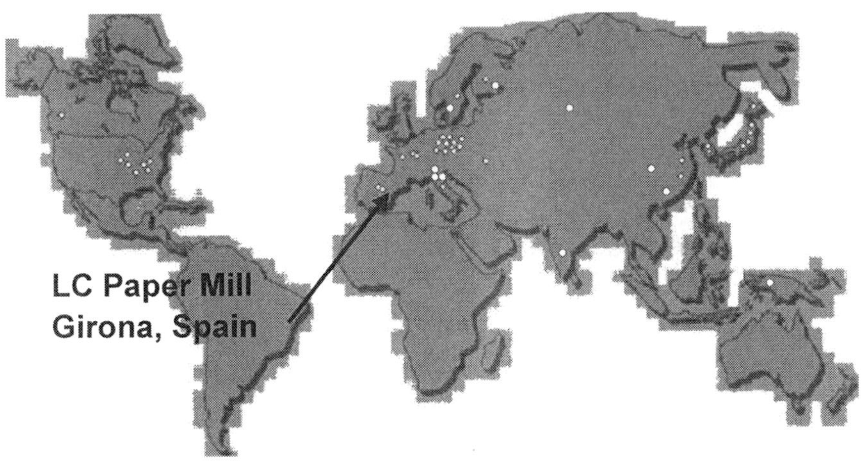

AFT
AIKAWA GROUP

POM Compact Wet End Principles

- Effectice gas removal
  - gas removal right after wire section
  - air/gasless short circulation, no flotation of dirt and pitch on air bubbles, less microbiological activity, less flocs, less pulsations
- Closed process
  - No tanks with open surfaces to collect pitch and dirt
- Compact dimensioning
  - Minimal space requirement and fast flow speeds, which will clean up the pipes
  - Small system volume which leads to fast grade changes
- As a result **stable, fast** and **clean** process

AFT
AIKAWA GROUP

LC Paper Mill in Spain

- Greenfield mill
- Colored tissue paper, 35 000 t/a ie. 39 000 ston/a
- Machine is mainly made by Italian Acelli
- Approach flow system volume is less than one third of a conventional approach flow system volume
- Complete POM system with POMp 730 degasser, Cyclopipe and two POMixes with premixing chambers
- No wire pit. Possible to do without basement => savings in construction costs. In this installation there is a basement.
- Target was to have a small system volume to have fast color changes and minimum change over losses – this has been achieved.

AFT
AIKAWA GROUP

POM System on a Crescent former tissue paper machine

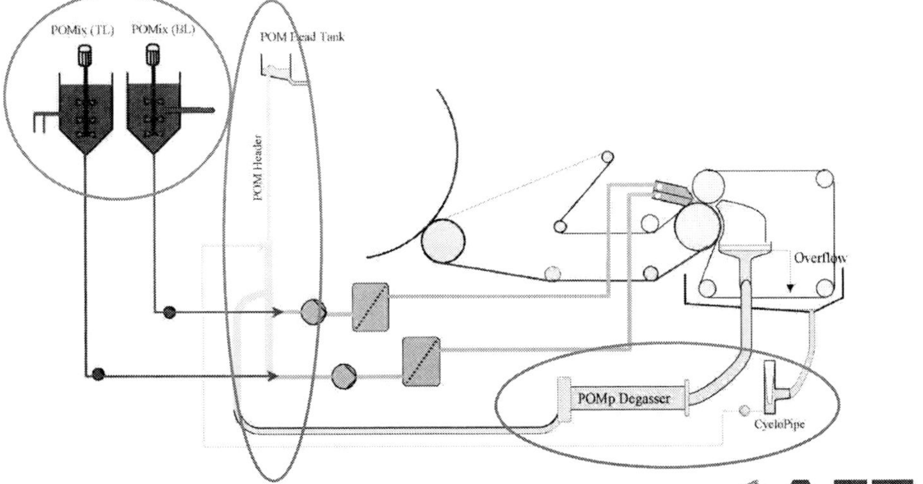

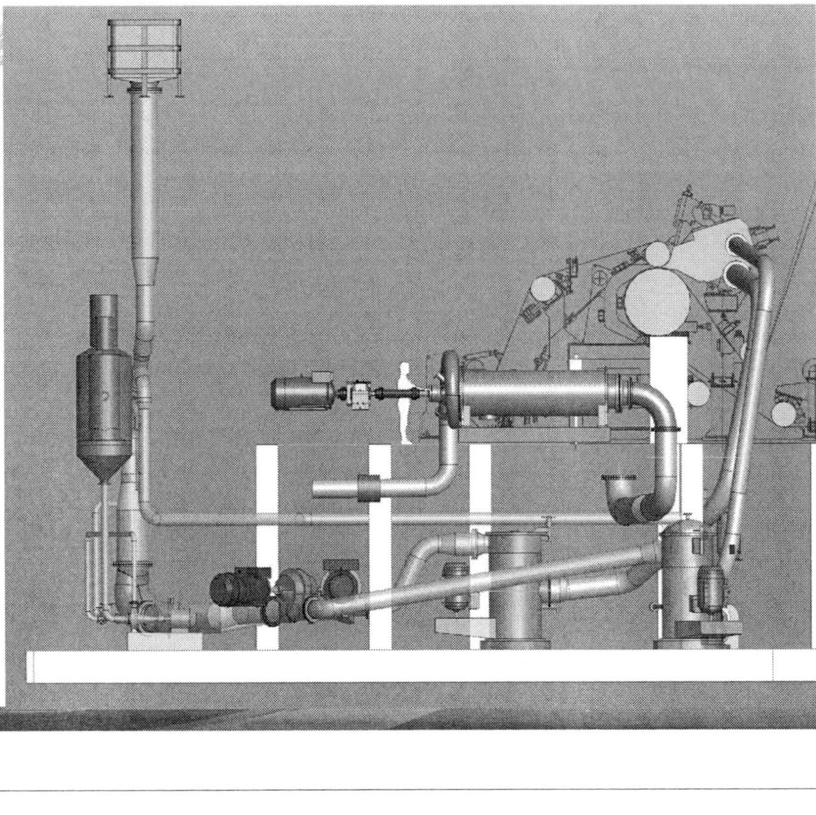

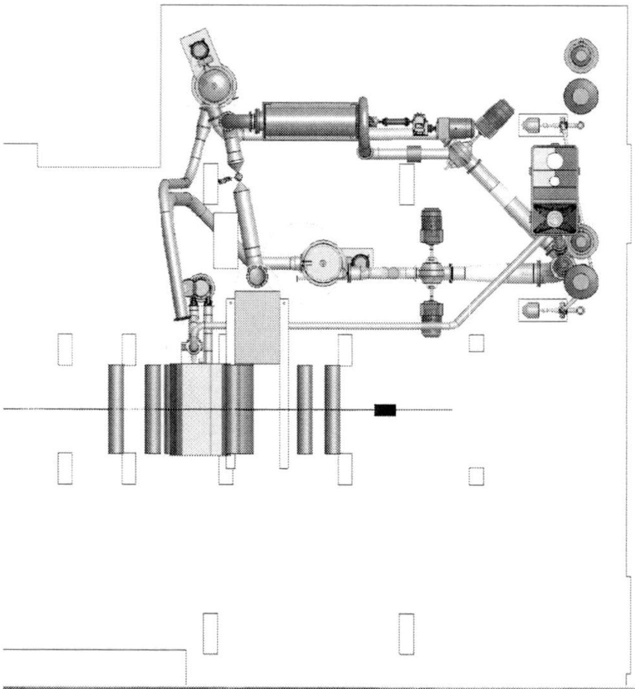

POMp degasser at machine level. Behind is the Crescent former tissue machine.

AFT
AIKAWA GROUP

AFT
AIKAWA GROUP

The POMp degasser

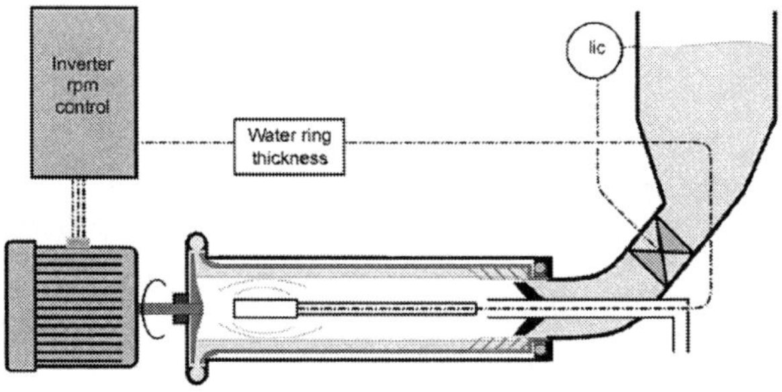

AFT AIKAWA GROUP

Cyclopipe for the suction waters and POMp degassers pumping head

AFT AIKAWA GROUP

POMix stock processor with premixing chamber and POM Header

AFT
AIKAWA GROUP

Mill trial

- White towel paper, 19 g/m2 basis weight
- All the machine set ups were kept constant, the target was to change only the entrained air content
- Air content was increased and compared to the situation with degassing
- Free air ie. entrained air content in the headbox is below 0,1% with degassing and it was increased to 0,4% and to 1,0%.

AFT
AIKAWA GROUP

Rotation speed of short fibre fan pump, when the air content was reduced back to normal (from 1,0% to below 0,1%*)

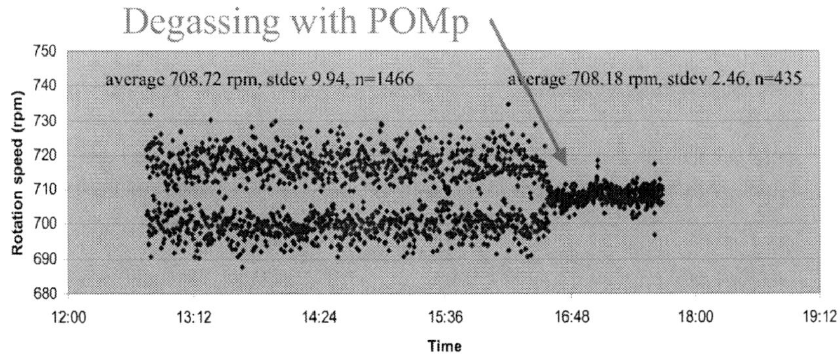

* Free air content measured from headbox with expansion method.

AFT AIKAWA GROUP

Pressure inside the headbox, when the air content was first increased and the decreased back to normal (from 0,4% to below 0,1%*)

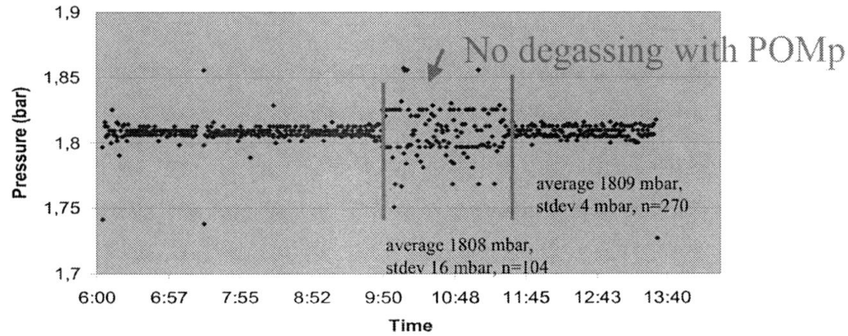

* Free air content measured from headbox with expansion method.

AFT AIKAWA GROUP

Collapsing bubble and attachment of pitch and dirt on bubble -> spots and deposits on paper

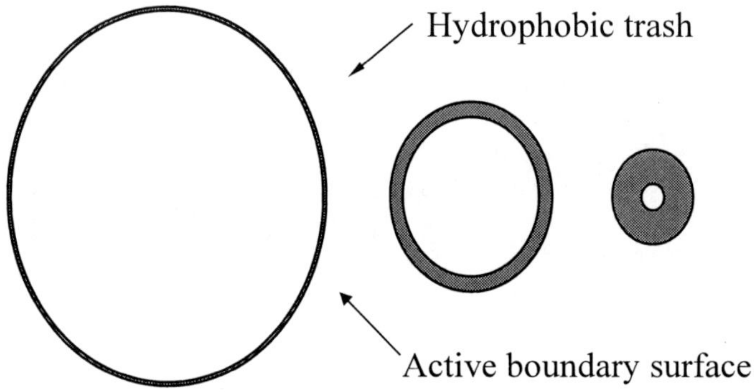

AFT
AIKAWA GROUP

Paper properties

- White towel paper, 19 g/m2 basis weight
- In tensile strengh and formation we could see some increased variation
- There were no other clear differencies in the paper properties with higher air contents
- The headbox consistency was only 0,2% and in addition tissue paper is very porous, which makes the paper more tolerant to large variations that could be caused by air in standard paper and board products

AFT
AIKAWA GROUP

Tensile strength of paper samples in three different air contents. The average values were not significantly changed. However the variation was increased with higher air contents.

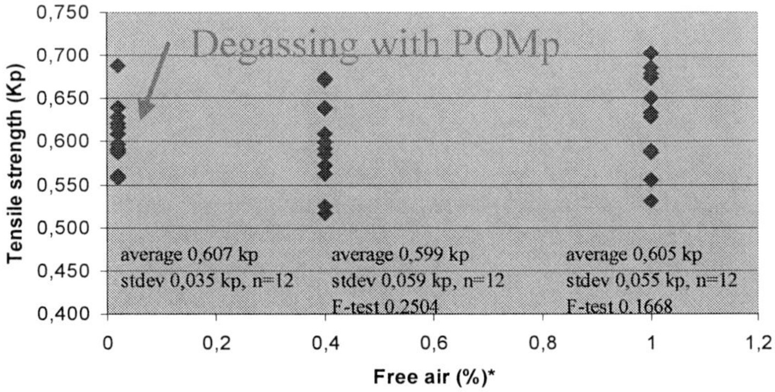

* Free air content measured from headbox with expansion method.

AFT AIKAWA GROUP

The formation variation was increased with high air contents.

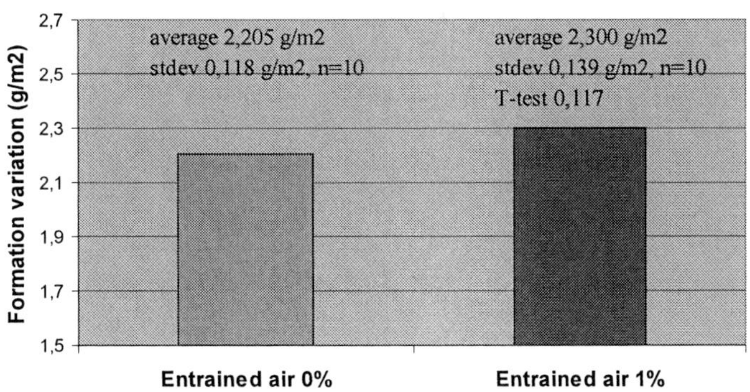

AFT AIKAWA GROUP

Water consumption

- Fresh water consumption has been 2,5m3/t (=600 gal/ston). Estimated consumption with conventional wet end was 4,0m3/t (=960 gal/ston)
- The main reasons could be:
  1. The approach flow system is very clean due to low air content, fast flow speeds and no open tanks
  2. High first past retention
  3. Vacuum fans use 1 m3/t (=240 gal/ston) less water
- First pass retention ie. FPR has been 89% with below 0,1% free air content with POMp degasser. With 1,0% free air content the FPR was 78% and with 0,4% free air content the FPR was 84%

AFT AIKAWA GROUP

Consistency after white water tray, when the air content was first increased and then decreased back to normal (air contents is in location A 1,0%, B 0,4% and C 0,1%*)

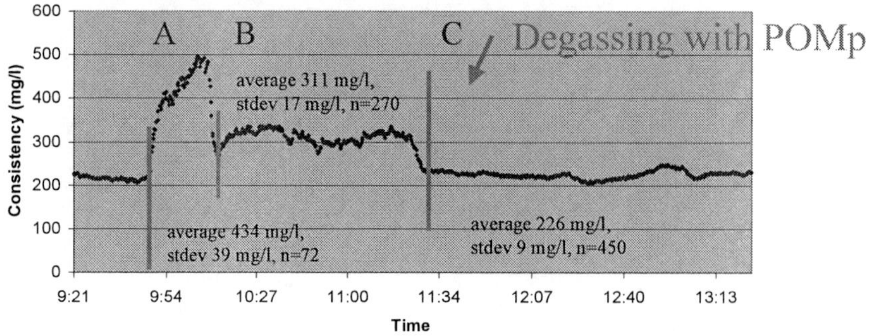

* Free air content measured from headbox with expansion method.

AFT AIKAWA GROUP

Mill results from LC Paper in Spain

- Color and basis weight changes are three times **faster** than in an conventional system
- More **stable** than conventional wet end. For example the feed pump rotation speed is more constant and there are less pressure variations in the headbox
- Improved first past retention by over 10%
- With improved retention, the headbox flow was decreased 12%, from 1403 l/s to 1249 l/s ie. 370 gal/s to 330 gal/s. This saves pumping electricity.
- Decreased fresh water consumption (0,5m3/t ie. 120 gal/ston)
- Less material losses
- **Cleaner** system and improved runnability

AFT AIKAWA GROUP

Additional effects of air in tissue

- Reduced defoamer consumption. Many defoamers contain hydrofobic components, which will effect to coating development on the yankee cylinder
- Improved FPR can lead to reduction of yankee adhesive use, because of increased levels of natural coating components contributing to the coating
- Even small increase in FPR can lead into large improvements in the fines retention. This will improve the felt cleanliness, DAF efficiency and reduced defoamer use.
- All above will help coating development in the pressure roll nip and furthermore into better creping.

AFT AIKAWA GROUP

Stable, fast and clean process

Development trends in tissue pressing

Ingvar Klerelid, VP Product Management, Tissue Mills, Metso

Anders Leandersson, Senior Process Engineer, Tissue Mills, Metso

Abstract:

Drying is the most energy consuming process in tissue making. The most effective way to reduce energy is to increase the ingoing dryness to the Yankee dryer. For a long time, most new Advantage DCT machines were supplied with a single press nip using a suction roll with linear load of less than 100 kN/m. At that time, there was more focus on sheet quality compared to dryness. Today the focus is more on energy as converting capabilities have increased to make premium products.

Tissue pressing is mainly a pressure controlled dewatering in the nip, but as heat is transferred from the hot Yankee to the web, the time in the nip also affects the dryness. Longer press nip increases residence time in the nip. This increases web temperature and reduces water viscosity, which improves dewatering. There are of course other parameters affecting the dryness such as choice of fiber, refining, felt design, etc.

Yankee dryers can today be designed for linear loads up to 170 kN/m. This has opened up the possibility for using different press concepts.

This paper compares different press concepts for both Advantage DCT and NTT machines. Interest for 2-press roll configuration has increased and it is an interesting alternative for improving dryness. Single press machines with a large diameter suction pressure roll at elevated linear load provide both improved dryness and bulk. The ViscoNip press running against the Yankee dryer produces the highest dryness and it is possible to reach 46 – 47% for low basis weights and up to 50% for higher basis weights. It has also the flexibility to run linear loads from 80-170 kN/m.

Advantage NTT has a single nip pre-press before the Yankee using a SymBelt shoe press running high linear load. This concept works with high peak pressure, which helps to achieve a dryness of 48-52% depending on the basis weight.

Background

The pressing technique used on conventional tissue machines has for many years been based on standard press rolls. Up until the early 90's it was common to use two press rolls. The first press was a suction press roll and the second was a blind drilled press roll. The first press was typically loaded at 80 kN/m and the second at 90 kN/m.

During the 90'-s the trend changed toward using only one press roll instead of two press rolls, mainly to improve the paper quality. By running only a suction press roll the compaction of the sheet was reduced, which increased the caliper in the creped product. A single press machine increases the bulk by about 10%. The downside is reduced press dryness by 4%. With only one press roll it was possible to increase hood wrap which helped to maintain production capacity. The disadvantage was higher energy cost, but as energy cost has been relatively low, the advantages of better tissue product and lower investment cost made the single press roll machine to be the most common, which is still the case.

The shoe press was introduced for tissue about 10 years ago. The main reason was to increase bulk and improve softness. The gain in bulk was up to 10% compared to a single press roll machine. This was possible to achieve with lower peak pressure in the nip and maintaining the same dryness after the press as with a roll press.

During the last few years energy costs have escalated and there is much more focus on this fact when producing tissue. Improved press performance is the most effective way to reduce energy consumption.

Press theory

Pressing in conventional tissue machines is different to pressing in most other paper machines. In paper machines the press section has typically two or three press nips. The press nips are double felted and normally single felted in last press nip, which also can be equipped with a process belt wrapping the counter roll. The temperature of the web is higher than room temperature. Web temperature is increased by temperature of white water and by steam shower if used. Increasing the temperature of the web, reduces viscosity, which improves dewatering in the press nip.

In conventional tissue machines the press nip is single felted with a hot counter roll, which is the Yankee dryer. In the press nip, heat is transferred from the hot Yankee to the web. This heat transfer increases the web temperature, which lowers the viscosity, resulting in a higher dryness compared to room temperature pressing.

A trial was made on the Metso pilot machine comparing pressing with a "cold" Yankee with a normal operating Yankee. The machine speed was changed from 500 m/min up to 1200 m/min. The basis weight was 16,5 gsm on the Yankee dryer. The white water temperature was 38°C. No steam shower was used.

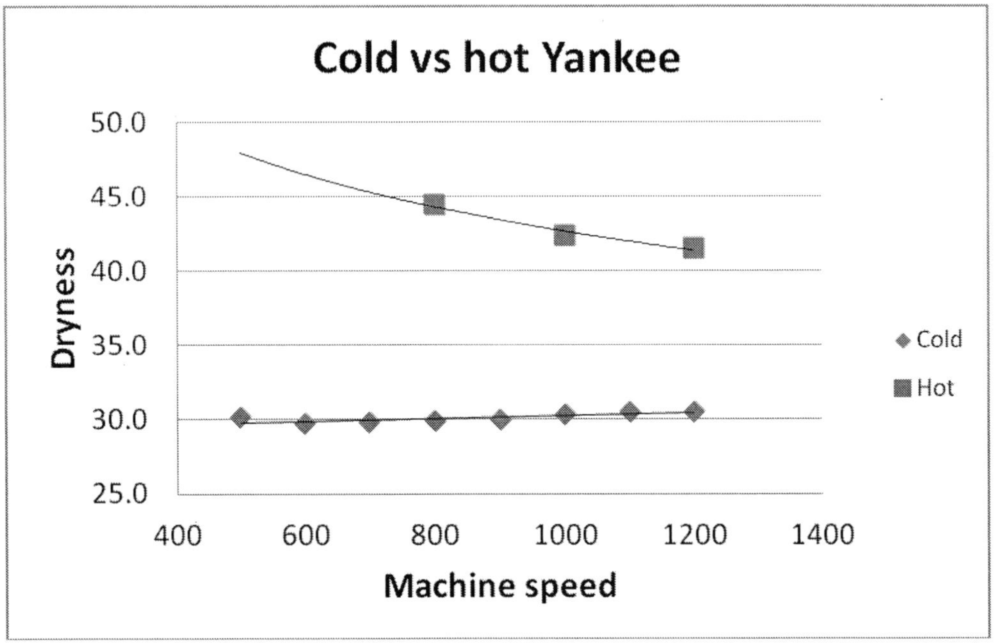

Figure 1. Post pressure roll dryness.

The trial showed that the dryness was much lower with cold Yankee compared to hot Yankee. This was not a surprise. More interesting is the fact that dryness with cold Yankee was not affected by the machine speed. The press nip profile remains the same and is independent of machine speed, but time in the nip is shorter with increased machine speed. This means that tissue of low basis weight can be regarded as a compression controlled nip. However, when the Yankee heats web, the time in the nip is important. Longer time means higher heat transfer and also a higher dryness as the higher temperature reduces viscosity.

By definition, a compression controlled nip will not be affected by the time in the nip, but this is only true when no heat is transferred to the web in the nip. Pressing against the hot Yankee results in higher dryness, but also declines in dryness when machine speed is increased, due to less time in the press nip.

In a compression controlled nip, the peak pressure is the most important parameter for increasing dryness. Second most important is the length of the nip or the time heat can be transferred to the web to reduce viscosity.

Rewetting is a very important mechanism. Rewetting is independent of basis weight for low basis weights. This means that the effect of rewetting is less for high basis weights resulting in higher dryness for high basis weight tissue.

The relationship seems to be linear for low basis weight.

$k_p = k_n + R/W$

k_p is the post pressure roll dryness

k_n is the dryness in the nip

R is rewetting, g/m²

W is basis weight, g/m²

With the assumption that rewetting is constant for low basis weight sheets such tissue (11-36 gsm) it is possible to estimate the dryness in the nip. Rewetting is in the range of 4-5 gsm. Rewetting in a Yankee press nip is mainly internal rewetting as the web follow the Yankee after the press nip. Rewetting outside the nip is, of course, an issue in a double felted press nip. This rewetting depends very much on the contact time. The longer the it is, the lower dryness.

Other parameters affecting press dryness result are

- Fiber
- WRV
- Refining
- Felt design
- Water content in felt

The difference in dryness for bath products with different amounts of hardwood is small. Increased refining reduces dryness. Recycled fibers normally produce higher dryness, while slush pulp reduces dryness.

The water retention value, WRV, is a measurement of fibers water holding capacity. Lower WRV results in a higher dryness.

The felt design is very important for water removal. Water removal can be increased by improving the uniformity of the application of available nip pressure. This is important for lighter weight grades such as tissue and is evidenced by use of fine batt fibers which improves micro pressure uniformity.

The water content in the felt has small effect on dryness. However, the felt has to be saturated in the nip. A new felt has to be compacted before it performs well.

Yankee Dryer

Yankee dryers for use in conventional one or two press roll configurations are traditionally designed for linear loads up to 90 kN/m. With the development of the extended nip press and the desire to improve sheet dryness after press, the so called "high load Yankee dryers" were developed to run linear loads ranging from 120 up to 170 kN/m. The casting of the thick walled shell is more challenging due to higher casting weight and high quality demands. However, the basic design of the ribbed shell remains the same.

It is important to achieve both a high and even heat flow in these dryers with thicker shells. The thermal capacity of the shell is a function of its thickness, the conductivity of the shell material and heat transfer from steam on the inside of the shell. The steam mainly condenses on the rib flanks and the condensate is collected on the bottom of the grooves by centrifugal force, from where it is removed. In the bottom of the grooves the heat transfer depends on the condensate level and its movement.

When dimensioning the shell, the loads from internal pressure, temperature gradients, rotation and press nips are considered. The shell is cast homogenous (plain). Maximum loading is, of course, restricted by shell thickness. The thicker shell the higher loading. For the same thickness, a plain shell will always is stronger than a ribbed regardless of its geometry. The purpose of the ribbed shell is solely to increase thermal efficiency.

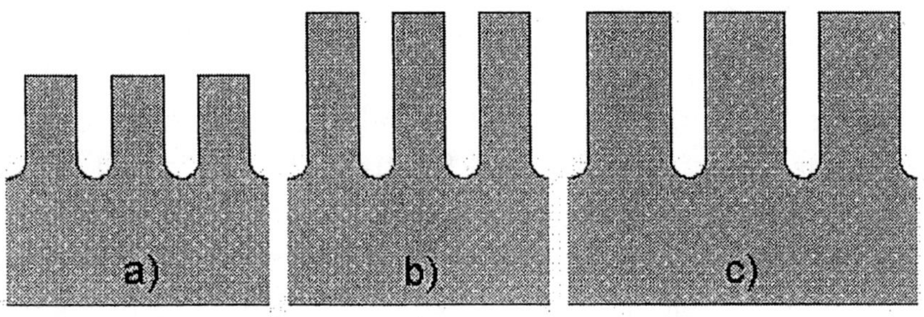

Fig. 2 Different shell designs

The width of the grooves is determined by the size of the soda straws and needed room for collecting the condensate. Given the groove width and number of grooves, the pitch is optimized with respect to thermal efficiency, thermal uniformity and structural strength. The strength can be increased by increasing the rib height making it possible to increase press loads.

A cast iron shell normally has a thickness of around 60-90 mm, in some high press load applications up to 100 mm. This to obtain the necessary strength and the higher Safety Factors stipulated, but also to provide sufficient amount of grinding allowance and thus service life before it needs to be metalized. The shell is thicker, but at the same time the grooves are deeper to ensure a high drying capacity.

In case a high drying capacity is desired in combination with a hard, thermally coated surface already from the start-up of the dryer a thinner root thickness can be used and strength be maintained by using even higher ribs. The rib and groove design decides the heat flow. With a deep grooved dryer the heat flow increases and can be higher than for a steel dryer.

There are some practical limitations. If the shell is too thin there might be problems with marking in the sheet from the ribs. A high heat flow that is too high may result in runnability problems such as sheet blistering and problems of achieving a uniform coating.

Press concepts

Most new tissue machines are still installed with only a single press roll, suction roll operating at linear load of 90 or 100 kN/m. But there are machines being installed with press configurations making it possible to save energy and the saving is huge. This trend will increase as the pay-back time on the additional investment is relatively short. In fact, there are currently several press configurations available, which can be used depending the need for top product quality or high dryness or both with the same press configuration.

Metso offers several press configurations for saving energy. Saving energy is today very important, but for the tissue producer there are also other parameters to consider. Tissue quality, softness and bulk, are important for the finished product. Machine performance and operating cost are also important to understand when selecting press configuration for a new machine or a rebuild of an existing machine. The following table describes the press concepts available from Metso. The press roll dimensions is for a 5,5 m wide DCT 200 high speed machine.

Press concept	Suction roll	Suction roll	2-press	2-press	ViscoNip	ViscoNip
Diameter, mm	1200	1465	1200/1100	1200/1100	1100	1100
Linear Load, kN/m	100	120	80 / 90	100 / 120	90-170	90-170
Nip length, mm	63,6	70,8	61,7 / 60,4	63,6 / 63,5	130	80
Peak pressure, MPa	2,4	2,6	2,2 – 2,5	2,4 – 2,8	1,8 - 4	2,7 - 5,2
Hardness	25	25	25 / 35	25 / 35	NA	NA

Rubber/PU	VacuFox TIS	VacuFox TIS	VacuFox TIS PressGat.ZL	VacuFox TIS PressGat.ZL	NA	NA

Single press roll configuration

Today the main difference is the possibility to run higher linear loads with the development of the Yankee dryers. The linear load for a suction press roll has for a long time been 90 kN/m, but currently it is today possible to supply a suction roll with linear load of at least 120 kN/m. For higher linear loads, Metso recommends installing a larger diameter suction roll. The benefits are a higher peak pressure and a longer press nip. Both have a positive effect on dryness as the longer time in the nip improves dryness. The longer nip is also advantageous with respect to sheet softness and bulk.

The gain in dryness is shown in the diagram below. The dryness increase is about 1.5%. This dryness is the expected when working with virgin fiber, medium refining and an acceptable felt. Important is also the temperature of white water. Machine speed, as discussed earlier, also has an effect on dryness. Therefore, it is important to use the numbers more for comparison than for absolute dryness.

Longer time in the nip also has an advantage with respect to sheet quality. Improved sheet adhesion has resulted in improved softness and bulk.

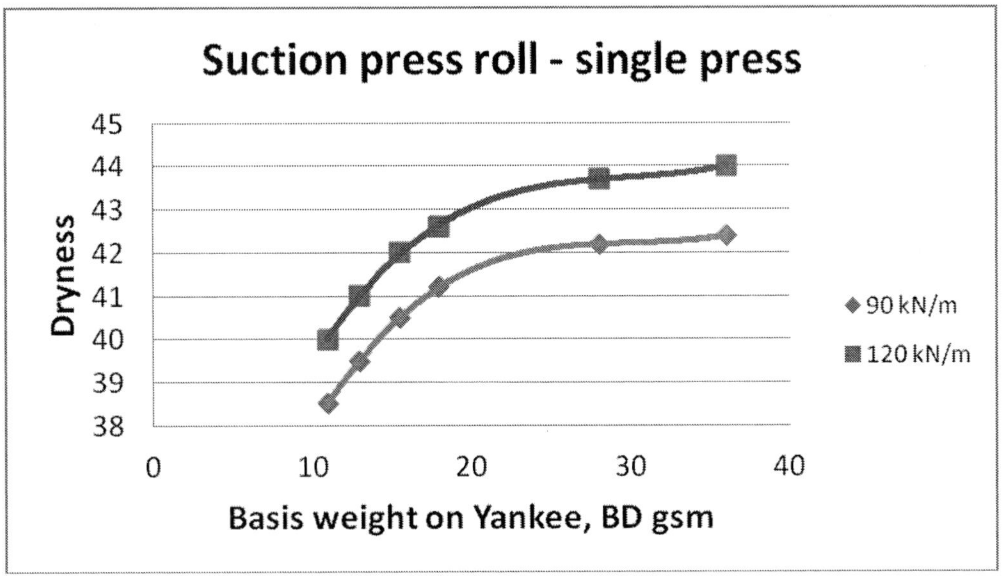

Figure 3. Dryness vs BW for suction press roll at different linear load.

2-press roll configuration

The 2-roll press configuration was earlier the most common. It is, of course, also possible to increase the nip load in this configuration. The recommended increase would be from 80 to 100 kN/m in the first press and from 90 to 120 kN/m in the second press. The first press is a suction press roll, while the second press is a bind drilled roll or combination of blind drilled and grooved. When comparing it to the single press roll, there are some disadvantages, but also some interesting possibility improving dryness.

Advantages of the single press are:

- Less equipment in the machine.
- Easier felt change.
- Longer felt life due to fewer nip cycles.
- Improved tissue quality with respect to bulk and softness.

Advantages of 2-roll press configuration are:

- Higher dryness after the press provides for lower production or higher drying costs. In the diagram below, the 2-press roll at two different load levels is compared to a standard single press.
- A second press evens out moisture profile variations in the CD direction. This means that the machine is less sensitive to any non-uniformity in the moisture profile.

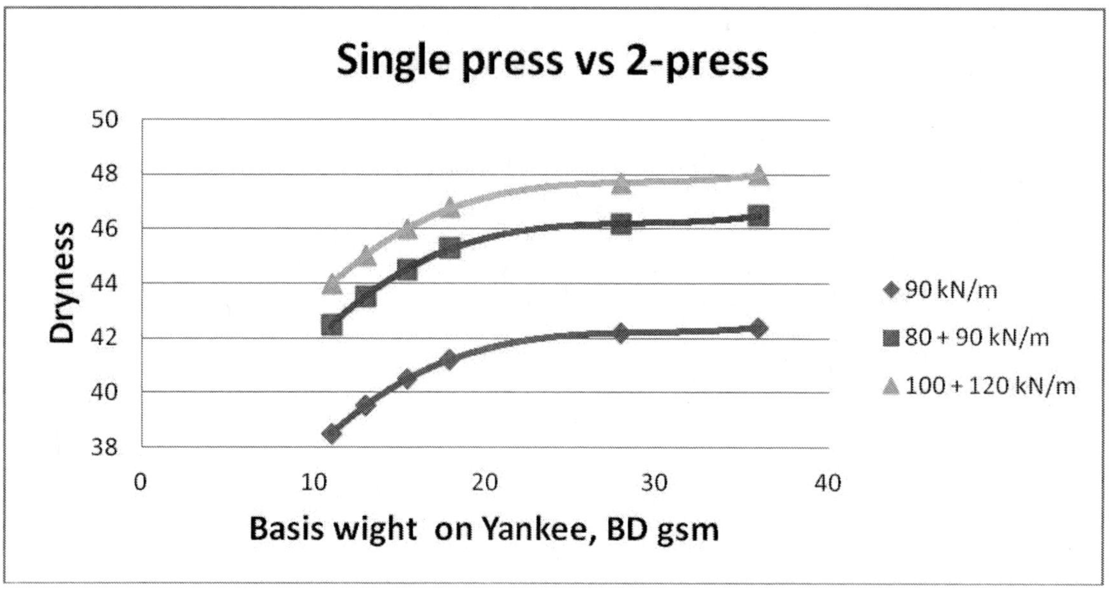

Figure 4. Dryness vs Basis weight on Yankee

The dryness increase adding by a second press is about 4%. This gain in dryness is achieved by a combination of slightly higher peak pressure but also the fact that the web temperature increases quickly between the two press rolls. The increased web temperature reduces viscosity which improves the dewatering. The sheet temperature before the second press is high, about 90°C. This is the main reason for the high dryness gain.

A significant difference in production capacity might be expected with a difference of 4% in dryness. The loss in dryness is to some extent compensated with higher hood wrap. Drying simulation shows the hood wrap compensates for about 2%. .

The peak pressure is determined by four parameters:

- Linear load
- Press roll diameter
- Press roll cover thickness
- Press roll cover hardness,

The easiest way to increase the peak pressure in order to increase dryness is to increase the hardness of the cover. The problem with a harder cover is that the pressure uniformity in the nip is much more dependent on being correctly crowned.

In general a one-press machine cannot use hard covers to achieve high peak pressures because it is very sensitive to crown issues.

ViscoNip press

The ViscoNip press was introduced on to the market in 2006. There are today 13 ViscoNip presses sold for different tissue applications. The press is designed to run a wide range of linear loads while maintaining a uniform nip load independent of counter roll deformation.

When designing an extended nip press or shoe press for use against the Yankee it is important to have a good understanding of how the Yankee deforms. The same four loads considered when dimensioning the shell are also the ones that contribute to the deformations. The Yankee is relatively stiff with respect to bending. Most of the deformation is caused by the shell deflection. A shoe press with steel shoe is stiffer than a rubber covered press roll. The ViscoNip press was developed to solve this problem. The press body consists of two or three chambers, which stretch over the width of the machine. This makes it possible to achieve a uniform CD press load and also over a wide range of linear loads. The figure below shows the ViscoNip body with two chambers unloaded and loaded to the Yankee dryer.

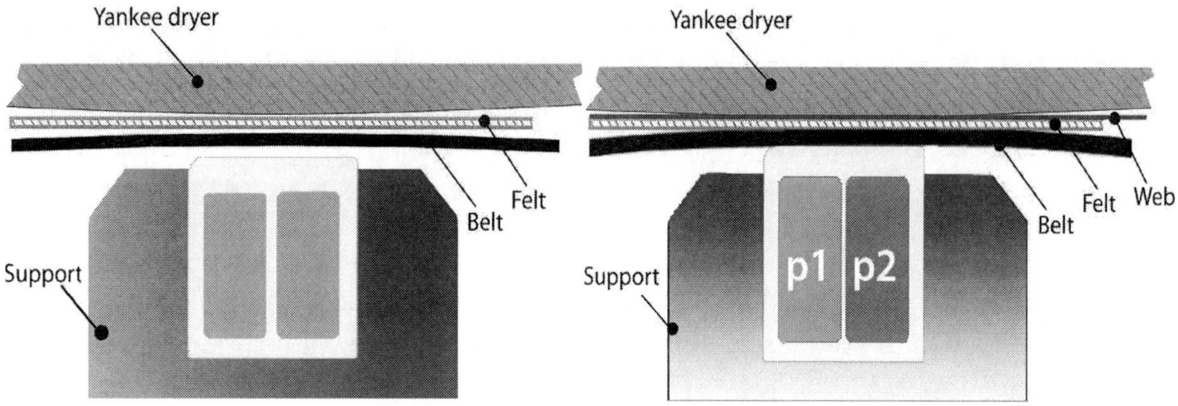

Figure 5. *Not pressurized* *Pressurized*

The linear load is the total load applied against the Yankee. The linear load is also the total load under the press nip curve. With the new press, it is possible to achieve a high peak pressure at the end of the curve, and it is also possible to have a very sharp drop in pressure to minimize rewetting. For lightweight sheets such as tissue the peak pressure in the nip is the dominating parameter for the achieved high sheet dryness after press. However, as discussed previously, the time in the hot nip and the amount of heat transferred also affects the final dryness. It is not only the sheet dryness that increases with increased load, but it has also an effect on the sheet quality. It is well known for the tissue maker, that bulk decreases with increased nip load. It is also well known that tensile strength increases with increased nip load.

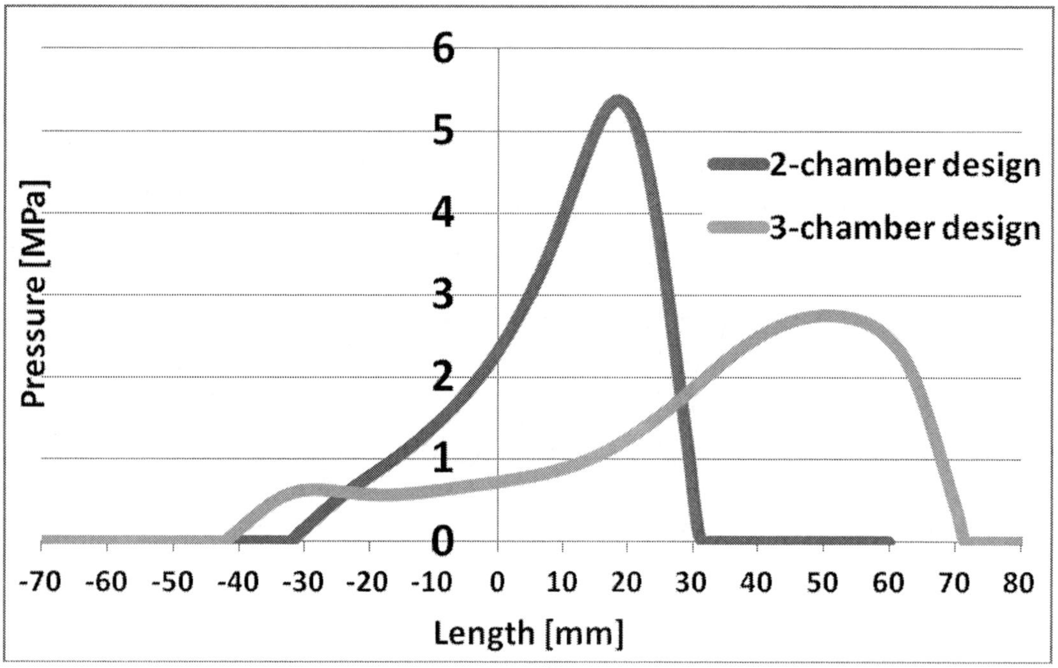

Figure 6. Pressure curve comparison 2-chamber vs 3-chamber design @ 150 kN.

Press dryness mainly depends on the peak pressure in the nip. High peak pressure is achieved with high total linear load and higher hydraulic oil pressure in the second chamber. The difference in dryness between low and high linear load is shown in the diagram below. There is an expected dryness increase of about 6% between running the 2-chamber body at 150 kN/m compared to 90 kN/m. The difference in peak pressure is 1.8 MPa at 90 kN/m compared to 5 MPa at 150 kN/m. This demonstrates the flexibility in operation of the ViscoNip press. The pressure ratio between the chambers can be varied, which adds flexibility when optimizing dryness and tissue properties.

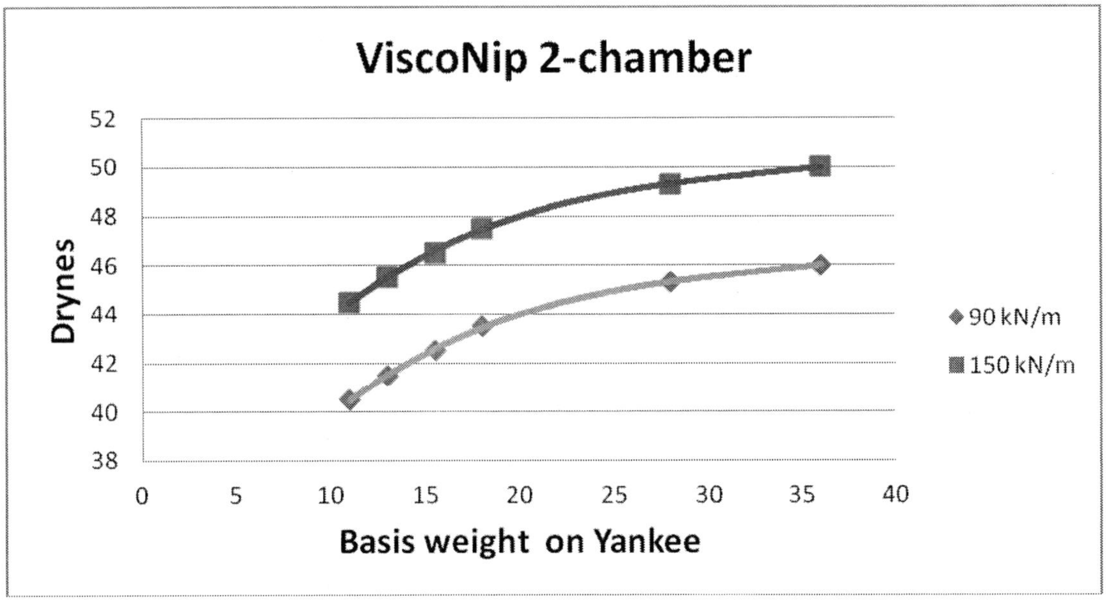

Figure 7. Expected dryness vs basis weight

When the shoe press for tissue was introduced it was claimed to be able to produce tissue quality that was close to TAD. This is not the case. Both in pilot scale and commercial experience the gain in bulk is about 10%, but it is at relatively low peak pressure and a dryness level that is the same as for a single roll press. The main advantage today for running a ViscoNip press is improved dryness. The press dryness CD profile has also improved as well as sheet softness. An additional advantage is the possibility to run different peak pressures by changing the ratio in oil pressure between the chambers.

High value grades, which require high bulk and softness, the ViscoNip press needs to operate at low nip pressure. Low value grades can be made with high nip load resulting in lower caliper, but a high production. The ViscoNip press has this flexibility without roll crowning issues.

Figure 8. ViscoNip installation

High peak pressure will has a negative effect on sheet caliper. However, both tensile strength and sheet adhesion to Yankee is improved.

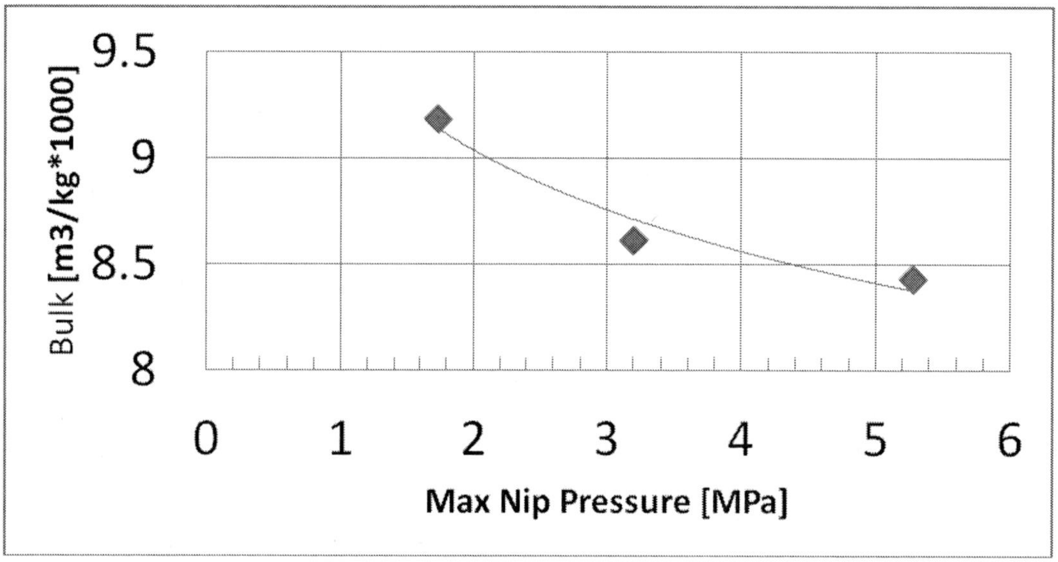

Figure 9. ViscoNip. Bulk vs Max Peak Pressure.

This improvement in sheet adhesion affects sheet softness. The crepe structure is finer, which can be measured as crepe waves per MD length unit. The Emtec TSA (Tissue Softness Analyzer) is becoming a more frequently used method for measuring sheet softness. A softness comparison was made in same mill between tissue produced on a machine with suction pressure roll and a machine with ViscoNip press. On both machines product specification was the same as well as the furnish mix used. The tissue produced with ViscoNip press had both finer crepe and higher softness.

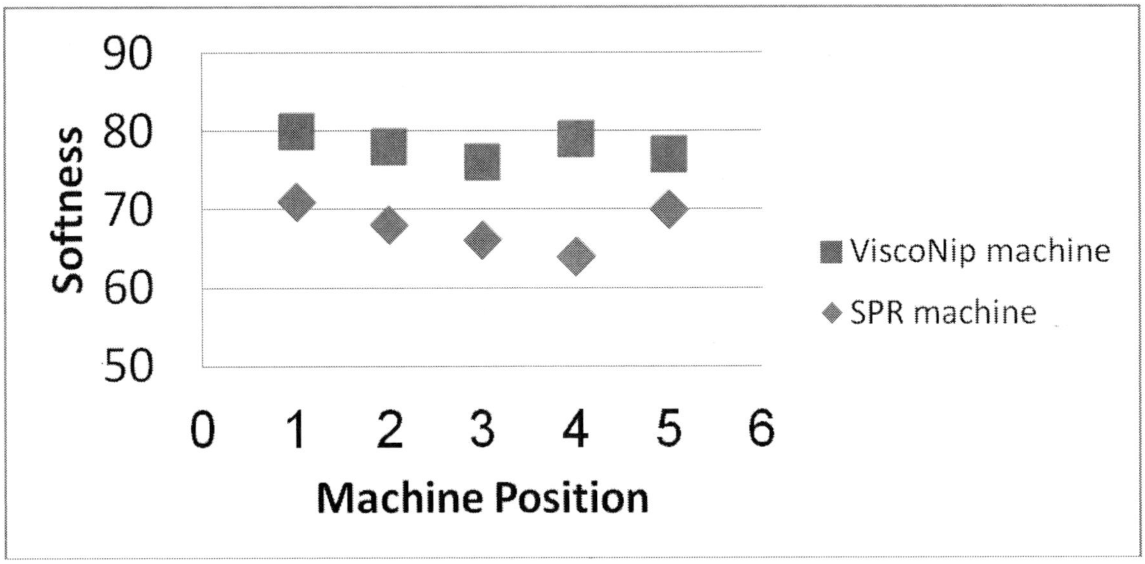

Figure 10. Softness measurement on tissue with BW 16 gsm

In the graph below there are dryness numbers from both the pilot machine and from commercial machines running with ViscoNip press. All the data is from the ViscoNip press with three chambers. The maximum peak pressure with this press body is 4.5 MPa. The dryness numbers from the pilot machine are higher compared to the commercial machine data. The difference is about 1 percentage point. The pilot data is mainly from a machine speed of 1500 m/min, while the commercial data is from 1900 m/min. The difference in machine speed explains the higher dryness obtained on the pilot machine. Pilot test comparing 1500 and 1800 m/min has showed a drop in dryness of about 1%.

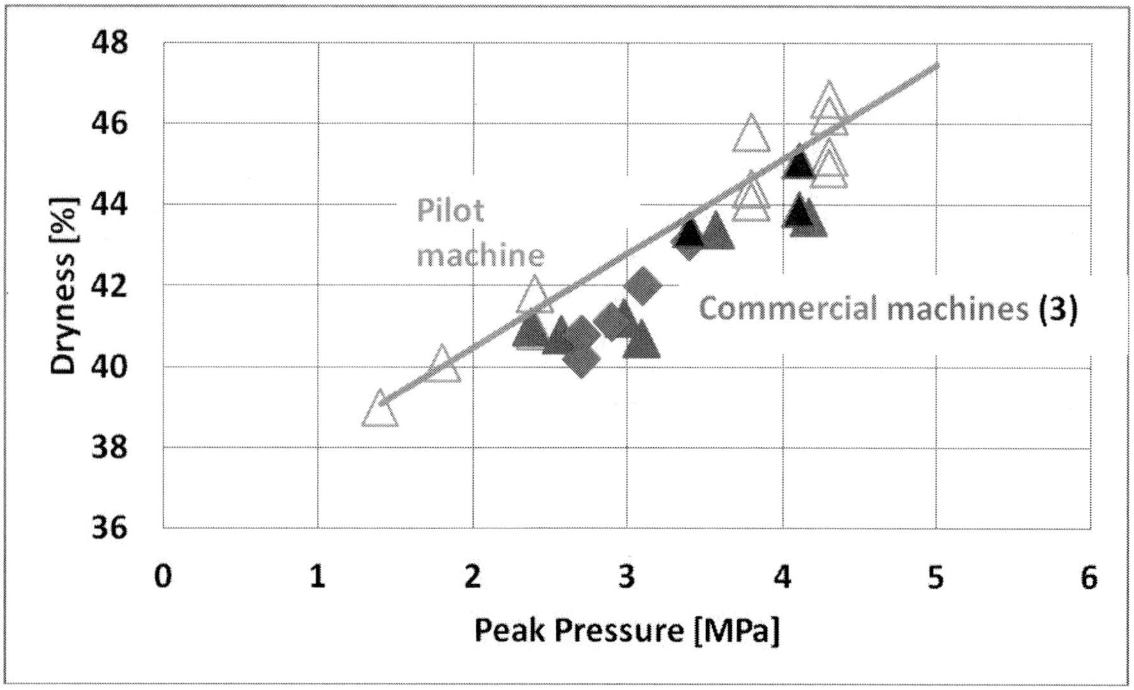

Figure 11. Dryness vs Peak Pressure, basis weight 16 gsm

Advantage NTT concept and press configuration

NTT is a new technology for making premium tissue and it can operate in two different modes. One mode, called the “Textured” produces high quality textured products of high bulk. The second mode, called the ”Conventional mode” produces products similar to a conventional dry crepe machine. The same machine configuration is used for both running modes. Changing from one to the other requires only a short clothing change.

From the forming section the web is carried on the press fabric to the pre-press. Prior to the pre-press nip there is a suction turning roll to dewater the paper web and the felt Dewatering of the felt and the web is very important for the performance of the press. For moisture profiling and to reach highest possible dryness it is recommended to use a steam box at the suction turning roll.

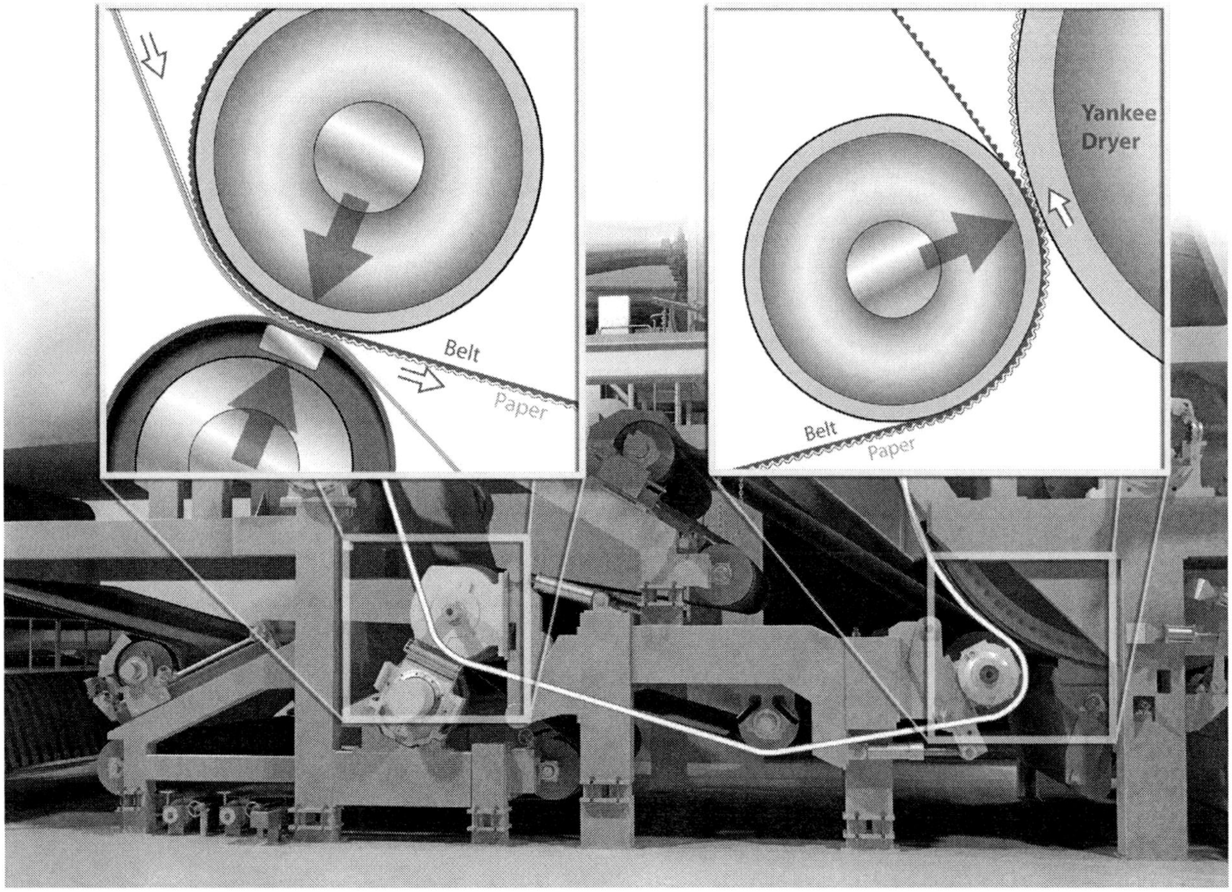

Figure 12. NTT press configuration.

The NTT press can be regarded a single felted press nip as the press dewatering is between a felt carrying the web and a belt that is wrapping the counter roll. The press can run with high linear loads and high peak pressure as the counter roll can tolerate much higher loads compared to a Yankee dryer. In the pre-press nip the dryness of the paper web is increased to a level of 44 - 47% using a textured belt depending on the basis weight. In the conventional mode the dryness is up in the range of 46 – 50%, depending on basis weight. A steam box is used in both cases. In the nip the web is also transferred to the belt and the belt then carries the web to the Yankee dryer. Rewetting of the web is minimal during the transportation to the Yankee dryer, and because of this, it is possible to achieve and maintain high dryness when web is transferred to the Yankee dryer.

A shoe press, SymBelt press, is used in the pre-press and the operating linear load is in the range 400 - 600 kN/m. For tissue producers this seems to be a high load, but the peak pressure in the nip at 500 kN/m is about twice as high as the Yankee press in a conventional machine. Pilot machine trials with a plain belt have actuallyproduced better dryness numbers than expected. Figure 9 shows the response in dryness for two basis weights when the pre-press load was varied from 400 to 600 kN/m. It is possible to reach a dryness of more than 50%. The steam box was used in both cases and the steam added increases the dryness by about 2%. In the NTT process, when compared to the conventional dry crepe, the dewatering of the web and the sheet transfer to the Yankee are separated. This results in both higher bulk and sheet dryness after transfer to Yankee.

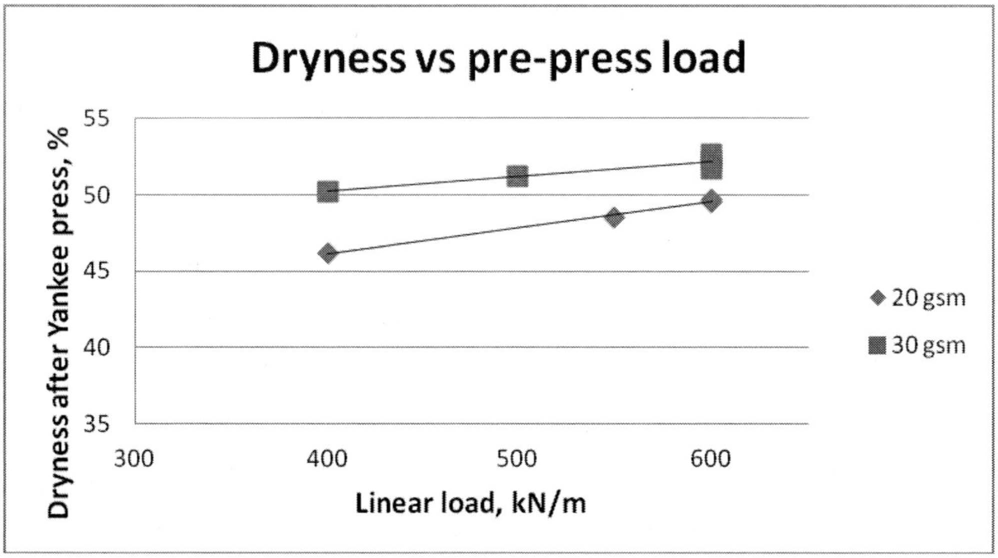

Figure 13. NTT in plain mode. Dryness vs linear press load.

The pressing in NTT is made in a pre-press with high loads, but not high temperature as when pressing against the Yankee in a DCT machine. The above dryness data is taken after the Yankee press nip. The Yankee press is actually only a transfer press operating at low load, about of 45 kN/m. The dryness on the belt compared to after the Yankee press shows a difference of about 1%.

Sheet quality and operating cost

As mentioned before, the dryness is only one parameter although a very important one for the tissue producer. The table below shows how various qualities gain or lose hen using different press concepts.

Press concept	Suction roll	Suction roll	2-press	2-press	ViscoNip 3-chamber	ViscoNip 2-chamber
Bulk index	100	103	90	90	110	105
Softness	Ref	+	-	-	+++	++
Fiber saving	Ref	+	-	-	++	+

When selecting a press concept it is also important to understand the operating cost as well as the investment cost. The operating cost includes fiber, energy, chemicals, clothing, maintenance, etc. The investment cost for a ViscoNip press is lower than for a 2-press roll configuration. The cost for clothing and maintenance is lowest for the single press, followed by the 2-press roll machine. Looking at the total operating cost the ViscoNip press is the lowest.

Is 50% dryness possible to achieve at low basis weights in a DCT machine?

The press dewatering in a DCT machine is made in the press nip(s) against the Yankee dryer. The applied load is limited by the design of the Yankee dryer. The peak pressure applied determines the dryness. However, in the press nip as discussed previously heat is transferred to the web. This heat transfer affects the dewatering, and with shorter time in the nip, the dryness will be lower. The highest peak pressure can be obtained with ViscoNip press, but the peak is limited in machine direction.

The 2-press roll configuration has limitations as regards the linear load that be applied. The peak pressure is limited by roll diameter, rubber cover hardness and material. The advantage with the configuration is high dryness without running high peak pressure. This is a result of heating the web between the first and second press nip.

By combining the suction press roll and the ViscoNip roll it will possible to achieve higher dryness is currently today. The suction press gives a dryness of about 40% for low basis weight tissue. The web is then heated up to the second press nip. In this press high peak pressure can be applied to maximize dryness. With such a concept it is most likely possible to reach dryness levels of 50% for low basis weight tissue. It may require further development of felt uniformity.

Conclusions

With the development of high load Yankee dryers and the extended nip press like ViscoNip there are more options for tissue pressing, which also offer advantages and disadvantages. Highest possible dryness is one important criterion for choosing a press section. There are several other issues to consider such investment cost, machine performance, operating cost, flexibility, tissue quality with respect to softness and bulk.

Metso can offer a wide range of press concepts, and has supplied machines with all of these different press concepts worldwide. The ViscoNip is the most flexible press concept making it is possible to run a wide range of linear loads, without having any crowning issues. This is not possible with roll presses. They have to operate at the linear load they are designed for.

The ViscoNip press produces tissue with higher bulk and softness compared any other concept at a given dryness. The absolute value of higher bulk and softness can always be discussed, but all tissue producers target for high quality.

The single press roll concept has the lowest investment cost, and as long as investment cost is more important than energy savings, there will still be new machines with only a suction press roll. The 2-press roll concept is attractive with respect to energy saving but produces tissue of lower bulk and softness.

Seamless Tissue Plant

Marco Pescantin
Comer SpA

DO MORE WITH LESS

EFFICIENCY OF ENGINEERING AND MANUFACTURING PROCESS
Smart hardware technology: the equipment can adjust itself to the process conditions

FAST PACE ENGINEERING AND INSTALLATION
Smart layout: pipes and piping are designed with the aim to minimize energy “First the pipes then the hardware”.

TIMELESS SOLUTIONS WITH MODULAR DESIGN
Designing equipment and hardware with multiple functions and same spare parts

INTEGRATIVE DESIGN

SMART DATA

High quality, timely and measured data are strategic especially in linking technical with economic performance.

MINIMIZE THE EQUIPMENT AND AUXILIARIES

Using modular technology (mini-process phases combined together) can simplify the final output and algorithms that anticipate and optimize what's about to happen.

REDUCE THE LOSSES IN THE SYSTEM

Most of the lucrative savings are not in the devices but in the piping design.

IMPROVE THE EFFICIENCY

Simply adding variable speed motor controls can cut energy by 15-20% and improve the final yield of the plant by 3 to 5%.

PAY ATTENTION TO THE WATER AND WASTE

Plants process huge quantities of water and produce significant amount of waste. Improving water and waste management can add value to the final product .

MDT deinking technology

Combines technologies in novel ways

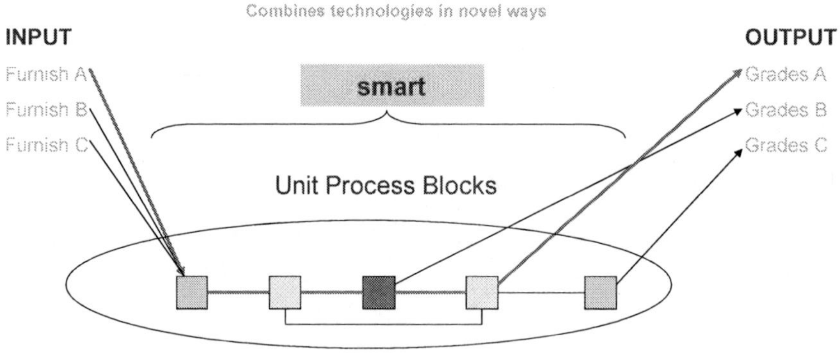

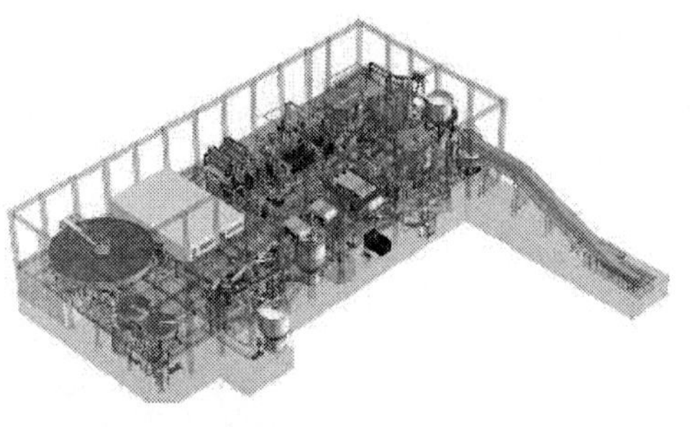

- Saving Energy
- Specific functions
- Process flexibility and selectivity
- Specific cost defined

CTM Tissue machines

Combines technologies in novel ways

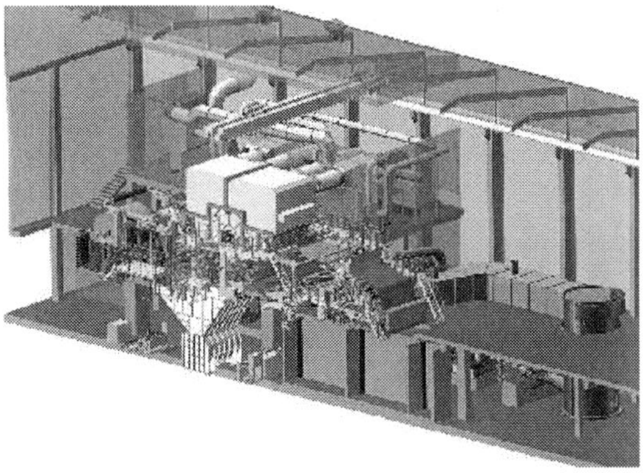

SMART DATA

High quality, timely and measured data are strategic especially in linking technical with economic performance.

MINIMIZE THE EQUIPMENT AND AUXILIARIES

Using modular technology (mini-process phases combined together) can simplify the final output and algorithms that anticipate and optimize what's about to happen.

CTM 12 & 15

The entry level family

	CTM12		CTM15	
Speed	1600	5250	1800	5900
Former (diam.)	1200	47"	1500	59"
Press (diam.)	1050	41"	1200	47"
Yankee (diam.)	3660	12'	4880	16'
Production (100% E)	75-80	84-90	100-110	110-122

36,000 MTpy tissue paper and facial tissue

What you need to know

PAY ATTENTION TO THE WATER AND WASTE

Plants process huge quantities of water and produce significant amount of waste. Improving water and waste management can add value to the final product .

Availability of clean and sorted RCF is quite uncommon today.
The trend is toward high contamination and high content of fillers and ONP

The quantity of sludge form the DIP process can provide opportunity for new energy.

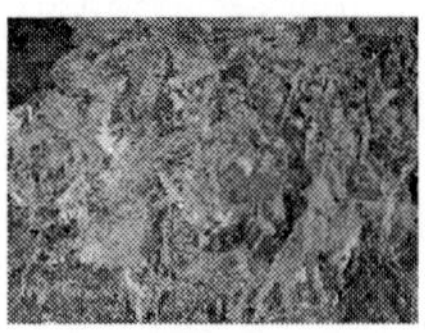

A large volume of plastics and contaminants result.

MOW to tissue paper and facial tissue

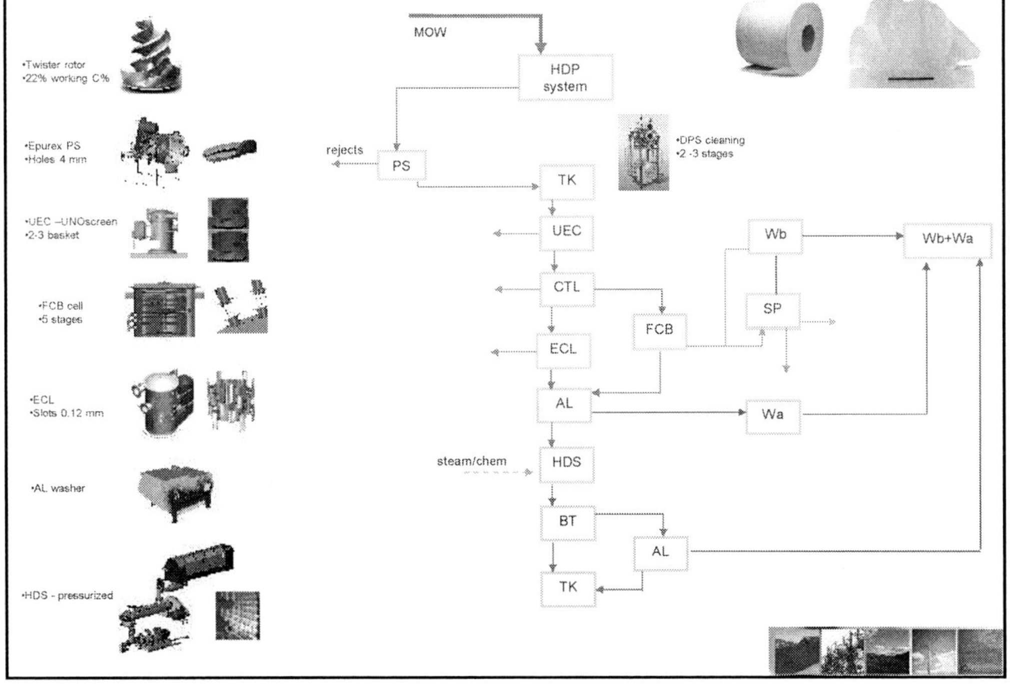

CTM 12 – Entry Level Machine

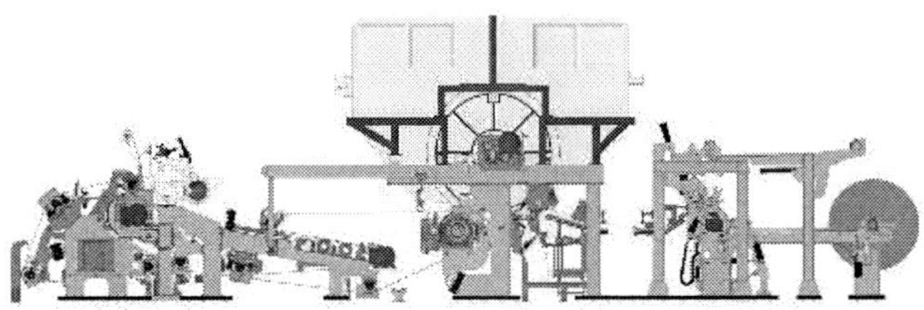

Production	75-80 MTPD	84-90 MTPD	
Grade	Toilet & Facial		
BW	11-34 gr/m2		
Speed	1600	5250	
Furnish	100%RCF 100% VF		
Steam	1.04 t/t		
Energy	750-800 kwh/t		(See refining)
Water	6-10 m3/t		

CTM12 Features

2 Turbulence Generator
Variable Geometry

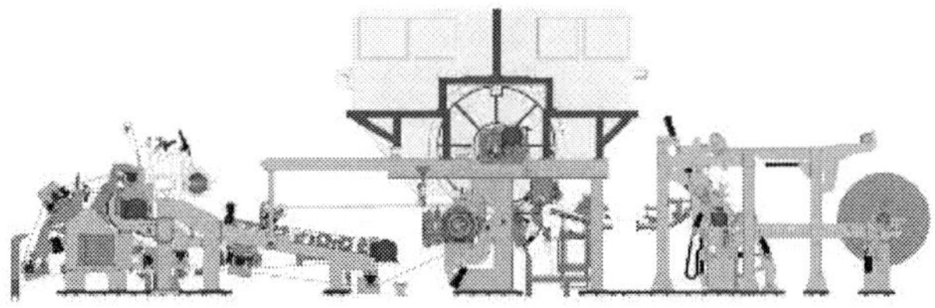

Jet impingement control

100% Wire impingement

Nominal impingement geometry

100% Felt impingement

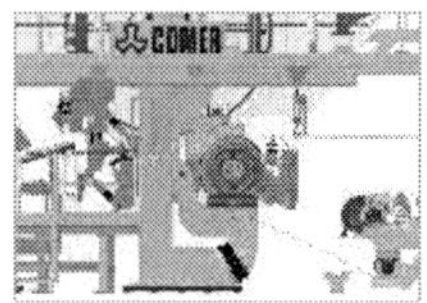

Rear side vacuum connection (No DRIVEN roll).
Rear Side and Tending Side Bearing identical (No internal bearing; only one spare.)
Trim and internal box angle adjustable during machine operation.

CyberDry-SteelShell

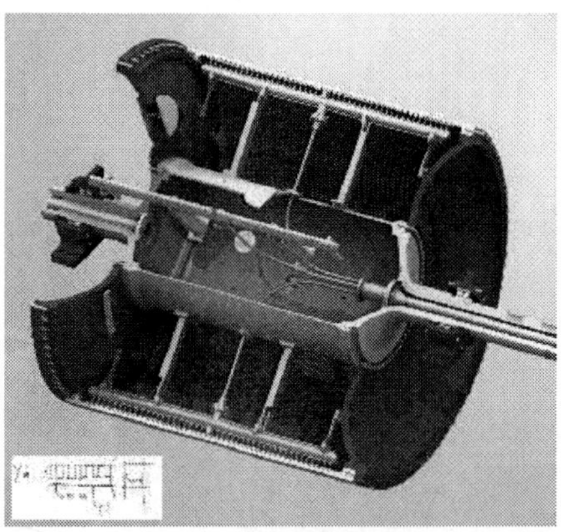

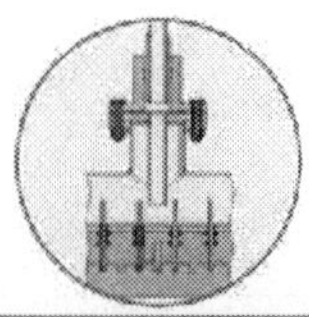

Condensate velocity control

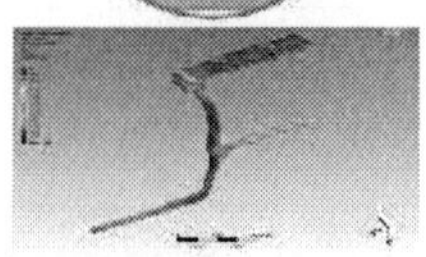

Finite elements stress analysis

	Y12		Y15	
Material	Steel shell Metalized		Steel shell Metalized	
Pressure (ASME stamp)	10	150	10	150
Hardness (HRC)	55	55	55	55

Hood and condensate

(partner supplier)

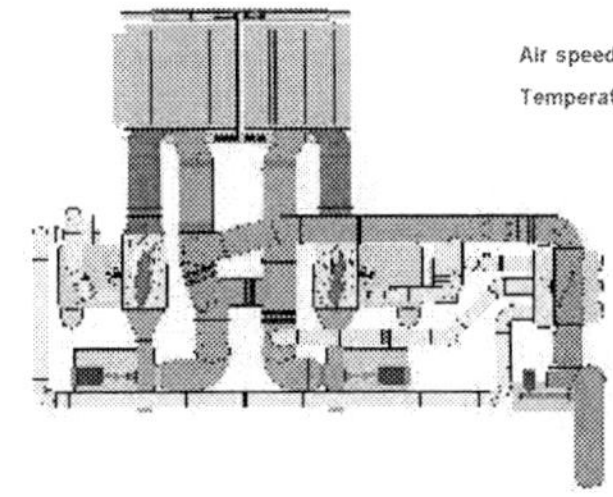

Air speed 100 m/sec

Temperature 500 (530) C

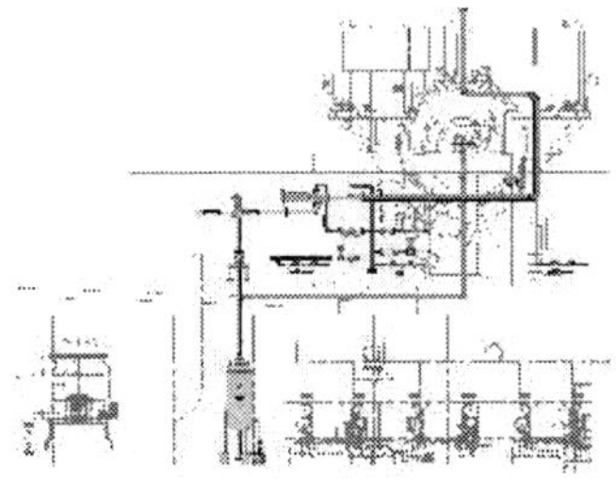

Results & Schedule

		TP	F
SR	SR	21	21
Brightness	D65/10 UVEX	75	78
L*-value	ISO	78	82
a*-value		0.5	0.5
b*-value		4	4
Stickies>20 micron	mm2/kg	900	500
Ash 560C		3	3
COD (filtrate accept pulp)	Kg/t	10	10
Brightness	D65/10 UVEX	82	88

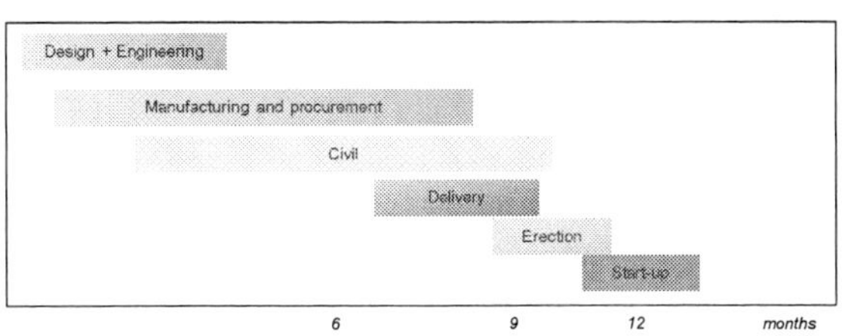

40,000 MTpy Kitchen Roll and Towel

MOW & Curbside to Kitchen Towel and Towel Rolls

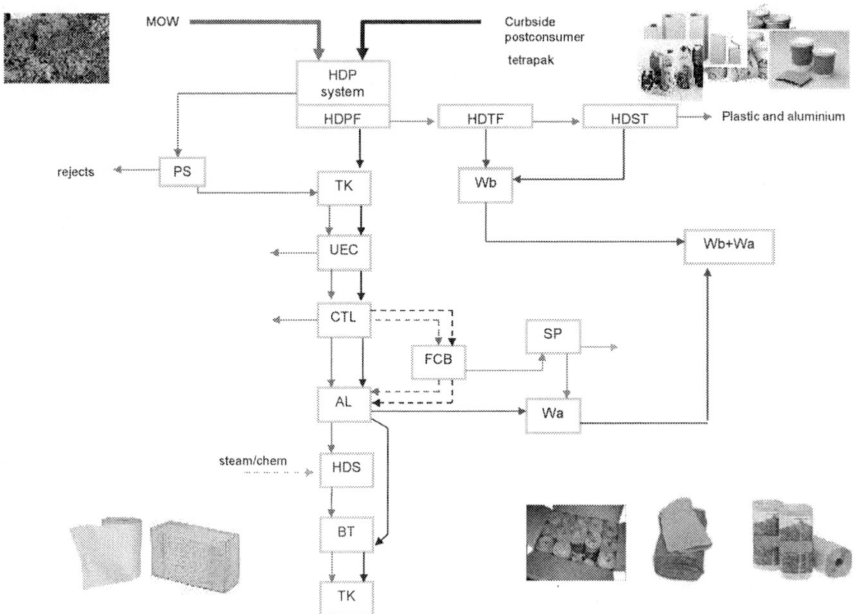

CTM 15 & ADT (Wet and Dry Crepe)

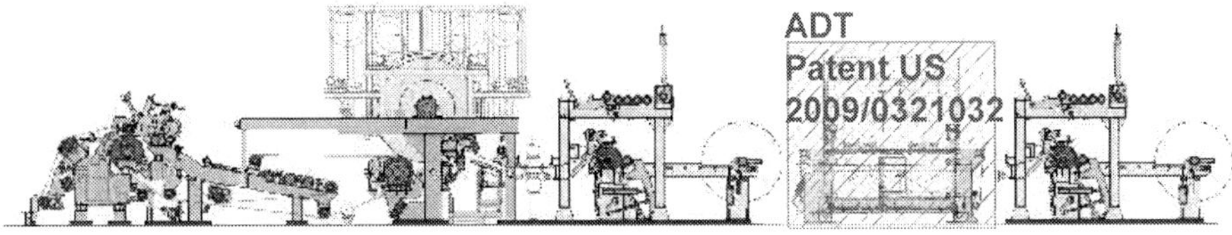

Production	100-110		
Grade	Towel and kitchen roll		
BW	18-32		
Speed	1800		
Furnish	100%RCF 100% VF		
Steam	1.04		
Energy	600-800		See refining
Vacuum			
Water	5-8		

ADT (Air Flotation Dryer)

Innovative technology for WET CREPE

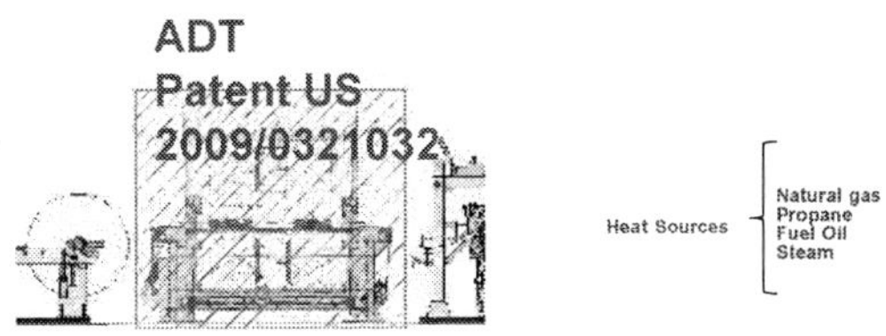

Production	Any capacity		
Grade	Towel and kitchen roll		
BW	High BW		
Speed	2000		
Furnish	100%RCF 100% VF		
Air speed	11,000 fpm T air bars 5,000 fpm B hole bars		
Gas requirement	25% less than TAD		

Energy Savings

MDT+CTM15+ADT

MDT+CTM	NRL Kwh/MT $	NRL daily avg. cost $	NRL Year savings $
Standard configuration @80 MTPD	580	2790	
@ 100 MTPD – phase A	405	2450	130,000
@ 100 MTPD – phase B	320	1915	312,000

IMPROVE THE EFFICIENCY

Simply adding variable speed motor controls can cut energy by 15-20% and improve the final yield of the plant by 3 to 5%.

CUSTOMIZING

...and beyond

Private labels are family business with few decision-makers.

For this specific customer we have developed a seamless concept tailored to provide first class turn-key projects. Simply stated we want to make the life of our customer easy.

The principle of INTEGRATIVE DESIGN have been applied for several phases of any plant's design: engineering, equipment manufacturing, spare parts and services.

Two examples were shown where MDT technology and the new CTM tissue machines have been applied to process low quality and critical RCF fibres to produce tissue products.

VOITH

Voith - New Pressing technology for Higher Bulk and Dryness on Conventional Tissue

Tissue World Americas, March 2012

VOITH

Tissue World

Current market situation

Tissue Consumption Per Capita kg/hab.year - 2010

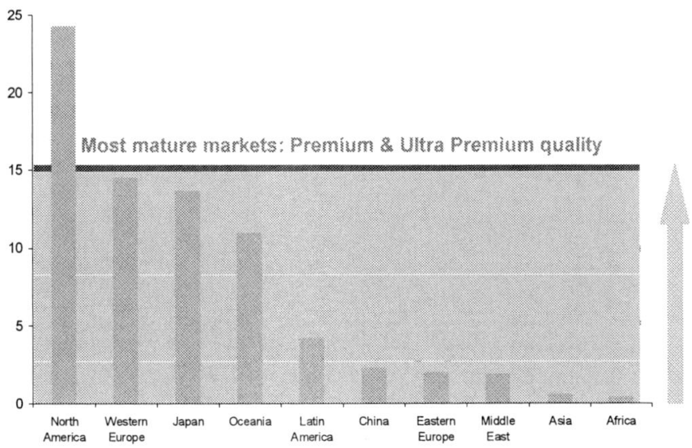

Productivity

- Speed
- Efficiency
- Quality

Less Production Cost

- Energy
- Fiber

2

VOITH

Conventional Tissue Machine (Dry crepe)

Current proven technologies

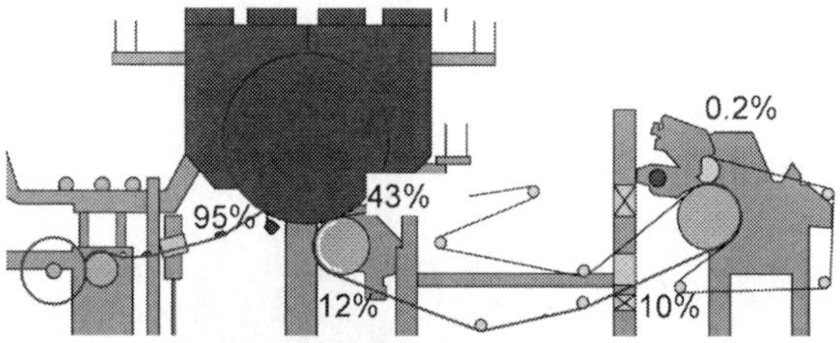

Suction press roll (SPR)

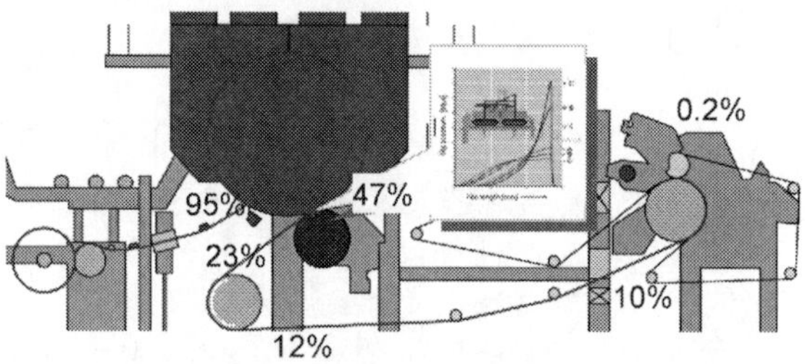

Voith NipcoFlex-T (NFT)

3

VOITH

Voith Goals

New developments

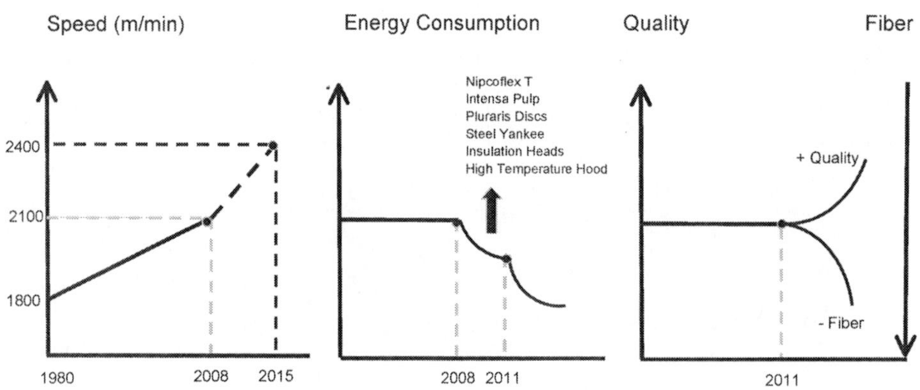

Higher Speed Machines

Less Energy Consumption

Higher Quality
or Less Fiber Consumption

4

VOITH

TissueLev pressing concept

Highlights

Low press load, hot air & vacuum

- Hot air to decrease water viscosity and improve drainage by vacuum
- Low pressing load applied when sheet is less resistant to mechanical pressing: from 12 to > 30% dryness
- Mechanical Press nip applied only at higher dryness: > 30%

Increase dryness and protect bulk on conventional machines (Dry Crepe)

6

VOITH

TissueLev pressing concept

Combination of Low press load, hot air & vacuum

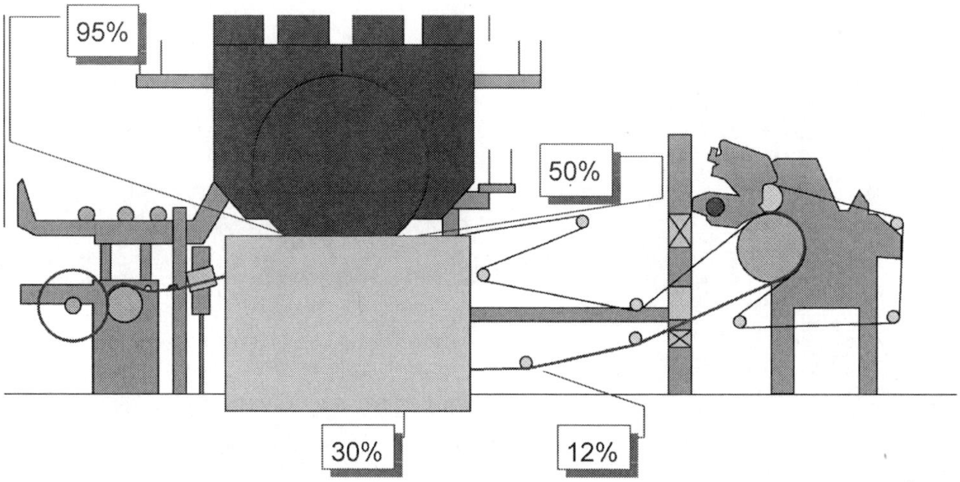

7

VOITH

TissueLev pressing concept

Pilot Trial results

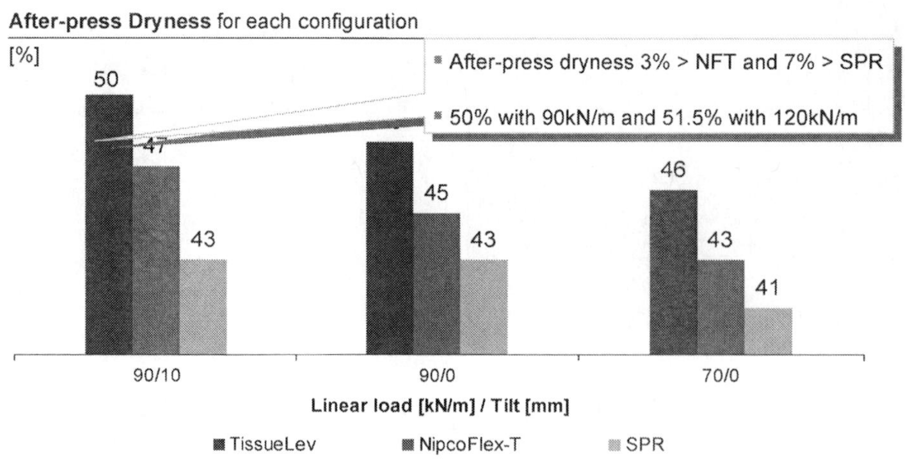

8

VOITH

TissueLev pressing concept

Pilot Trial results

Bulk vs. After-press Dryness for each configuration

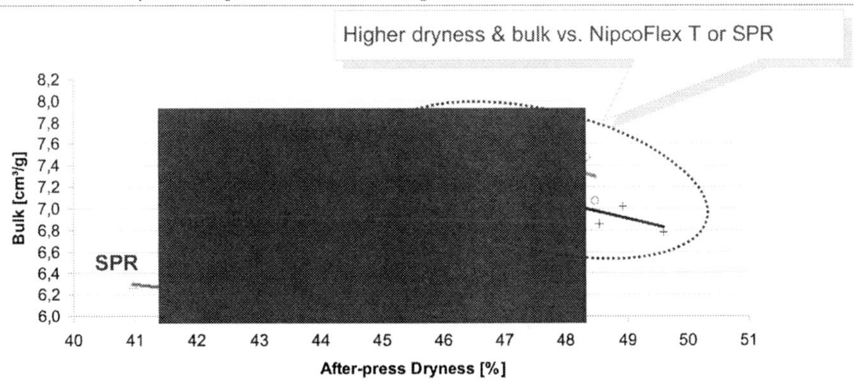

VOITH

TissueLev pressing concept

Energy comparative – Theoretical Calculation

Energy Consumption for each configuration

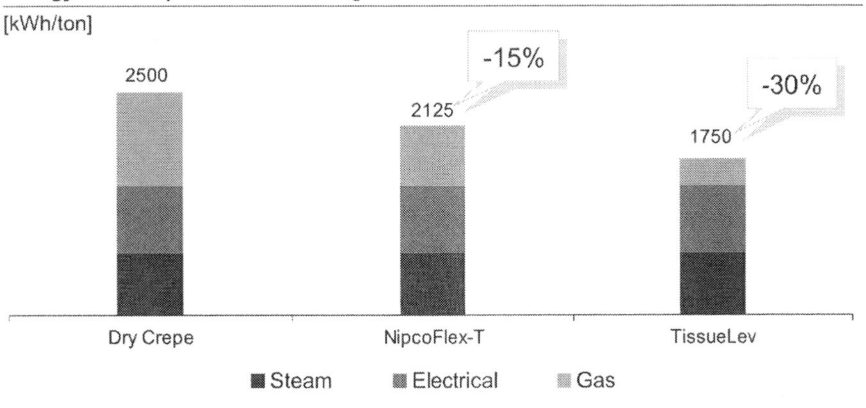

VOITH

TissueLev pressing concept

Benefits

- To be installed on Dry Crepe machines

Energy savings & higher production:

- 3% dryness gain = 15% higher production or 15% energy savings vs. NF-T
- 7% dryness gain = 30% higher production or 30% energy savings vs. SPR

Paper quality & fiber savings:

- Up to 15% bulk gain vs. NipcoFlex-T
- Up to 25% bulk gain vs. SPR

Final development stage - commercially available soon

11

VOITH

Engineered Reliability

12

Yankee Dryer System: Critical Asset Protection, Monitoring and Control utilizing 3D TRASAR® Technology by Nalco

By Jeff Fox, Industry Technical Consultant and Steve Hoefs, Senior Industry Development Manager, Nalco An Ecolab Company

ABSTRACT

Reliable Yankee dryer performance is a key contributor to the smooth operation of every tissue machine. This paper will explore how through the use of 3D TRASAR Technology for Yankee Dryer Systems from Nalco, tissue manufacturers are able to protect this vital mill asset by monitoring/controlling critical parameters within the Yankee dryer condensate system.

When applied, 3D TRASAR Technology for Yankee dryer systems is able to monitor/control the return condensate system to ensure that appropriate chemical treatment programs are applied in response to the dynamic nature of the Yankee dryer steam system. Through examination of the chemical reactions occurring within the return condensate system and customer case histories, this paper reviews the actual performance results achieved through use of this technology including:

- *Minimized corrosion in the Yankee dryer. The Nalco Corrosion Stress Monitor (NCSM) and pH measures the corrosion potential of the condensate in the Yankee dryer system as opposed to the bulk condensate. A supplemental amine feed is proportionally controlled based upon changes in the measured corrosion potential.*
- *Increased machine runnability by minimizing soda straw plugging caused by iron deposition.*
- *24/7 monitoring and alarming of the Yankee dryer operational conditions and condensate system to quickly respond to system upsets and provide diagnostic support.*
- *Analyze the effect of Mechanical, Operational and Chemical (MOC) changes in the steam production and Yankee dryer condensate systems in real time, reducing the production of corrosion by-products.*

INTRODUCTION

In the tissue manufacturing process, the performance of the Yankee dryer has a direct impact on the quality of the finished product. Yankee dryer performance is directly related to the continuous delivery of a high quality steam supply to the dryer and an efficient evacuation system to remove the condensed steam. Both of these elements are critical in maintaining a consistent skin temperature across the dryer surface. These key elements are influenced by the operation of the boiler pretreatment system and control / monitoring of the return condensate chemical treatment program.

When the Yankee dryer performance declines, the tissue machine may experience reduced drying capacity, increased pressure drop at constant steam blowthrough, wet streaks and random coating patchiness. Eventually the crepe quality and uniformity of the sheet will degrade and chatter can occur. When these symptoms become apparent, it is too late, quality and production have already been adversely affected!

To maintain Yankee dryer system performance, the tissue manufacturer must develop a comprehensive plan which addresses three specific areas; steam/condensate chemistry, corrosion / deposit control, and system monitoring.

Steam / Condensate Chemistry – Effect of Alkalinity and Oxygen

The boiler pretreatment system has a direct effect on the amount of carbonic acid and oxygen that can enter the Yankee dryer system and negatively affect its performance. In sodium zeolite softened water systems all of the raw water alkalinity is allowed to enter into the boiler feedwater system. Under boiler water temperature and pressure conditions the alkalinity breaks down and carbon dioxide gas is generated.

The reaction is shown below.

Breakdown of feedwater alkalinity

$$2\,HCO_3^- \xrightarrow{\text{Heat}} CO_3^{=} + H_2O + CO_2$$

Bicarbonate — Carbonate — Carbon Dioxide Gas

$$CO_3^{=} + H_2O \xrightarrow{\text{Heat}} 2OH^- + CO_2$$

Carbonate — Hydroxide — Carbon Dioxide Gas

The carbon dioxide gas travels out with the steam and possesses no cause for operational concern until the steam is condensed back into a liquid form. At that point, the condensing water reacts with the CO_2 gas forming carbonic acid which will equilibrate in solution as shown by the reaction below.

$$CO_2 + H_2O \longleftrightarrow H_2CO_3$$

Carbon Dioxide — Water — Carbonic Acid

$$H_2CO_3 \longleftrightarrow H^+ + HCO_3^-$$

Carbonic Acid — Hydrogen — Bicarbonate

Systems operating with higher boiler feed water alkalinity will generate more carbonic acid in the return condensate system and, if not continuously monitored and adjusted for, will produce periods of depressed condensate pH and a more corrosive environment to the return condensate internals including the Yankee dryer. It should be noted that the remaining bicarbonate alkalinity will recycle with the return condensate back to the boiler where it will once again breakdown to form additional carbon dioxide that will leave with the steam.

One ppm of carbon dioxide in a 140°F condensate stream will drop the pH an entire pH unit, from 6.5 to 5.5. A pH of 5.5 is 10,000 times more aggressive to mild steel than a pH of 7.5 and 1,000,000 times more aggressive than a pH of 8.5. Since corrosion rates increase with temperature, the aggressiveness of a low pH condensate stream cannot be ignored. Figure 1 shows how carbon dioxide can depress the pH of pure water at varying concentrations and temperatures.

Alkalinity levels in the boiler feedwater can be reduced through enhancements to the boiler pretreatment equipment. In most tissue mills, which operate with low pressure boiler systems, the boiler feedwater make-up system is typically based around sodium zeolite softeners alone. By adding a reverse osmosis train to the pre-treatment strategy, mills have shown a reduction in make-up water alkalinity

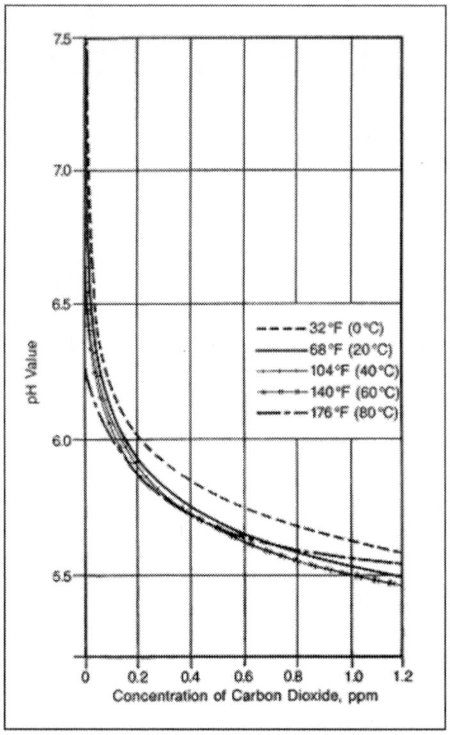

Figure 1 – *pH values of solutions of carbon dioxide in pure water at various concentrations*[1]

by up to 95%. Note: Care must be taken to consider operational changes that will be required to the boiler system during reverse osmosis outages. The sudden increase in alkalinity with no operational changes to the boiler can charge the system with large amounts of carbon dioxide, causing significant pH depression across the Yankee dryer.

For tissue mills which operate at higher boiler pressures demineralizers will be utilized in their pretreatment system. In demineralized make-up systems, the make-up water alkalinity concern has already been addressed as this process reduces alkalinity levels to near zero.

Most tissue mills indirectly monitor/control the effects of carbonic acid present in the condensate system by checking the pH of the main return condensate receiver in the power plant. Through the use of 3D TRASAR technology, we have seen that this practice is insufficient in maintaining consistent system protection of the Yankee dryer system leaving it vulnerable to low pH excursions.

Another element that must be taken into consideration within the pre-treatment and return condensate system is oxygen. Oxygen can enter in the condensate system and the Yankee dryer through ineffective operation of the feedwater deareator or air inleakage from vacuum seals. Oxygen dissolved in 140°F pure condensate is 6 to 10 times more corrosive to iron than its carbon dioxide molar equivalent.

When both carbon dioxide and oxygen are present the resulting corrosion rate is 10 to 40 times greater than the sum of the corrosion rates of the two gases acting separately.

Corrosion/Deposit Control

Yankee dryer performance is directly related to the effective removal of the condensed steam from within the dryer. A common Yankee dryer condensate removal design utilizes "soda straws" to evacuate the condensed steam. If flow through these straws is restricted, the condensate can build up within the dryer causing inefficient operation and poor sheet quality. The most common cause of soda straw flow restriction is plugging of the straw orifice with accumulated iron. The tip of the soda straw is very susceptible to deposit formation due to the large pressure drop experienced in that section of the pipe. This iron deposit can be caused by the formation of active corrosion by-products within the dryer system or by the rapid release of the passive magnetite layer which forms on the drum wall and internals. Both of these events can occur when a corrosive atmosphere is present within the Yankee dryer system.

A corrosive atmosphere will be present whenever there is an imbalance between the level of carbonic acid and oxygen present within the condensate stream and the chemical treatment program utilized to neutralize/scavenge these elements. Because of the dynamic nature of the tissue machine operation (grade changes, breaks, etc.), this balance is continuously shifting making it difficult to maintain a non-corrosive atmosphere at all times.

A critical element effecting Yankee dryer performance and reliability is in the establishment and maintenance of a thin passive layer of iron on the internal drum surfaces referred to as magnetite. Metal passivation is a critical process that occurs and provides protection of the Yankee dryer system. Passivation is the formation of an insoluble nonporous protective oxide on a metal surface. Cast iron and steel will self-passivate in water if no contaminants are present. The reaction is very slow under 212°F. Passivation can be accelerated by the use of chemical treatment (amines), however, under the operating conditions present within the Yankee dryer system, the passivated layer is not very robust and is easily damaged, both mechanically and chemically if the system is not properly controlled. Damage to the magnetite layer will release particulate iron into the bulk condensate stream and, depending on severity, cause a threat of plugging at the tips of the Yankee dryer soda straws. The reactions that drive the passivation process are shown below.

1) $Fe^{0} \rightarrow Fe^{+2} + 2e^{-}$ (oxidation)

2) $Fe^{+2} + 6H_2O \rightarrow [Fe(H_2O)_6]^{+2}$ (hydrated)

3) $[Fe(H_2O)_6]^{+2} \rightarrow Fe(OH)_2 + 2H^{+} + 4H_2O$ (hydrolyzed)

4) $Fe^{+2}(OH)_2 + H_2O \rightarrow Fe^{+3}(OH)_3 + 0.5H_2$ (oxidation)

5) $n\,Fe^{+3}(OH)_3 \rightarrow [Fe^{+3}O(OH)]_n + n\,H_2O$ (polycondensation)

6) $2[Fe^{+3}O(OH)]_n + n\,Fe^{+2}(OH)_2 \rightarrow n\,Fe_3O_4 + 2\,n\,H_2O$

Note in step 1) above, two electrons are released as the base metal is ionized. The reaction must be allowed to passivate the base metal, but limited, to prevent excess magnetite formation. In the presence of carbonic acid and/or oxygen, the passivation process becomes a corrosion process. Figure 2 shows how oxygen impacts the corrosion process and how hydrogen is released by the process.

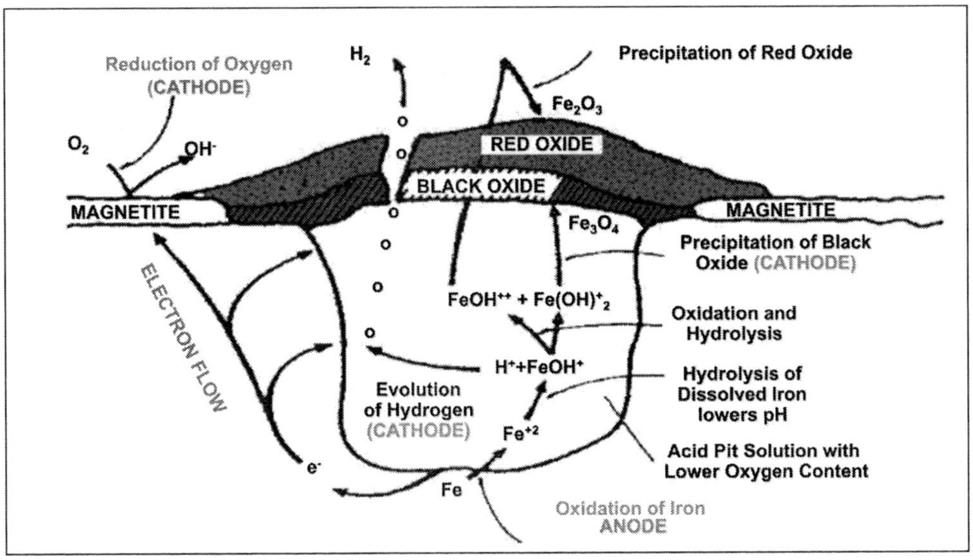

Figure 2 – *Electrochemical oxygen corrosion reaction*[2]

Please note the electron flow from the anode of the corrosion cell to the cathode. Measuring this electron flow is a direct indication of the corrosion rate within the corrosion cell.

MONITORING

Because of the variability in boiler pre-treatment system and tissue machine operation, it is imperative that a monitoring and chemical control program be established that maintains a non-corrosive atmosphere within the Yankee dryer and return condensate system at all times.

Table 1 lists common techniques used to monitor the return condensate systems within the tissue industry today. The table also highlights some of the limitations these monitoring techniques face when trying to assure/confirm proper corrosion control within the Yankee dryer system.

None of the tests listed above provide a direct indication of the atmosphere present within the return condensate in the Yankee dryer. All of these techniques utilize indirect measurements in an attempt to quantify the corrosive nature of the Yankee condensate and to provide assurance that the appropriate chemical treatment program is being applied in response to the changing operational conditions.

3D TRASAR Technology for Yankee Dryer Systems

Realizing how dependent Yankee dryer system performance is on a comprehensive plan that effectively addresses the steam/condensate chemistry, corrosion/deposit control and monitoring, 3D TRASAR technology for Yankee dryer systems from Nalco was developed to continuously monitor and control those critical parameters associated with maintaining dryer performance. The technology incorporates sensors which monitor Yankee dryer condensate system pH, conductivity, presence of boiler water carryover (via inert TRASAR), and water corrosivity in real time. 3D TRASAR technology collects these system measurements along with other Yankee dryer data (steam flow, dryer pressure, etc.) every 6 seconds and utilizes this information to determine the appropriate chemical treatment adjustments necessary to maintain a stable Yankee dryer condensate environment. The level of responsiveness achieved through this technology to the changes in system demand minimizes system upsets and limits the production of corrosion by-products and soda straw deposit formation within the Yankee dryer.

A critical component of the 3D TRASAR technology is Nalco's innovative way of directly monitoring the corrosive nature of the return condensate stream. Instead of relying on indirect measurement techniques, the patented Nalco Corrosion Stress Monitor (NCSM) was developed to measure the oxidation reduction potentials of hot water systems at system temperatures and pressures. When combining the NCSM readings with Fe^{+2}, soluble iron, wet lab testing, an optimal control range is able to be determined for the Yankee dryer operation which minimizes system corrosion potential and the subsequent soda straw plugging concerns that may result. Once this range is determined, the 3D TRASAR technology will continuously monitor and re-adjust the chemical treatment program to maintain system control within the specified range.

The 3D TRASAR technology also offers the ability to monitor remotely on a 24/7 basis generating alarm notifications when the system is operating outside of control limits that are established by the mill. This provides the mill and Nalco the ability to quickly respond to upsets that are outside of normal operating conditions and provide diagnostic support to determine their cause. Alarm conditions are communicated immediately to persons chosen by the customer and Nalco.

Table 1 – *Common methods of return condensate monitoring and their limitations*

Instrument or Test	Limitation
pH	Indirect measurement of only one of the components of corrosion. Does not measure effect of oxygen in the return stream
Dissolved Oxygen	Oxygen will preferentially flash off with the steam through the condensate flash tank and will not remain with the condensate.
Conductivity	Single component measurement to identify boiler water carryover, amine overfeed, and/or raw water contamination.
Particle Monitor	Slow sample time that cannot be used to control amine.
Millipore Testing	Grab sample only. Requires 5 fps sample flow for measurement to be representative. No continuous monitoring of system conditions.

The following case studies highlight the use of 3D TRASAR technology for Yankee dryer systems as part of Nalco's comprehensive plan to improve Yankee dryer performance by addressing the steam/condensate chemistry, corrosion/deposit control and monitoring of this critical mill asset.

Case Study 1

A US mid-western tissue mill had experienced Yankee dryer soda straw plugging concerns in the past. As a result of these concerns, the mill established a satellite condensate treatment application ahead of the Yankee dryer system. After being introduced to 3D TRASAR technology, a decision was made by the mill to evaluate the use of this technology for the Yankee dryer systems to determine if improved treatment program efficiency and effectiveness could be gained through use of this technology.

Prior to the use of 3D TRASAR technology to automate the satellite condensate treatment, the program was adjusted by the operations staff based upon the pH of a grab sample taken off the Yankee flash tank. 3D TRASAR technology was installed utilizing this same sample point so that data could be compared to previous results obtained through manual control. The pH control range that the operators were targeting was 8.8 to 9.2 based on the metallurgies in the Yankee dryer and soluble iron wet tests that showed 0 ppb iron levels when operating within this control range.

Figure 3 below shows the results of the evaluation performed utilizing 3D TRASAR technology for Yankee dryer systems. The first portion of the graph shows when the 3D TRASAR system was in monitoring mode only. This initial evaluation established the operating benchmark for manual control in maintaining the desired pH control range. In manual control, the Yankee dryer system was experiencing regular pH swings from 7.5 to 9.5.

The second half of the graph shows when the 3D TRASAR for Yankee dryer systems was placed in control mode. The 3D TRASAR controller was able to maintain Yankee dryer pH between the desired control limits, 8.8 to 9.2 on a consistent basis. Based upon these results, the mill is proceeding with additional enhancements to the satellite chemical feed system to provide additional tightening of this control band.

Once this 3D TRASAR system control was put into place, an evaluation was performed to determine the resilience of the passive magnetite layer that had now been established on the Yankee dryer internals. This passive layer provides a protective barrier against active corrosion on the metal surface within the Yankee dryer. A test to simulate a low pH excursion was performed to determine the length of time this passive layer would remain intact before releasing iron into the return condensate stream. In addition to collecting the information from the 3D TRASAR system, soluble iron testing was performed throughout the test. Table 2 summarizes the relationship observed between the soluble iron levels, pH and the NCSM.

This data shows that the passivated magnetite iron layer in the Yankee dryer deteriorated in a short period of time and is not robust. It took just over 3 hours for the passivated layer to begin shedding measureable amounts of iron. This soluble iron once released into the bulk condensate stream could deposit at the tip of the Yankee dryer soda straws causing flow restriction.

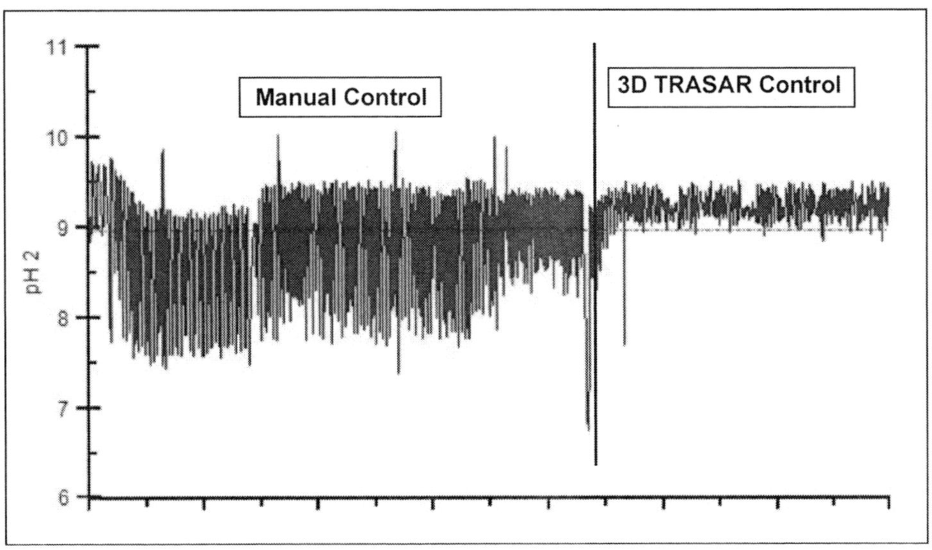

Figure 3 – *Condensate pH control obtained through 3D TRASAR for Yankee dryer systems*

Table 2 – *Effect of low pH excursion on metal passive layer and soluble iron throw*

Time	pH	NCSM reading (mv)	Soluble iron (Fe^{+2})
8:30 am	9.01	-702	0 ppb
Amine off 8:40 am	9.01	-702	0 ppb
9:40 am	6.95	-640	0 ppb
10:45 am	6.64	-625	1 ppb
11:45 am	6.54	-620	1 ppb
1:00 pm	6.2	-590	3 ppb
3:00 pm	6.01	-560	10 ppb
Amine on 3:15 pm			
5:00 pm	9.0	-702	0 ppb
6:30 pm	8.9	-700	0 ppb

Case Study 2

During the first seven months of operation of a greenfield tissue machine, the tissue machine experienced five unscheduled maintenance outages due to poor tissue quality created by concerns with the performance of the Yankee dryer system. Each unscheduled machine outage was preceded initially by a gradual decrease in blowthrough steam velocity within the Yankee and an increase in differential pressure across the system. Just prior to the tissue machine being taken out of service, both of these operational conditions accelerated in their rate of degradation, as seen in Figure 4. The longest period between these outages was 42 days.

In each incident, soda straw plugging caused by iron deposits was determined to be the cause of the poor Yankee dryer performance. After each outage, operational changes were made to try and eliminate soda straw plugging – increased venting off the condensate flash tank, increased pH control range in main return condensate sample, new soda straws installed, installation of multiple sample points around the Yankee dryer, etc. With each operational change made, the time period between unscheduled outages remained unchanged.

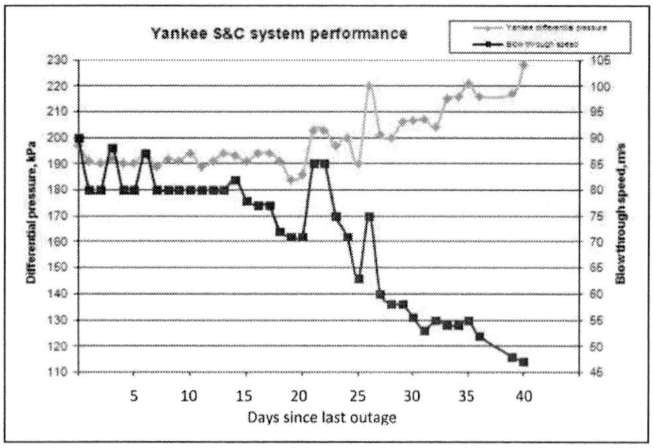

Figure 4 – *Yankee S&C System Performance – Plugged Soda Straws layer and soluble iron throw*

Nalco was asked to evaluate the problem and to offer recommendations for operational changes to minimize the soda straw plugging and eliminate these unscheduled outages. Because of the speed and nature of the deposit formation, it was believed that the plugging occurring in the soda straws was due to excessive corrosion rates present within the Yankee dryer system. Corrosion rates of this magnitude indicated that a combination of elevated carbon dioxide and oxygen levels were likely present in the return condensate system.

A comprehensive plan was developed to improve the mechanical removal of oxygen and carbon dioxide within the steam supply system (increased dearator venting and RO installation) and to improve the level of system monitoring and control in the boiler feedwater and Yankee dryer condensate systems via 3D TRASAR technology.

3D TRASAR Technology for Boilers was installed in the feedwater system to monitor deaerator mechanical performance and to optimize the feed of chemical oxygen scavenger resulting in reduced feedwater system corrosion. Nalco 360 Service was utilized to monitor system performance around the clock in analyzing system alarms through human interaction, not static computer monitoring.

To improve the monitoring/control of the Yankee dryer, the mill incorporated 3D TRASAR technology for Yankee dryer systems in the return condensate line between the Yankee dryer and the condensate flash tank. This system was also tied to Nalco360 service so that immediate analysis and action could be taken in case of system alarm, Figure 5.

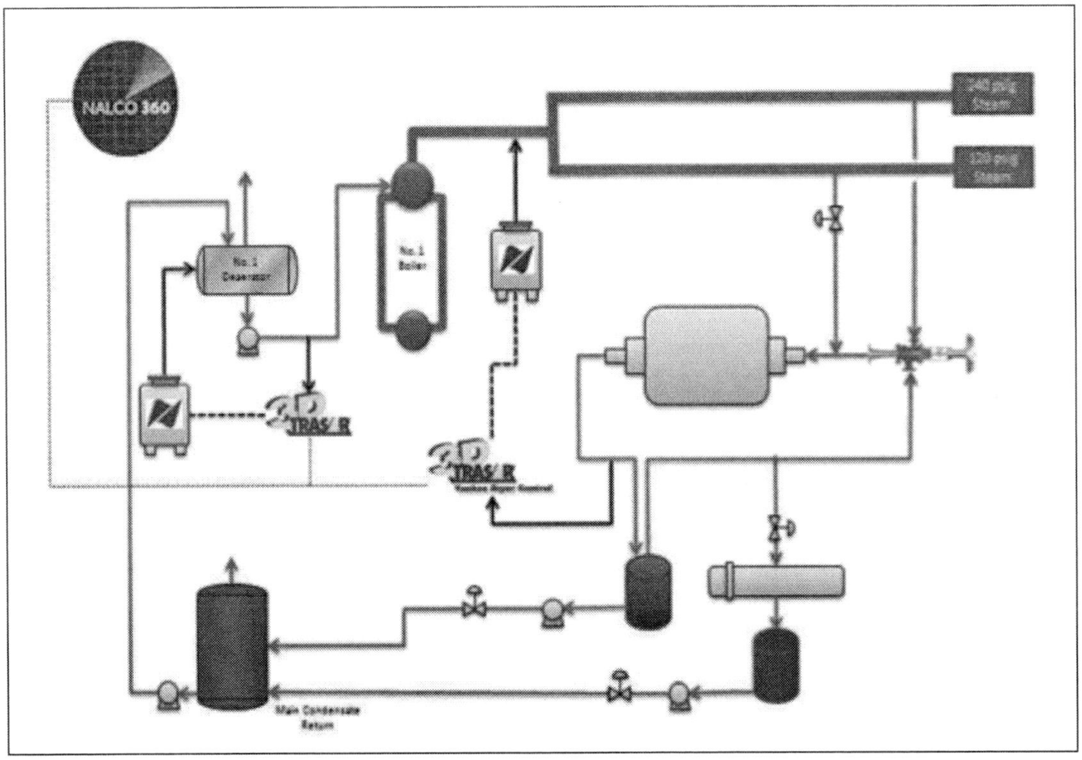

Figure 5 – *Layout of 3D TRASAR Technology application points and Nalco 360 Service*

Through the combination of mechanical changes made to reduce oxygen and carbon dioxide levels, and the implementation of 3D TRASAR technology in the boiler feedwater and Yankee dryer systems, condensate corrosion rates have been significantly reduced and soda straw plugging concerns eliminated. This tissue machine is now running consistently on the mill's scheduled six month maintenance outage plan. There has been no sign of soda straw plugging as evidenced by a consistent Yankee dryer differential pressure and blow through velocity between outages (Figure 6) and through internal inspection performed during the six month outage.

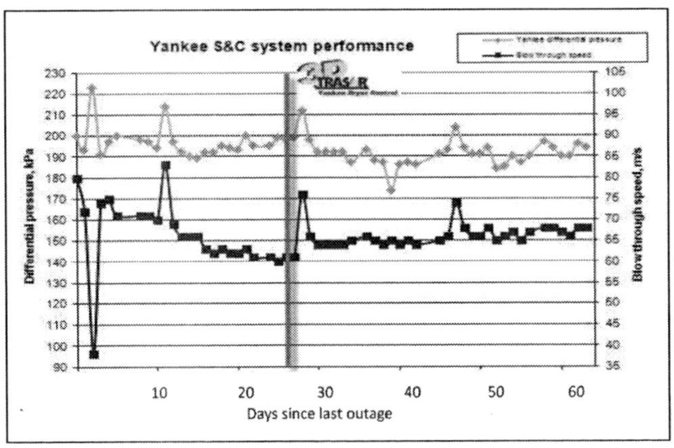

Figure 6 – *Yankee S&C System Performance – Clear Soda Straws*

SUMMARY

Reliable Yankee dryer performance is a key contributor to the smooth operation of every tissue machine. Through application of 3D TRASAR Technology for Yankee Dryer Systems from Nalco, tissue manufacturers are able to maintain the performance of and protect the internal surfaces of this vital mill asset by monitoring/controlling critical parameters within the Yankee dryer condensate system.

REFERENCES

1. P. Hamer, J. Jackson, and E.F. Thurston 'Industrial Water Treatment Practice' Butterworth, 1961
2. I. Cotton 'Oxygen scavengers – the chemistry of sulfite under hydrothermal conditions' National Association of Corrosion Engineers, March 1987

Nalco Company 1601 West Diehl Road • Naperville, Illinois 60563-1198
SUBSIDIARIES AND AFFILIATES IN PRINCIPAL LOCATIONS AROUND THE WORLD

New TAD Simulator Provides Tissue Makers a Competitive Edge

Doeung David Choi

March 21, 2012

ASHLAND.

Outline

- Structured sheet
- Release dynamics
- TAD simulator
- Product development
- Summary

ASHLAND.

Formatting Issue - Best Available Paper/Graphic

Structured Sheet Process

- Sheet transfer (rush transfer) from forming to structured fabric
- Through air drying
- Sheet transfer to Yankee dryer

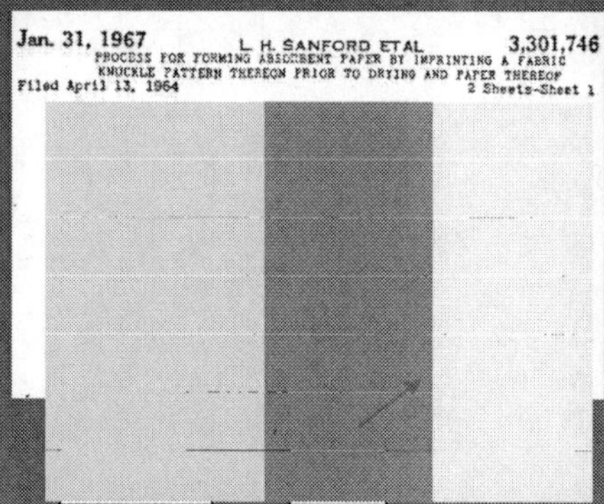

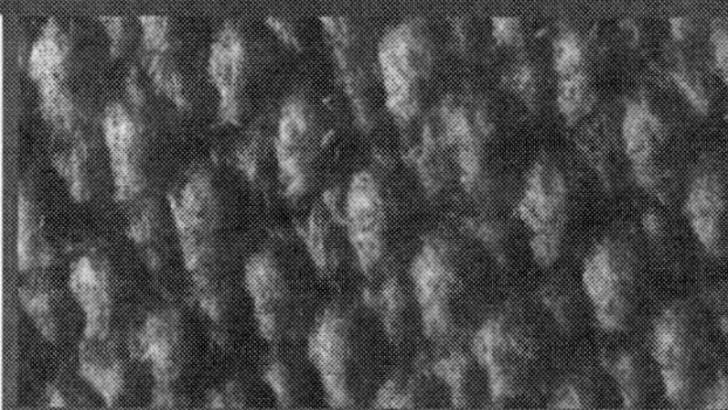

ASHLAND.

Potential Issues with Poor Release

- Poor structural definition

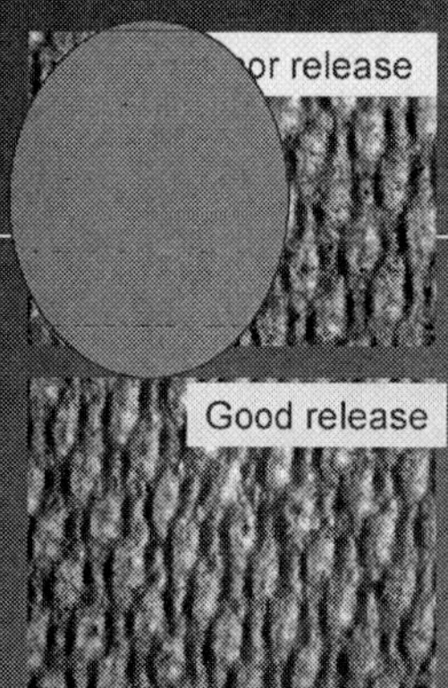

- Poor sheet transfer to Yankee

- Contaminants

- Thermal degradation
- Deposit
- Negative effect on absorbency

ASHLAND.

Product Developmental Plan

- TAD Process review
  - Sheet transfer from forming fabric to TAD fabric
  - Moulding
  - Through air drying
  - Sheet transfer from TAD fabric to Yankee dryer
- Test method development
  - Release force measurement
- Sheet structure qualification
- Product development
  - Benchmark
  - New product development

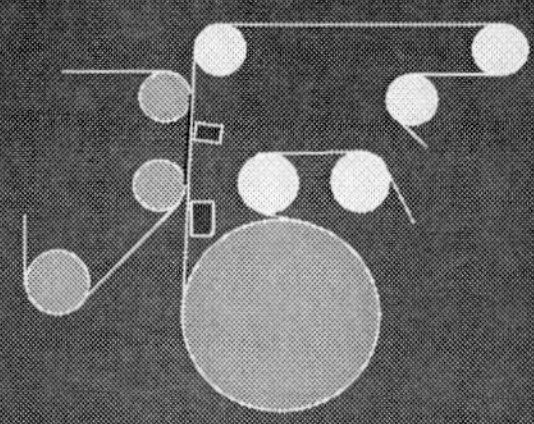

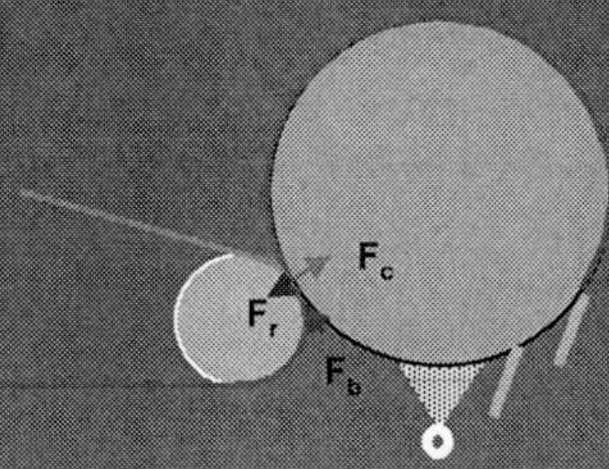

ASHLAND.

Release Dynamics

- Forces at work
  - Fc : Adhesion force between wet web and coating
  - Fr : Adhesion force between wet web and the fabric
  - Fb : Wet web bond strength
- Factors affecting forces
  - Sheet moisture level
  - Wet end chemical additives
  - Fabric design
  - Fabric degree of wear
  - Adhesion level of coating
  - Fabric release aids

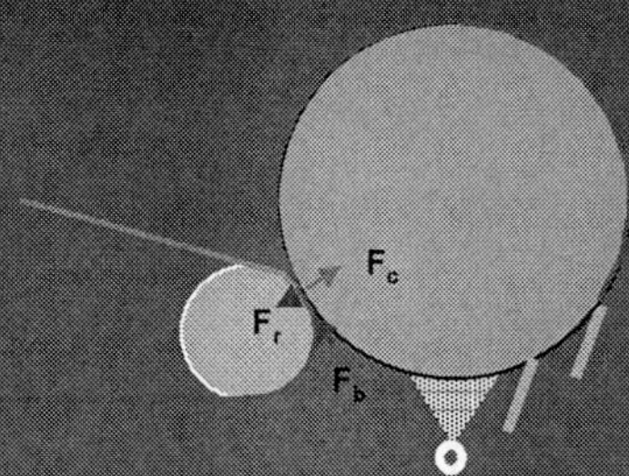

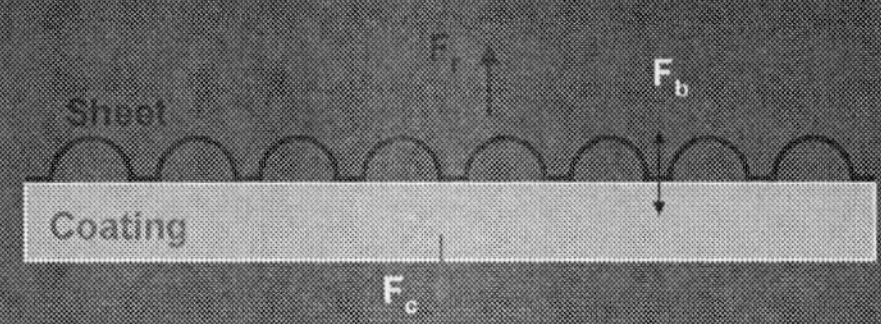

ASHLAND.

Release Performance

- Good release
  - If $F_c > F_r$ and $F_r < F_b$
  - Well defined sheet structure
  - No negative impact on absorbency
  - No deposit, contaminant

- Poor release
  - If $F_c < F_r$ or $F_b < F_r$ or F_c
  - Poorly defined sheet structure
  - Deposit, contaminant
  - Negative impact on absorbency

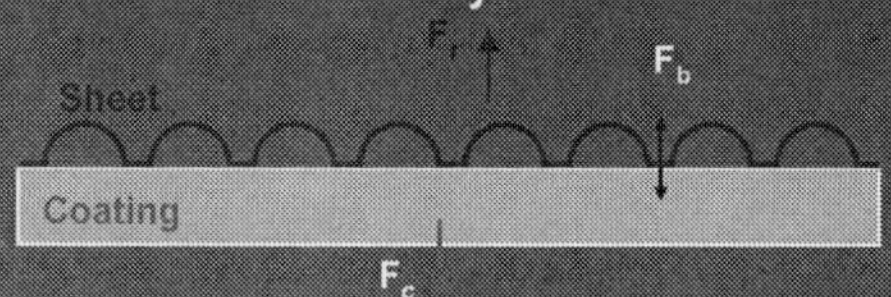

ASHLAND.

TAD Fabric Release Evaluation Process

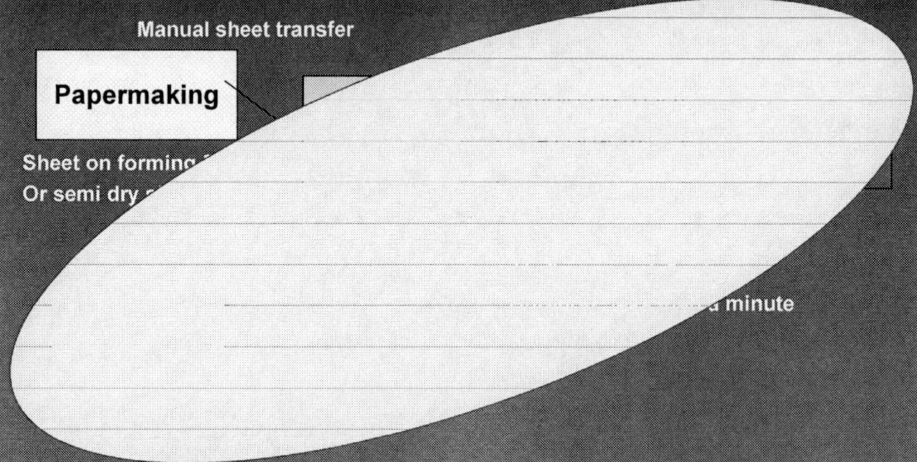

ASHLAND.

TAD Fabric Release Test

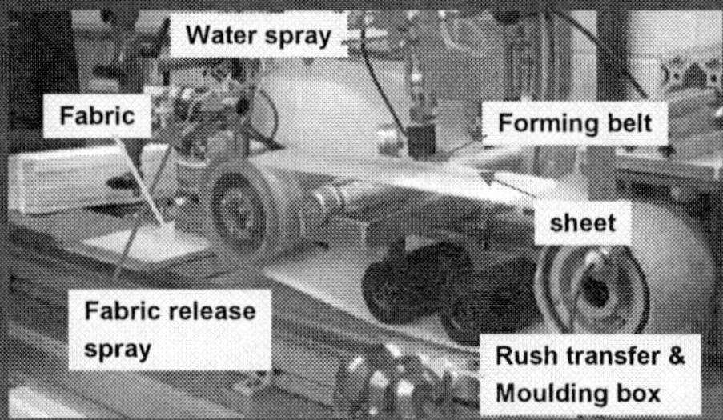

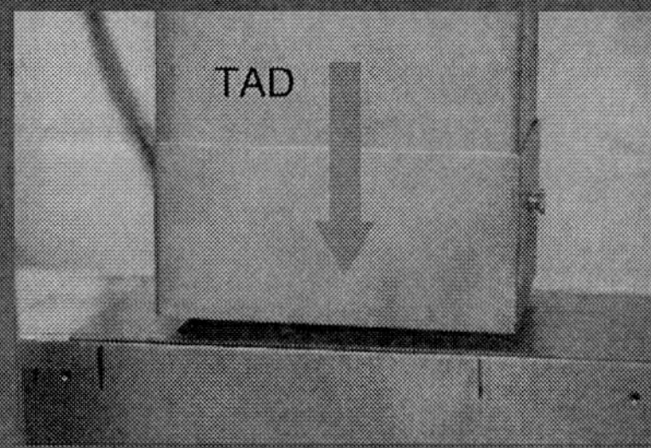

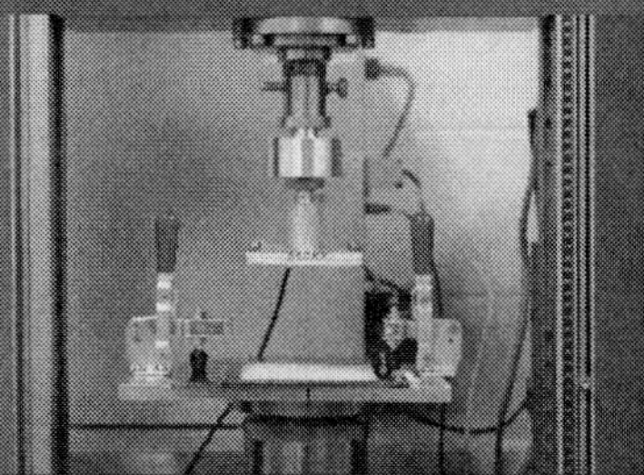

- Design target speed: 500ft/min
- Rush transfer up to 20% speed differential
- Through air velocity of up to 2m/sec

ASHLAND.

Benefits of the Simulator

- Fabric release product development is fast
- Grade development could be fast, less expensive, and less risky.
  - Realistic pre-screening of fabric release and operational conditions.
- Optimization for TAD and hybrid drying processes that is faster, less risky and less costly.
  - Furnish utilization, internal chemical usage, fabric release agents and structured fabric design, etc

ASHLAND.

Product Development

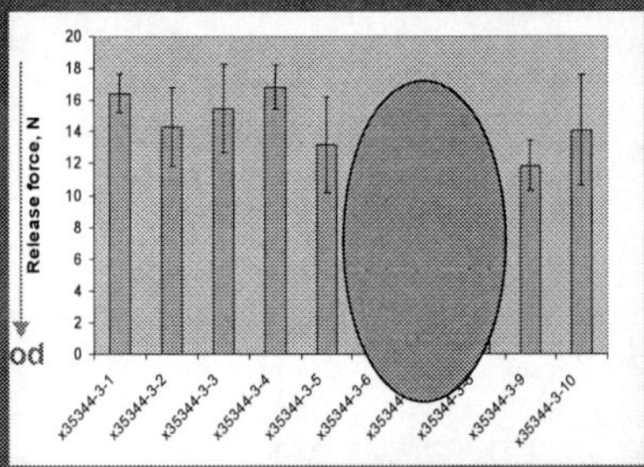

Formulation	VOC, % Neat
1	0.625
2	0.620
3	0.636
4	0.460
5	0.716
6	0.631
7	0.540
8	0.721
9	0.812
10	0.545

VOC
Thermal degradation

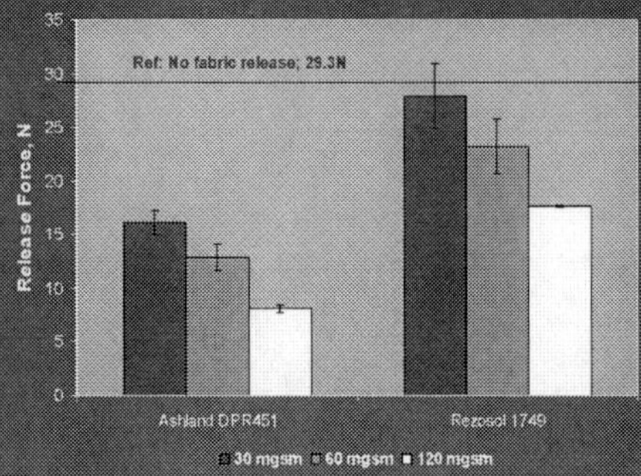

New product significantly reduced release force.

ASHLAND.

Summary

- Unique laboratory instrument has been developed to simulate structured sheet formation and through air drying.
- Realistic pre-screening of fabric release and operational conditions.
  - Fast product development
  - Fast grade development
- A new TAD fabric release product has been developed
  - High performance, low VOC (>1%), low thermal degradation

ASHLAND.

Formatting Issue - Best Available Paper/Graphic

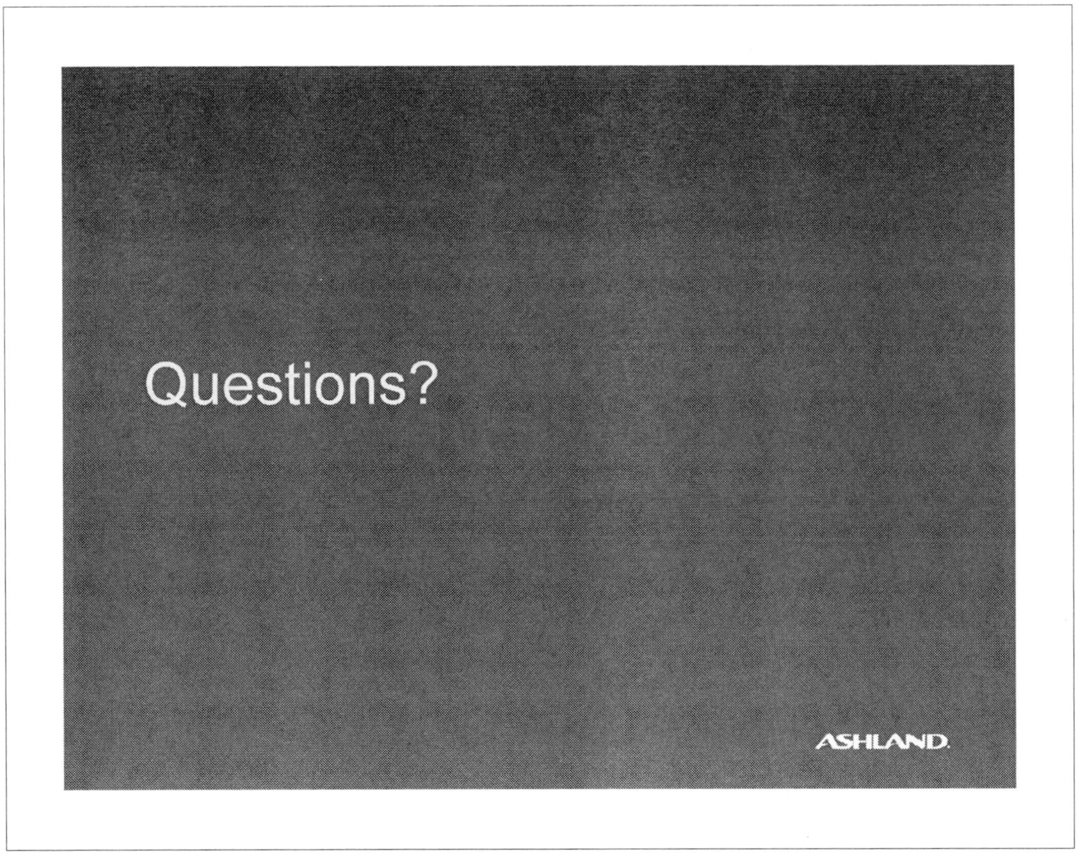

Towel Wet End Optimization:
Utilizing Strength Additives and Functional Promoters

Clayton Campbell, Harold Goldsberry
Kemira

Kemira

Tissue Industry Challenges

- This presentation will focus on improving wet and dry strength performance efficiency, maximizing process operational efficiency, while meeting the desired towel sheet quality attributes.
- Effective wet end process management is vital in towel grades in order to:
  - Achieve the desired sheet quality attributes
  - Optimize machine runnability and productivity
  - Realize the lowest chemical treatment spend
- Wet end processes can be enhanced and optimized by utilizing the most cost effective dry strength resins and anionic functional promoters.
- PAE wet strength additives are the most commonly used process chemical to obtain wet tensile sheet strength but these have challenges that can impact machine runnability and productivity.
  - High dosages of insoluble cross-linked wet strength, can bind to fines and fillers and cycle up in the short loop process.
  - Unretained PAE has the potential to cause a variety of operational and sheet quality issues.

 Kemira

Wet Strength Performance Efficiency (WSPE)

- The key to optimum towel production is maintain the highest wet strength performance efficiency
  - The wet strength performance efficiency (WSPE) is determined by the amount of wet strength required to achieve the desired wet tensile target. For example is 5 kg/t was required to achieve a 330 gF/in CD wet tensile, then the WSPE is 66.
- Increasing WSPE allows the manufacturer to increase the wet-over-dry tensile ratio and/or use lower wet strength dosage costs.
- Some process factors that can reduce WSPE are:
  - Absorbency aids
  - Improper addition point (i.e. before the refiner)
  - Higher chlorine residual levels
  - Lower operating pH
  - High ash furnish
  - High short fiber ratio
  - Insufficient anionic sites on the fiber
  - Poor sheet retention
  - Performance variability in the wet strength chemical

Lower Wet and Dry Strength Performance Efficiency Leads to Process Problems

- Poor sheet floc structure
- Poor sheet drainage – slower machine speeds
- Wire and felt filling (unreacted wet strength or fines / filler)
- Sheet holes and breaks – increased downtime due to felt picking
- High defoamer usage
- Poor creping performance - Hard, insoluble wet strength deposits building z directional thickness, pushing back the doctor blade, contributes to sheet stealing
- Plugged felts cause high moisture streaks; strip crepe coating
- Hypochlorite felt cleaning to remove wet strength; also strips crepe coating
- Inability to meet desired sheet tensile specs

Bottom Line Result :
Excessive Chemical Costs, Reduced Quality and Machine Productivity Losses

Dry Tensile Strength Additives

Towel Dry Strengths Additives

- •Cationic glyoxalated polyacrylamide (GPAM)*
- •Solution anionic polyacrylamide (SPAM)*
- •Polyvinyl amine (PVAM)
- •CMC
- •Enzyme

Novel Cationic Glyoxalated Polyacrylamide (GPAM) Dry Strength Technology

- Utilization of lower cost furnish (fiber substitution)
- Reduced basis weight
- Optimized refiner(s) use
- Reduced energy consumption
- Improved paper machine productivity
- Meeting grade strength requirements
- Reduced linting and dusting
- Improved converting efficiency

Dry Strength Performance

- Advanced glyoxalated PAM technology
- More efficient wet and dry strength resin
- 15-20% improved efficiency compared to conventional GPAM

Conditions:
- •Virgin furnish (50/ 50 hw/sw)
- •50 lb basis weight hand sheets
- •5kg/t active resin w/ stock pH @ 7.0

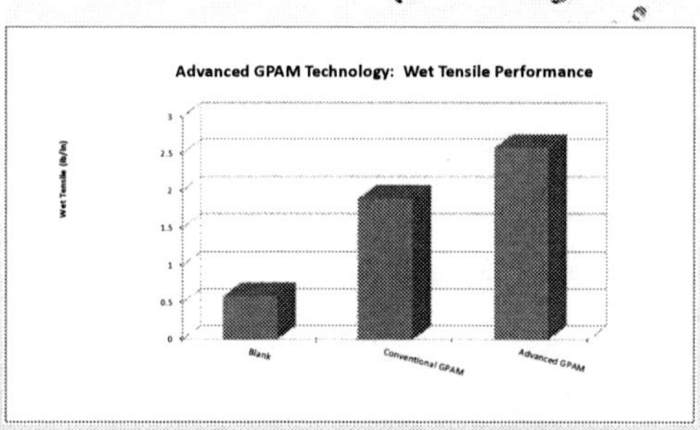

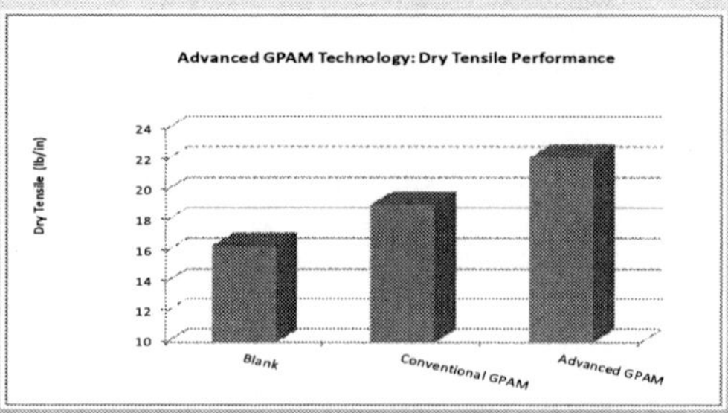

Case Study I – Advanced GPAM Technology

Machine Overview

Segment / Grade:	AFH Towel – bleached
Basis weight:	16 lb
Machine type:	1 layer , 200 TPD
Sheet moisture	3.8%
Furnish:	60% in-house MOW 40% bale softwood

Needs

- Increase machine speed
- Tensile limited on light weight grades.

Customer Benefits

- 1.5 kg/t solids of advanced GPAM dosage level resulted in equal to higher dry tensile while increasing machine speed by 250 fpm.
- FPR up from 80% to 86%
- Better save-all runnability

Functional Promoter (FP) Technology

- Synthesized products of acrylamide and acrylic acid copolymer technology are manufactured with varying molecular weights and charge densities:
  - act to improve the performance efficiency of wet end functional chemicals
  - distinct anionic promoters that vary in dry tensile contribution
  - provide higher levels of wet tensile beyond what PAE can deliver by itself
- Higher anionic charge density FP provide high degree of effectiveness at low dosages (0.25-0.50 kg/t solids) and minimum impact on dry tensile
- Lower anionic charge FP perform effectively to enhance WSPE and increase dry tensile strength
  - Easy to handle "Pump and Go" liquid

Potential Program Benefits:

- Increasing WSPE allows for:
- wet/dry tensile ratio
- reducing wet strength dosage cost
- Manage process charge
- Reduce foaming
- Reduce felt filling issues
- Improve retention of other wet end additives
- Improve sheet drainage

Functional Promoter (FP) Concept: Charge Control Laboratory Evaluation

Functional promoter charge control mechanism

PAE - Wet Strength Dosage (KG Solids/MT)	FP Dosage (KG SOLIDS/MT)	Zeta Potential (mV)
0.0		-29.1
		12.7
7.5		20.3
10.0		25.3
7.5	1.0	7.1
7.5	2.0	-2.9
7.5	3.0	-10.2

Functional Promoter (FP) Concept: Charge Control Laboratory Evaluation

Functional Promoter charge control mechanism leads to increased wet and dry tensile along with wet/dry tensile ratio

PAE Wet Strength (KG Solids/MT	FP Dosage (KG Solids/MT)	Dry Tensile (LBS/IN)	Impact on Dry Tensile % Improvement over 7.5KG PAE only	Wet Tensile (LBS/IN)	Impact on Wet Tensile % Improvement over 7.5KG PAE only	Impact on Wet /Dry Tensile ratio Improvement over 7.5KG PAE only
0	0	12.2		0		
5.0	0	16.5		3.7		
7.5	0	16.0		3.9		24%
10.0	0	16.8		4.5		
7.5	1.0	18.4	15%	4.8	23%	26%
7.5	2.0	19.5	21%	5.4	39%	28%

Case Study II: Functional Promoters

Machine Overview

- Bleached Towel / Napkin- Northeast US
- Wet crepe – Y speed at 1,128 mpm 35Ktpy
- Basis weights: 11-12.5lb
- Furnish: 100% DIP

Objectives

- Reduce wet strength cost – Improve WSPE
- Reduce foam in wire pit / defoamer usage

Customer Benefits

- Increased wet strength performance efficiency (WSPE) by 25-35% (grade dependent)
  - Reduced wet strength dosage by 30%
  - Increase CD wet /dry tensile ratio by 11%
- Increased first pass retention by 12% (absolute) or from 53% to 65%.
- Reduced defoamer usage by 34%

Summary

- Best practices: maintain the highest wet strength performance efficiency (WSPE) for each specific grade.
- Lab screening provides the most accurate means of developing an effective solution:
  - Grade specific process survey
  - Hand sheet tensile strength
  - Retention/ drainage study
  - Process chemistry and charge analysis testing
  - Advanced GPAM technology is a cost effective means to develop dry tensile strength, provide additional wet tensile strength, and enhance machine productivity.
  - Anionic functional promoters can significantly improve the wet and dry strength performance efficiency and be an effective tool to control process charge.
  - Wet end optimization is a required precursor to effective creping and maximizing a machines overall operational efficiency.

Forecasting and Validation of Fossil Fuel and Steam Savings from Tissue Machine Heat Recovery

Robert Triebe, COO, Thermal Energy International
Marc Desaulniers, Manager, Energy Conservation, Kruger Products L.P.
Gordon Buchanan, Kruger Products L.P.
Michel Francoeur, Engineering Manager, Thermal Energy International
Feikai Chu, Thermal Energy International

1.0 ABSTRACT

Accurate determination of fossil fuel reductions from proposed and implemented energy savings projects is paramount. Great confidence in projected and achieved savings is central to capital approval of major projects, and is also potentially needed to meet funding agency application and / or emission and GHG reduction reporting requirements.

When delivering recovered energy to heat air and water either directly or indirectly feeding a tissue machine, or the machine room, factors are used to determine the reduction in mmBtu fossil fuel burned relative to the mmBtu heat added. Poor understanding and estimation of these factors can result in misrepresenting an energy savings project, negatively or positively. These factors are often estimated and applied, but not frequently verified.

A condensing heat recovery system was installed at a Kruger Products facility, with a heat recovery capacity of 30 mmBtu/hr (31.5 GJ/hr).

Bump tests were completed to measure the reduction of steam at each user, and total gas / oil savings resulting from the heat recovery system. Measurements of mmBtu delivered, mmBtu steam saved, and mmBtu gas use reduction, due to each user of recovered heat, were completed. The factors served to:

1. Validate savings for Kruger Products;
2. Validate savings for the funding agency; and,
3. Forecast savings for future potential projects.

The methodology and reasoning for the pre-project savings estimates, and the experimental validation of these factors, are presented in this paper, to highlight considerations and pitfalls in forecasting and validating fossil fuel and steam savings from tissue machine heat recovery.

2.0 SITE AND SAMPLE PROJECT DESCRIPTION

The Kruger Products Gatineau mill, with a capacity of 80,000 metric tonnes /year, is comprised of a distribution center, a converting facility and a mill. It operates three tissue machines (#3, #4, and #5) and state-of-the-art converting equipment to produce well known products such as Scotties®, White Swan® facial tissues, White Swan® Napkins and SpongeTowels®, and White Swan® bathroom tissues. The mill shares facilities and services such as processed water, steam and effluent treatment with a neighboring forest products company, currently out of operation.

Thermal Energy International Inc. (TEI) designed and implemented a FLU-ACE® Heat Recovery Solution Project at the Kruger Products Gatineau Mill, and the system was made operational in October of 2010. The system is designed to recover waste energy from the combined exhaust of Machines #3 and #4. Waste

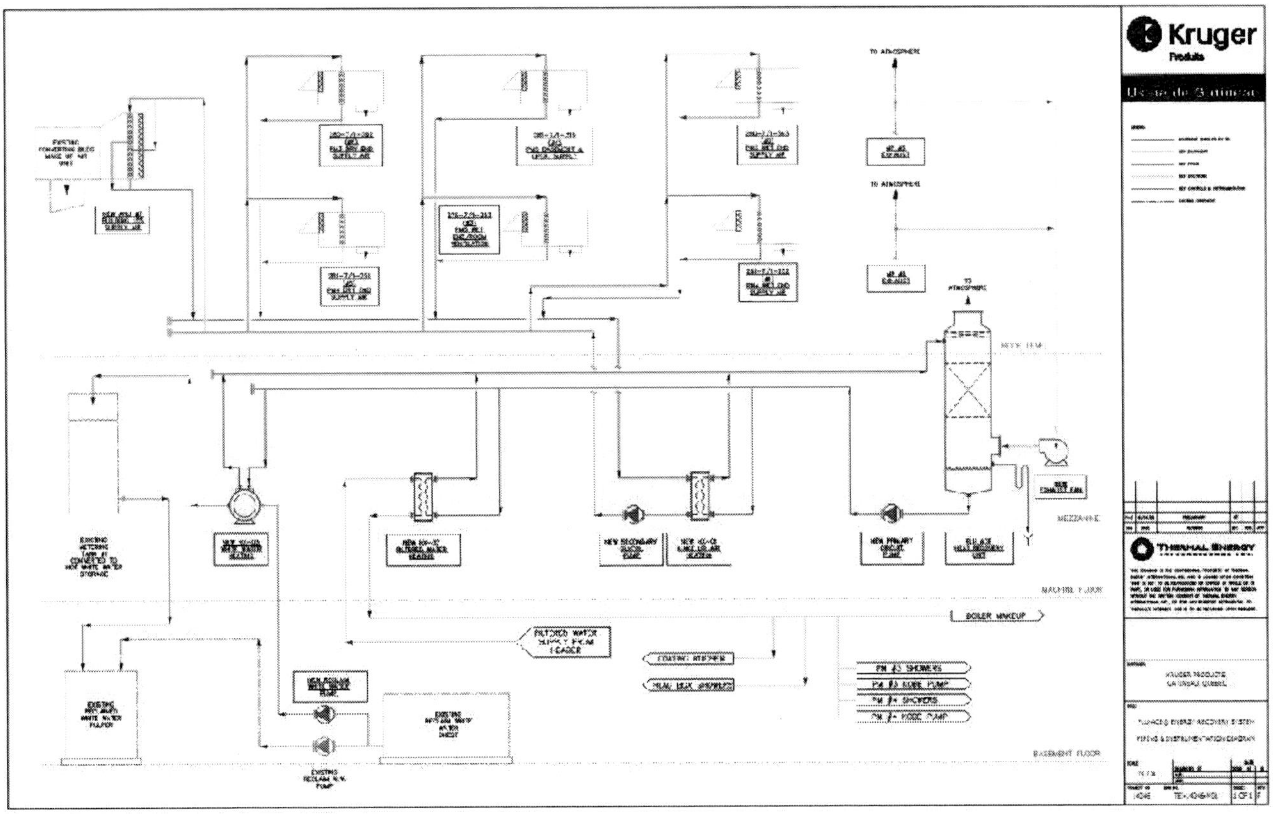

Kruger
Produits
TO ATMOSPHERE
NEW SECONDARY
BOILER MAKEUP
PM #3 SHOWERS
PM #3 KOBE PUMP
PM #4 SHOWERS
MEZZANINE
BASEMENT FLOOR
THERMAL ENERGY
KRUGER PRODUCTS

This is often the show stopper for many fantastic potential projects – the team knows that it will save a tremendous amount of energy, but without a reliable quantification of this number, the project will stop at financial approval. ***Often excellent projects cannot be approved, not because they have poor impact, but because the impact is difficult to firmly assess.*** It is key to focus on projects that have readily quantifiable and verifiable energy conservation impacts. Heat recovery projects often fall into this category.

For a heat recovery project delivering energy to a number of different end users, the impact of delivering energy to each individual energy user must be considered independently. As an example we will consider a common application for tissue mills, shower water heating, in order to save steam used to heat the water through a heat exchanger, or to displace direct-inject steam in the pit that is ultimately used to heat some or all of the shower water. This analysis must be completed individually for any location where waste heat is to be applied. Let us consider the following:

20 US gallons per minute (or 76 litre per minute) of shower water heated from 50°F to 150°F (10°C to 66°C), a heat load of 1 mmBtu/hr (1.05 GJ/hr). What is the impact (fuel and dollar savings) based on heating this stream?

3.1 Heating Currently Unheated Streams

If the stream was not previously heated, then determining savings can be difficult. As the shower water falls into the tissue machine, some will travel with the sheet / felt, some will drop out of the circuit, and some may go to a steam-heated wire pit. Clearly not all energy added will directly displace steam, and some modeling approach (CadSim or similar) can be used to estimate the impact, but confidence in such numbers is often not suitable for approval of major capital projects. This type of savings, although valid, loses value due to lack of confidence, and numbers between 0.5 and 1.0 (steam energy saved at end user / recovered heat delivered) have been suggested and modeled at various mills.

3.2 Heating Currently Heated Streams

On the other hand, if the shower water is currently heated with an existing heat exchanger before being sprayed onto the felt or a cylinder, and we heat the water directly upstream of this existing heat exchanger with a new heat exchanger using recovered energy, we can comfortably say that each unit of energy delivered to the water stream, up to its current setpoint, say 150°F (66°C), will displace one unit of steam used in the existing heat exchanger. However, a number of footnotes and factors must be applied to determine the amount of fuel used (and hence money spent) to deliver one unit of energy to the water using the existing steam heat exchanger.

$$Q_{Fuel\ Saved} = Q_{Energy\ Delivered} / E_{Heat\ Exchanger} / E_{Steam\ Distribution} / E_{Boiler\ Plant}$$

Where:

$Q_{Fuel\ Saved}$	Fuel Savings (mmBtu or GJ);
$Q_{Energy\ Delivered}$	Energy Delivered to end user from the heat recovery system (in this example case, 1 mmBtu/hr or 1.05 GJ/hr);
$E_{Heat\ Exchanger}$	Efficiency of Heat Exchanger, accounting for radiant losses and steam trap losses at the heat exchanger, effectively this is how much energy is delivered to the heat end user per unit of steam energy delivered to the heat exchanger (ratio);
$E_{Steam\ Distribution}$	Efficiency of Steam Distribution, accounting for radiant losses and steam trap losses in distribution system, effectively this is how much steam energy is delivered to the heat exchanger per unit of steam energy raised at the boiler plant (ratio);

$E_{Boiler\ Plant}$ Efficiency of Boiler Plant, accounting for useful energy delivered from the boiler plant (typically enthalpy of generated steam less enthalpy of feed water) per unit of fuel energy consumed in the boiler (ratio).

Note that if direct-inject steam is being displaced, another factor must be considered, specifically that the boiler plant must heat water from make-up water temperature to steam temperature, but energy delivered to the end user must be determined between the steam temperature and the end user temperature, so energy added at the boiler plant between make-up water temperature and end user temperature is wasted. In addition, new make-up water requires more steam in the deaerator, not just for heating up to feedwater temperature, but also passing through the deaerator to serve to remove non-condensable gases. This is an additional loss with direct inject steam, or an additional savings when displacing direct-inject steam. Each stream presents its own factors to be considered in such an analysis.

Clearly determining fuel savings is more than a case of simply measuring heat delivered. If heat delivered is measured, which is simply done with a water stream knowing the mass flow and temperature rise, this must go through a number of factors to determine actual fuel savings.

In this selected shower water heating case, we assumed for the purpose of conservative forecasting that:

$E_{Heat\ Exchanger}$ 90%, or 0.9, due to radiant losses and steam trap losses;

$E_{Steam\ Distribution}$ 100%, or 1.0 - although steam distribution systems are much less efficient, depending upon insulation and steam trap quality and maintenance, this factor is typically carried as between 80% and 90% (demonstration for every unit of steam reaching the end user that between 1.1 and 1.2 units of steam must be raised at the boiler plant), but this steam loss will continue unless the steam distribution system delivering this energy is shut off. In this case the steam system (distribution lines and steam traps) serving the user are left on and open as a back up, so this factor is not considered in determining savings, and is left as 1.

$E_{Boiler\ Plant}$ 79%, or 0.79, as determined by calculating useful energy delivered from the boiler plant (typically enthalpy of generated steam less enthalpy of feed water) per unit of fuel energy consumed in the boiler (ratio).

Therefore, for this case of shower water that is currently heated through an indirect steam heat exchanger:

$$Q_{Fuel\ Saved} = Q_{Energy\ Delivered} / E_{Heat\ Exchanger} / E_{Steam\ Distribution} / E_{Boiler\ Plant}$$

$$Q_{Fuel\ Saved} = 1\ \text{mmBtu/hr} / 0.9 / 1.0 / 0.79$$

$$Q_{Fuel\ Saved} = 1.41\ \text{mmBtu/hr}$$

And since the basis is 1 mmBtu/hr:

$R_{Fuel\ Saved}$ = Ratio of Fuel Saved to Energy Delivered = 1.41

So the ratio of fuel energy saved per unit of energy delivered to the end user is 1.41, for this sample project, for this specific end user.

3.3 Factor Used for Sample Project

Of course, the general rule here is to remain conservative but representative. If the technical team can agree that the savings are both real and likely conservative, the likelihood of management accepting the project and providing capital approval leading to successful energy conservation is heightened.

For this sample project, the factor of 1.41 was applied for all end users for the purpose of remaining accurate or conservative, as commented on below.

- **Machine #3 shower water and Plant make-up air** - the 1.41 factor should be relatively accurate and representative, as this stream is currently heated with an indirect steam heat exchanger, however if the steam distribution to this system can also be shut off, the savings would be higher than estimated with this factor;
- **Reclaim pulper make-up (white) water and Chemistry (kitchen) water** – the 1.41 factor is conservative as this end user is using direct inject steam, leading to some energy loss and additional steam use in the boiler plant deaerator – however the $E_{Heat\ Exchanger}$ factor should not strictly be applied, as there is no loss at the heat exchanger for a direct inject application if all steam is condensed at the end user, and if the steam distribution to this system can also be shut off, the savings would be higher than estimated with this factor;
- **Boiler plant make-up water** - the 1.41 factor is conservative as this end user is using direct inject steam, leading to some energy loss and additional steam use in the boiler plant deaerator – however the $E_{Heat\ Exchanger}$ factor should not strictly be applied, as there is no loss at the heat exchanger for a direct inject application if all steam is condensed at the end user.

Based on this applied factor, and based on historical trended flow and temperature data from the existing plant data acquisition systems, and from end-user flow and temperature measurements taken for the purpose of developing this project, and monthly profile of forecast fuel savings was developed and is presented in Figure 2 below.

Note that for the purpose of conservatively forecasting savings, basing flows and temperatures on existing equipment design data is not recommended, particularly for air handling / make-up units, the actual operational flow should be measured. Although water stream flows are often well known, air flows through make-up units should be measured to better represent actual savings, as flows are typically 60% to 80% of the design flow after a number of years of operation, and forecasting savings from design flows will inevitably lead to an aggressive savings estimate.

The profile developed and presented in Figure 2 was accepted and considered conservative for the reasons above, and capital approval was obtained, and the project was implemented.

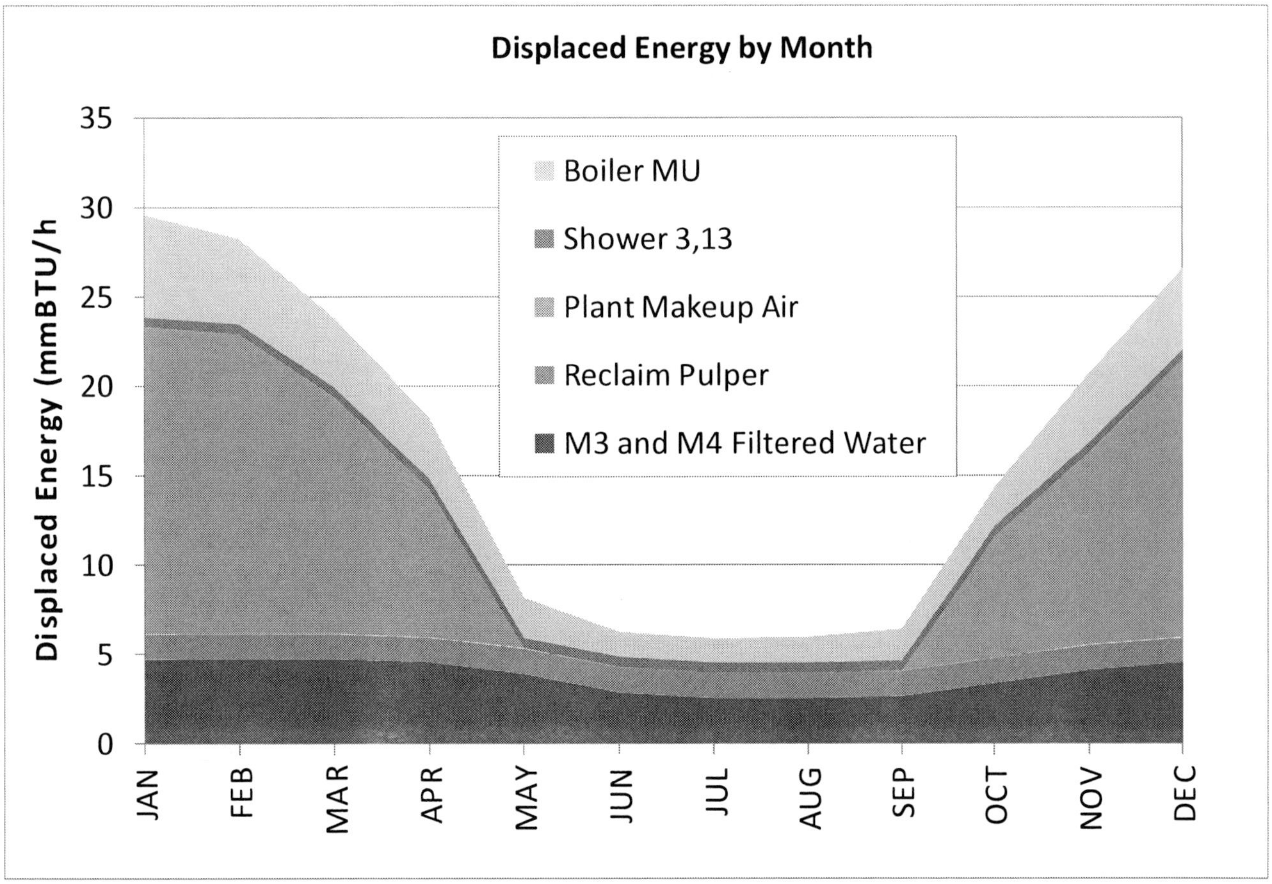

Displaced Energy by Month
Displaced Energy (mmBTU/h
35
30
25
20
15
10
5
0
Boiler MU
Shower 3,13
Plant Makeup Air
Reclaim Pulper
M3 and M4 Filtered Water
JAN
FEB
MAR
APR
MAY
JUN
JUL
AUG
SEP
OCT
NOV
DEC

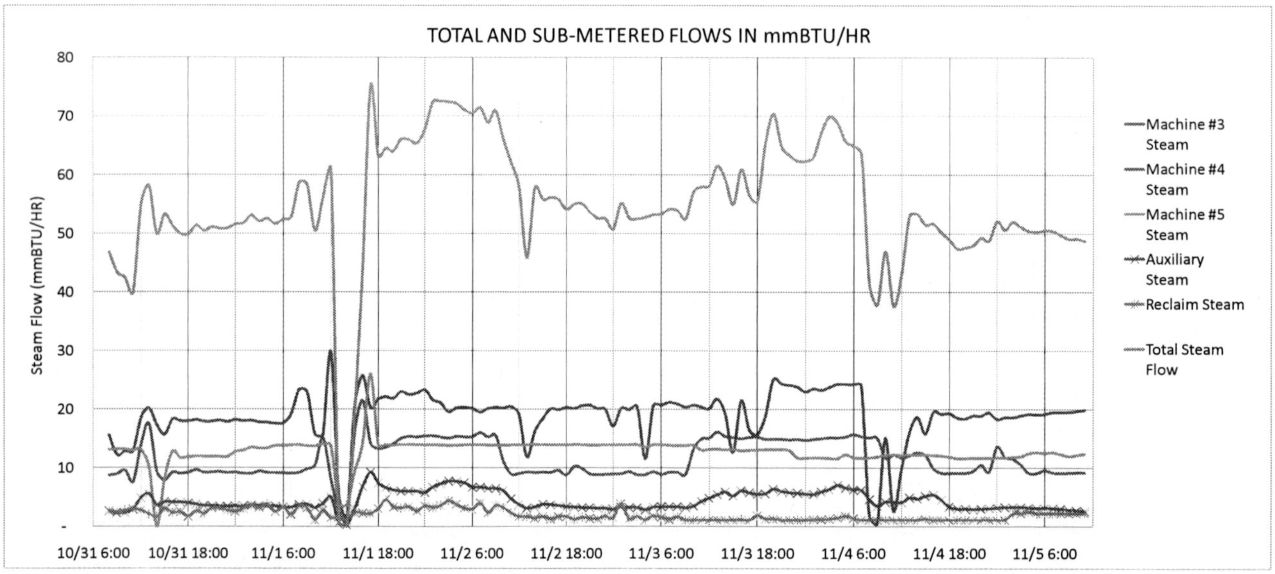
TOTAL AND SUB-METERED FLOWS IN mmBTU/HR
Steam Flow (mmBTU/HR)
80
70
60
50
40
30
20
10
-
10/31 6:00
10/31 18:00
11/1 6:00
11/1 18:00
11/2 6:00
11/2 18:00
11/3 6:00
11/3 18:00
11/4 6:00
11/4 18:00
11/5 6:00
Machine #3 Steam
Machine #4 Steam
Machine #5 Steam
Auxiliary Steam
Reclaim Steam
Total Steam Flow

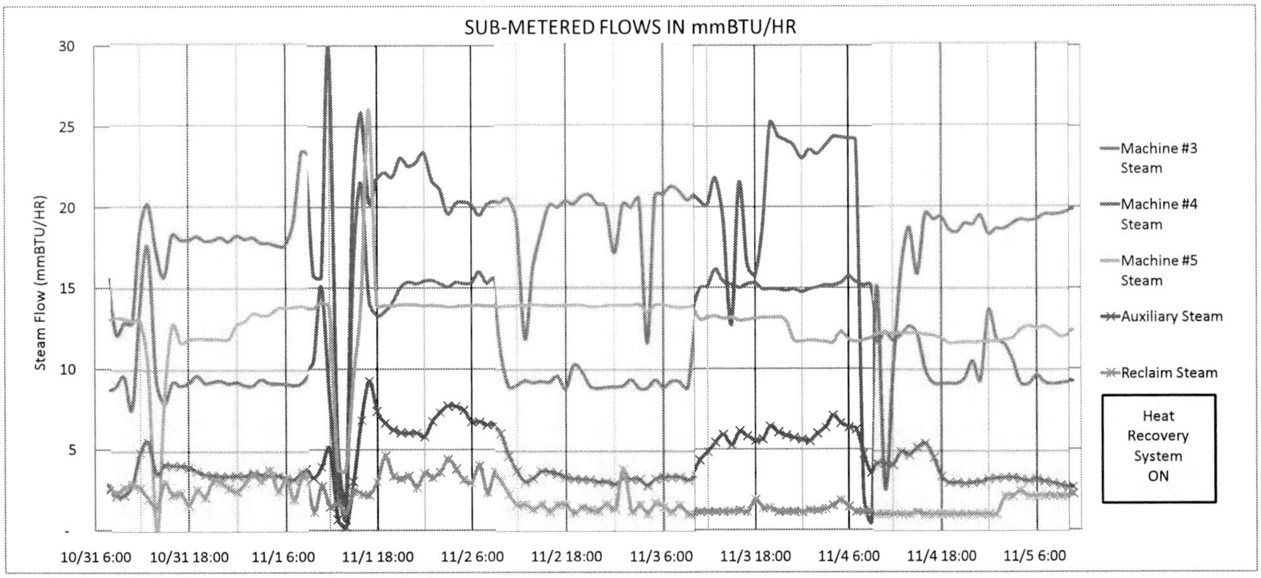
SUB-METERED FLOWS IN mmBTU/HR
Steam Flow (mmBTU/HR)
30
25
20
15
10
5
-
10/31 6:00
10/31 18:00
11/1 6:00
11/1 18:00
11/2 6:00
11/2 18:00
11/3 6:00
11/3 18:00
11/4 6:00
11/4 18:00
11/5 6:00
Machine #3 Steam
Machine #4 Steam
Machine #5 Steam
Auxiliary Steam
Reclaim Steam
Heat Recovery System ON

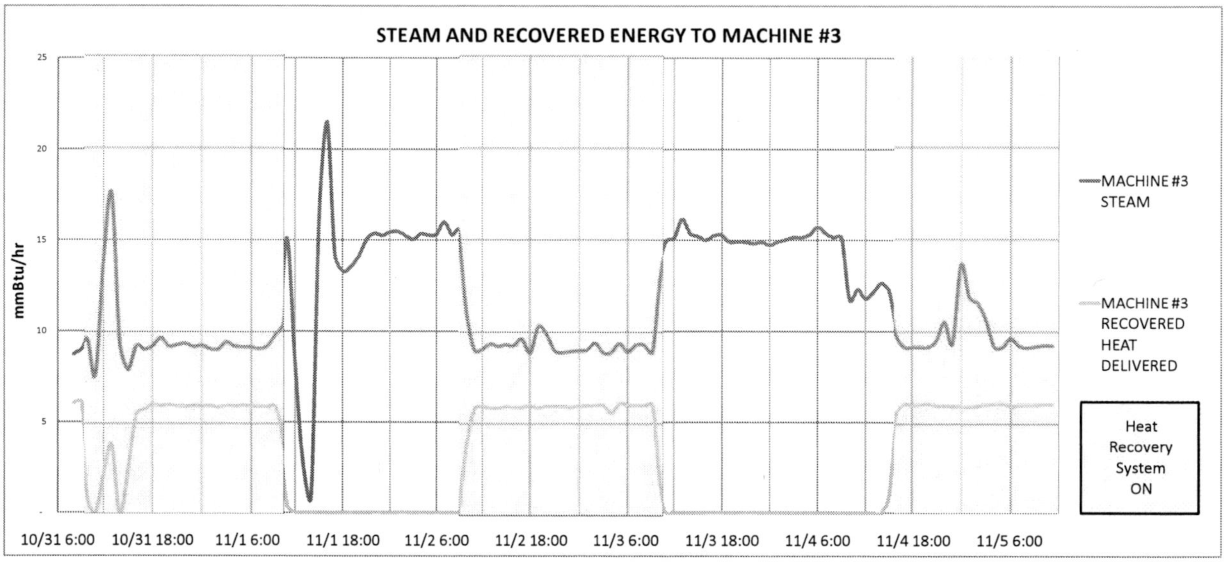
STEAM AND RECOVERED ENERGY TO MACHINE #3
mmBtu/hr
25
20
15
10
5
-
10/31 6:00
10/31 18:00
11/1 6:00
11/1 18:00
11/2 6:00
11/2 18:00
11/3 6:00
11/3 18:00
11/4 6:00
11/4 18:00
11/5 6:00
MACHINE #3 STEAM
MACHINE #3 RECOVERED HEAT DELIVERED
Heat Recovery System ON

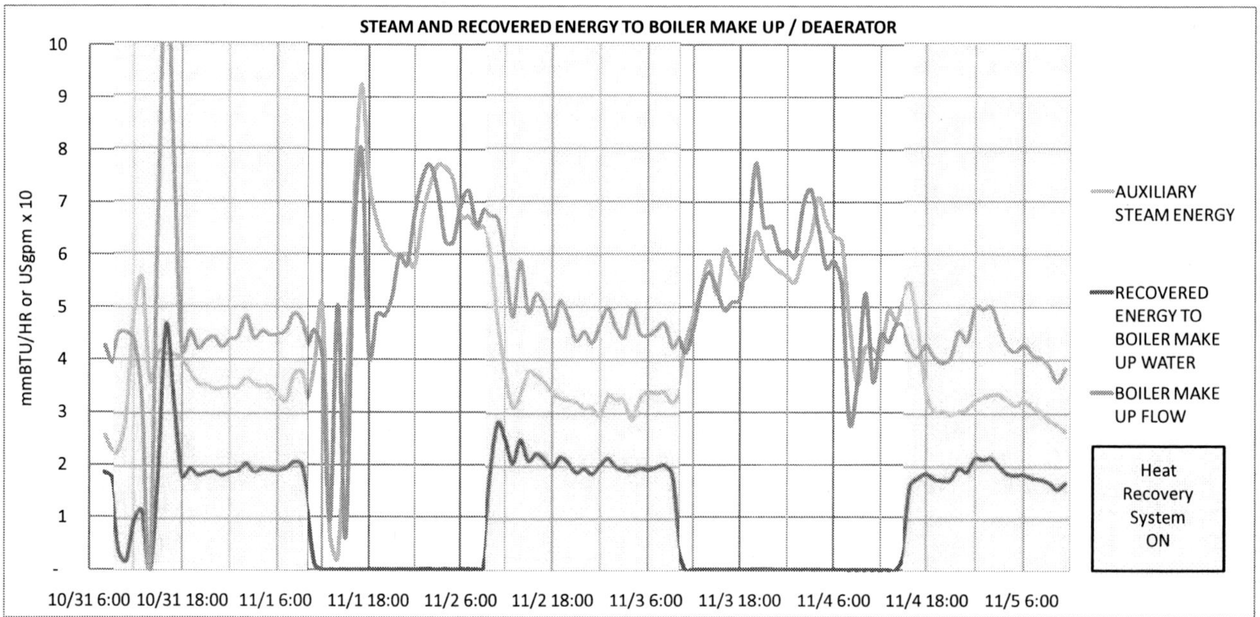
STEAM AND RECOVERED ENERGY TO BOILER MAKE UP / DEAERATOR
mmBTU/HR or USgpm x 10
10
9
8
7
6
5
4
3
2
1
-
10/31 6:00
10/31 18:00
11/1 6:00
11/1 18:00
11/2 6:00
11/2 18:00
11/3 6:00
11/3 18:00
11/4 6:00
11/4 18:00
11/5 6:00
AUXILIARY STEAM ENERGY
RECOVERED ENERGY TO BOILER MAKE UP WATER
BOILER MAKE UP FLOW
Heat Recovery System ON

$R_{Fuel\ Saved} = 1.29$

For the Boiler make-up water:

$R_{Fuel\ Saved} = 2.16$

4.2 Discussion of Results

Effectively the assessment revealed that the pre-project factor of 1.41 applied to all heat sinks was aggressive (by less than 10%) for the recovered heat users previously heated with indirect steam heat exchangers, but was very conservative for the boiler make-up heating, and likely for all recovered heat users previously heated with direct steam, due to the additional requirement for cold make-up water heating and deaeration of the additional flow of boiler make-up water.

Across the entire project and all recovered heat users, applying $R_{Fuel\ Saved}$ of 1.29 to all (former) indirect steam users (Machine #3 shower water and Plant make-up air) and $R_{Fuel\ Saved}$ of 2.16 to all (former) direct steam users (Reclaim pulper make-up (white) water, Chemistry (kitchen) water, and, Boiler plant make-up water) resulted in higher fuel savings per unit of recovered energy delivered than was initially forecast, and the effort to forecast with conservative accuracy was proven successful.

Another point of interest to consider during this type of investigation is the potential change in boiler efficiency that can result from a significant decrease in steam production at the boiler plant, potentially making the pre-project boiler plant efficiency different from the post-project boiler plant efficiency due to change in loading of the boiler.

In addition, ***given that the results were satisfactory, this in no way suggests that the results are perfectly accurate, nor are all-encompassing.*** Testing in winter and summer, or under other general high or low load conditions at the boiler plant, will impact the boiler plant efficiency, and the overall factors that arise. In addition, a process facility is not a stagnant, constant operation by any means, and although the validity and integrity of these results is excellent, conducting the same tests 6 months later with different outside air temperatures, different fresh water temperatures, different grades on the machines, and possible operator changes, will not yield precisely the same results. It is very important in this kind of endeavour to develop a strong validation program, execute the program, and accept the results, particularly if they are logical and acceptable. An attempt to find the "true" value is a wild goose chase – the dynamic process environment will not allow for determination of an overall, time-independent "true" value – the truth is that the values determined are for the conditions they were determined under, and if the conditions were well-selected, the factors will be applicable all year around, but will never be precise. Develop and execute a good program and accept the results – there is more energy to be saved in other areas of the plant, and this must be the new focus. Revisiting the process every few years might be advised, particularly since for a case like this the "bump" test approach is so simple, to ensure you continue to get the savings expected from the system in question.

5.0 CONCLUSIONS / RECOMMENDATIONS

Accurate determination of fossil fuel reductions from proposed and implemented energy savings projects is paramount. Great confidence in projected and achieved savings is central to capital approval of major projects, and is also potentially needed to meet funding agency application and / or emission and GHG reduction reporting requirements;

Projects (such as heat recovery) with savings that can be readily and accurately forecast and validated are more easy to validate than projects with more speculative and hard to validate impacts (such as steam trap replacement);

Providing waste heat to currently heated air or water streams provides more certain and verifiable savings than adding heat to air or water streams not currently being heated which would provide indirect savings;

Each end-user taking advantage of delivered waste heat will have a different factor ($R_{Fuel\ Saved}$) to translate the recovered heat delivered to the fuel energy saved, affected by factors including but not limited to use of direct or indirect steam, efficiency and operation of associated steam traps and steam distribution systems, and impacts on boiler efficiency – the factor can vary substantially and requires attention for accurate forecasting;

Accurate forecasting and validation requires accurate measurement, design data should not be used;

The simplest way to determine the true immediate impact of a heat recovery system is the "bump" test, or simply stopping and starting the system, and comparing the increase in steam and fuel at the boiler plant to the amount of recovered energy delivered before and after shutting the heat recovery system down;

A long period (no less than 8 hours, up to 24 hours) should be used to ensure that at least a good number of hours of steady operation can be observed with the system on and off – the 24 hour period also, to some degree, accounts for daily outdoor temperature variations; and,

Data obtained through a properly planned bump test can be relatively accurate and dependable, but the dynamic environment of a tissue mill will thwart attempts to find an all-encompassing, time-independent "true" value, so the outcomes of a good program should be accepted, and the attention is best spent on the next conservation measure.

ANDRITZ
Pulp & Paper

Andritz Paper Machinery
Steam Energy Reduction in Tissue Drying

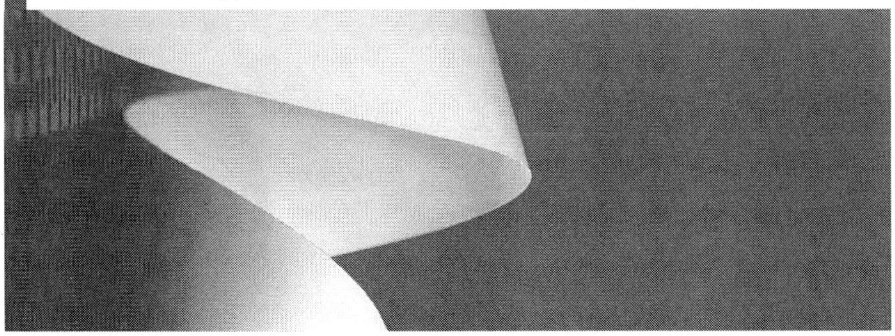

www.andritz.com

We accept the challenge!

Steam Energy Reduction in Tissue Drying
Focus Points

- **Drying Energy Overall Cost**
- **Steel Yankee Application**
- **Re Evaporation**
- **Yankee Heat Losses Reduction**

2 Andritz Paper Machinery – Energy Reduction in Tissue Drying Processes

ANDRITZ
Pulp & Paper

Steam Energy Reduction in Tissue Drying

Energy Consumption Cost in Tissue Production

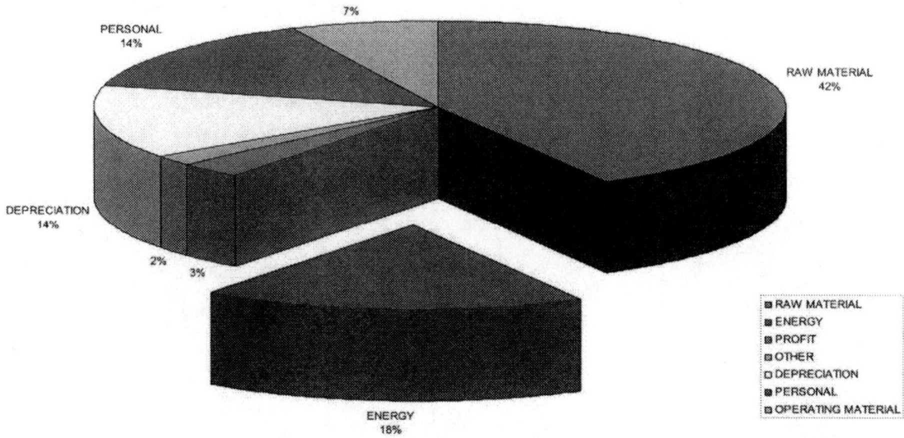

3 Andritz Paper Machinery – Energy Reduction in Tissue Drying Processes

ANDRITZ Pulp & Paper

Steam Energy Reduction in Tissue Drying

Major Energy Costs Distribution in Typical Tissue Machine

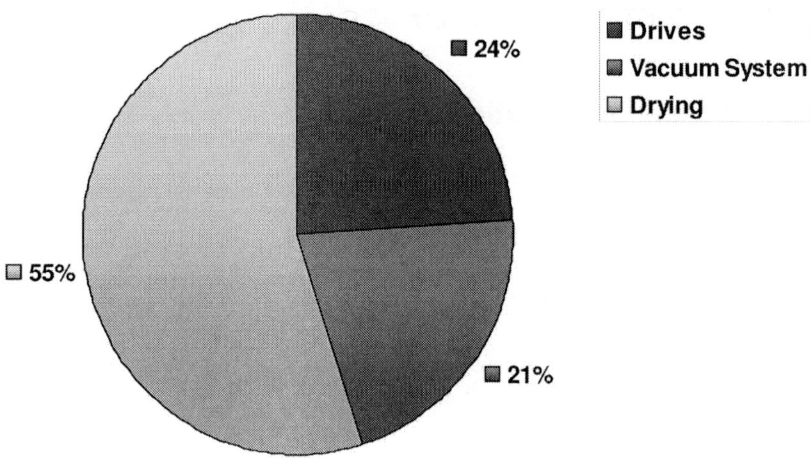

4 Andritz Paper Machinery – Energy Reduction in Tissue Drying Processes

ANDRITZ Pulp & Paper

Steam Energy Reduction in Tissue Drying

Cost of Natural Gas

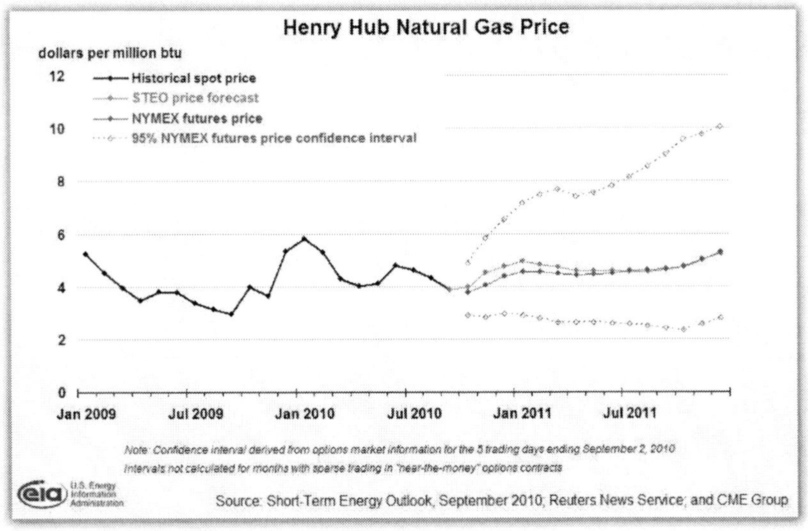

ANDRITZ Pulp & Paper

Steam Energy Reduction in Tissue Drying

Natural Gas Prices – Historical and Forecast

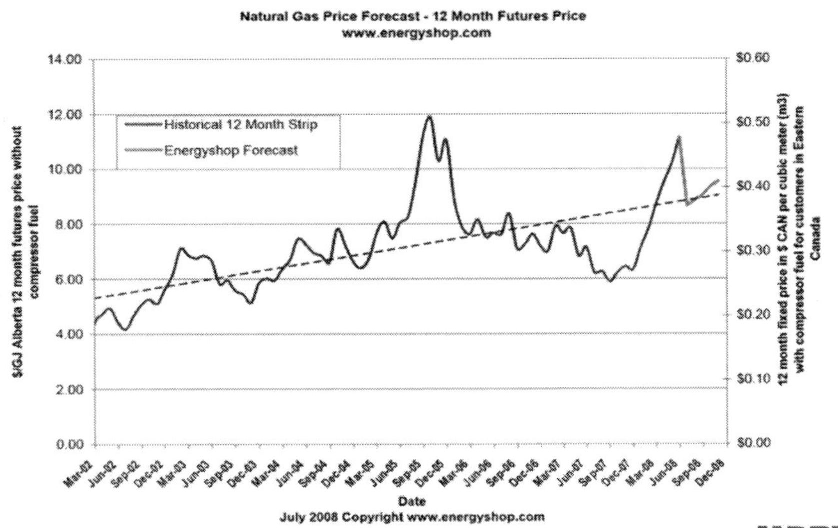

ANDRITZ Pulp & Paper

Steam Energy Reduction in Tissue Drying

Typical Steam Consumption

Machine Size	Small 2.8m; 12ft dia.	Medium 3.6m; 15ft dia.	Large 5.5m; 18ft dia.
Typical Production	70-90 T/day	110-150 T/day	200-230 T/day
Typical Steam Consumption, approx.	4,000 Kg/hr	6,500 Kg/hr	10,000 Kg/hr

Note: TAPPI Rw drying contribution split ,Yankee Dryer / Yankee Hood : 50/50

Steam Energy Reduction in Tissue Drying

Typical Drying Energy Costs

Machine size	Small	Medium	Large
Machine diameter	12 ft. (3.6m)	15 ft. (4.5m)	18 ft. (5.5m)
Sheet width	110" (2.8m)	140" (3.6m)	200" (5.5m)
Production	70 – 90 t/day	110 – 150 t/day	200 – 230 t/day
Typical electrical energy cost per year	US$230,000	US$320,000	US$480,000
Typical natural gas cost per year	US$1,145,000	US$1,650,000	US$2,800,000
Typical steam cost per year (gas heated boiler)	US$670,000	US$1,100,000	US$1,680,000

Natural gas: US$7.9/MBTU; electricity: US$0.065/kW; steam : US$ 20/ton

Steam Energy Reduction in Tissue Drying

Energy Cost Savings with Steel Yankee vs. Cast Iron Yankee

2,8 m working width | 17,5 gsm (toilet paper) | 15 ft Yankee | 1 press roll | 1800 m/min

	Steam	Gas	Electricity	Energy	Energy Savings	Energy Cost	Cost Savings
	kg/h	kg/h	kWh/h	kWh/h	kWh/h	€/h	€/year
Conventional Yankee	4 806	425	427	8 588	0	334	**0**
PrimeDry Steel Yankee	5 975	361	328	8 316	271	319	**121 500**

Annual Energy Cost Savings up to
€ 120.000.- (5 %)

or **8 % production increase**
due to higher Yankee performance

Rw TAPPI value for Steel Yankee **15% higher** than for Cast Iron Yankee

(Values shown are indicative only)

Steam	20	€/t
Gas	0,5	€/kg
Electric	0,06	€/kWh

Prices typical for Western Europe today, (changes from country to country and from mill to mill possible)

ANDRITZ Pulp & Paper

Steam Energy Reduction in Tissue Drying

Re Evaporation - Steam generation for the Yankee dryer

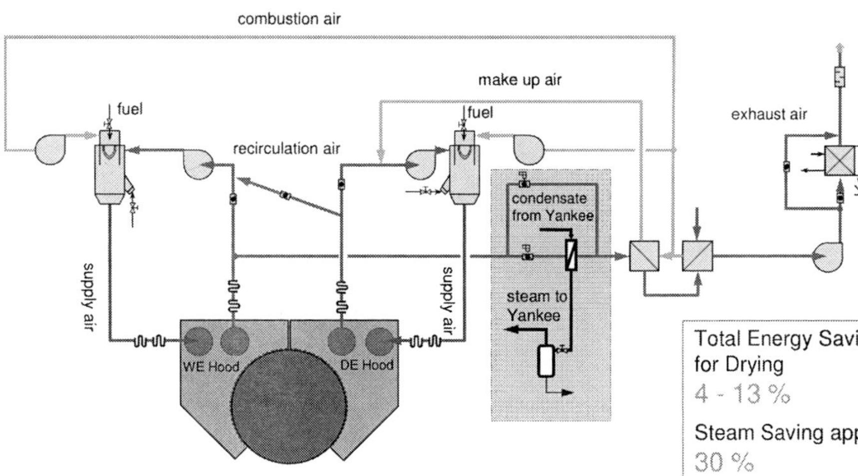

Total Energy Saving for Drying
4 - 13 %
Steam Saving approx.
30 %

ANDRITZ Pulp & Paper

Steam Energy Reduction in Tissue Drying

Re Evaporation - Steam generation for the Yankee dryer

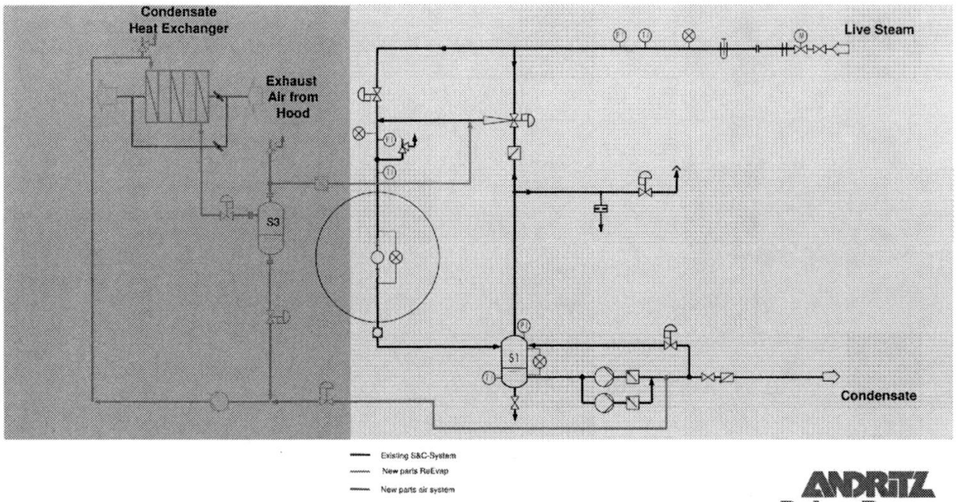

ANDRITZ Pulp & Paper

Steam Energy Reduction in Tissue Drying

Heat exchanger - arrangement

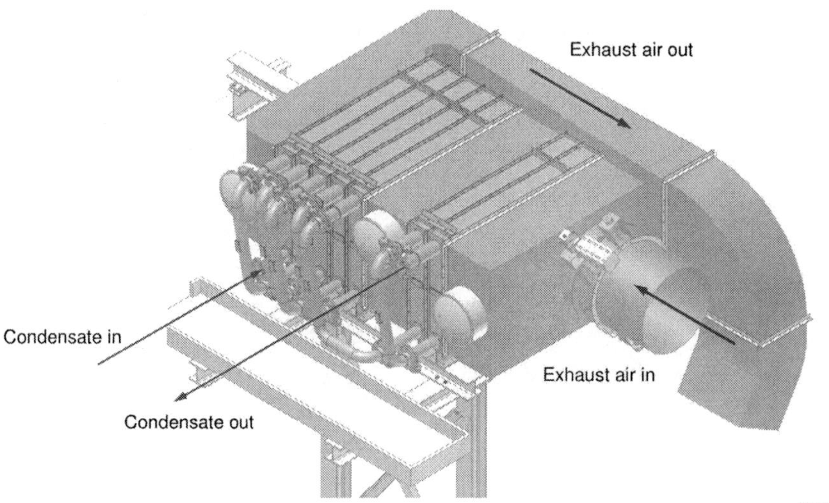

ANDRITZ Pulp & Paper

Steam Energy Reduction in Tissue Drying

References

Customer	Country	Delivery	Width [mm]	Speed [m/min]	generated steam [kg/h]	Start-up
Hengan-16	China	ReEvaporation for Yankee steam	5600	2000	2900	2012
Hengan-15	China	ReEvaporation for Yankee steam	5600	2000	2900	2012
Hengan-12	China	ReEvaporation for Yankee steam	5600	2000	2900	2012
Hengan-11	China	ReEvaporation for Yankee steam	5600	2000	2900	2011
Metsä Tissue Kreuzau	Germany	ReEvaporation for Disperger steam	5600	2000	2925	2010
Thüringer Hygiene Papier (Sofidel)	Germany	ReEvaporation for Yankee steam	2760	2200	1600	2010
Wepa Giershagen TM5	Germany	ReEvaporation for Yankee steam	2750	1750	2200	2009
Wepa Kriebstein TM9 (KRIEPA)	Germany	ReEvaporation for Yankee steam	5400	2000	3500	2003

ANDRITZ Pulp & Paper

Steam Energy Reduction in Tissue Drying

Steam Savings ,Confidential Projects, Germany

	Production			Yankee			Hood		Re- evaporated steam from hood	
	basis weight	operating speed	working width	diameter	steam pressure	steam consump	imping. velocity	imping. temp.	steam flow	portion of total steam consumption
	g/m²	m/min	m	ft	bar	kg/h	m/s	°C	kg/h	%
No.1	12,8	1700	5,4	15	5,5	5960	140	580	**1625**	**27**
No.2	22	1495	5,11	18	7,5	7470	130	600	**2460**	**33**
No.3	13,7	1800	5,55	15	8	9340	155	520	**2250**	**24**
No.4	18,3	1100	2,88	12	710	3218	150	500	**840**	**26**
					(smooth yankee)					

ANDRITZ Pulp & Paper

Steam Energy Reduction in Tissue Drying

Energy Savings – Yankee Head Insulation

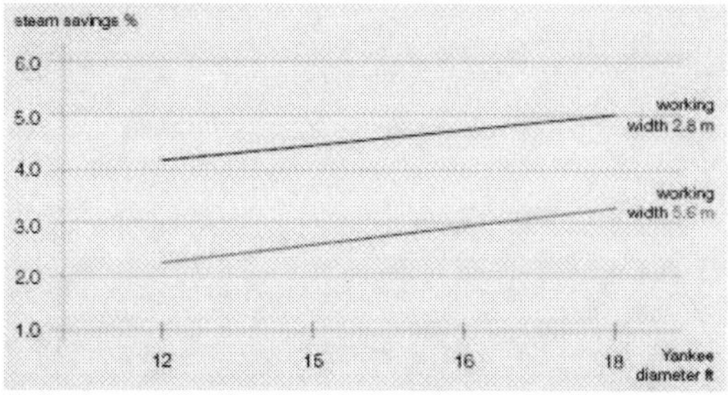

The chart shows the energy saving potential at 8 bar steam pressure for the PrimeDry Yankee head insulation for two different working widths.

2-5% potential savings

Crepe and MG Yankees of paper, board and tissue machines show significant heat losses at the head. Here, the saving potential – referring to the steam consumption – is at 2-5%.

The head insulation is not only applicable to new Yankees, but can also be retrofitted on existing units – no matter who the original manufacturer was.

Less contamination

In addition to energy savings, the Yankee head insulation produces less contamination due to its even surface. It prevents steam leakage, as well as sealing out dust that could enter the hot Yankee head.

Better sheet run

At tissue machines one main advantage is that the head insulation covers the head screws thus eliminating disturbing airflows and enabling a better sheet run.

Steam Energy Reduction in Tissue Drying

Total Potential Savings

- **Steel Yankee Application : 10% - 15%**
- **Re Evaporation : 25% - 30%**
- **Yankee Heat Losses Reduction : 3% - 5%**

Runtech Systems

ECOPUMP TURBO™ FOR TISSUE MACHINE

Tissue World 2012 - Jussi Lahtinen, Runtech Systems

More Profitability - Less Energy www.runtech.fi

Runtech Systems

ECOPUMP TURBO™

Capacity controlled turbo blower saves energy and water in vacuum system

- 30..60 % energy savings in vacuum system
- Automatic speed control to adapt to changes in speed, basis weight and felt condition
- Wide air capacity range eliminates bleed valves
- Easy to install so EP Turbo™ is an excellent option for vacuum system rebuilds and optimization
- Totally water free solution
- Speeds up to 10 500 rpm
- Impellers directly mounted on high speed motor shaft so no gearbox is needed
- Fast and easy maintenance

More Profitability - Less Energy www.runtech.fi

Runtech Systems

ECOPUMP TURBO™

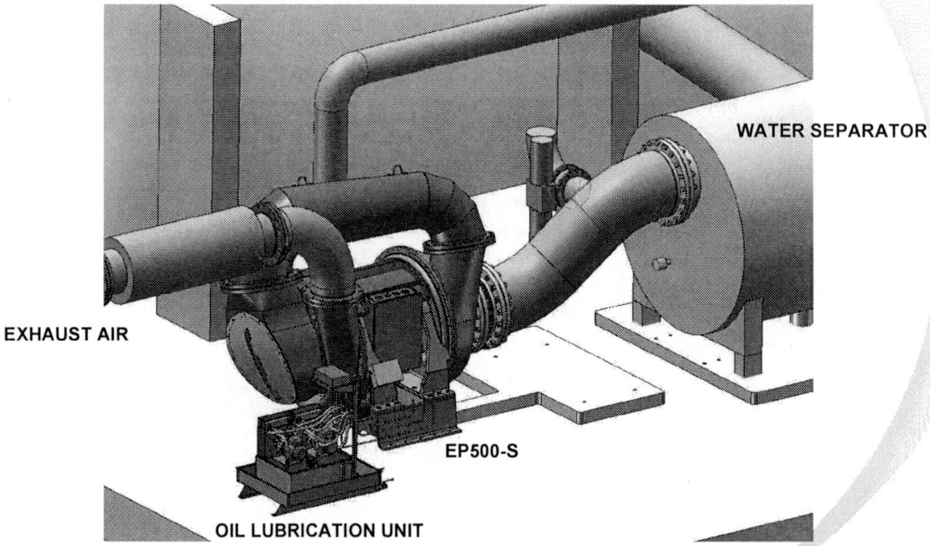

More Profitability - Less Energy www.runtech.fi

Runtech Systems

EXAMPLE SYSTEM CONNECTION WITH TWO UNITS

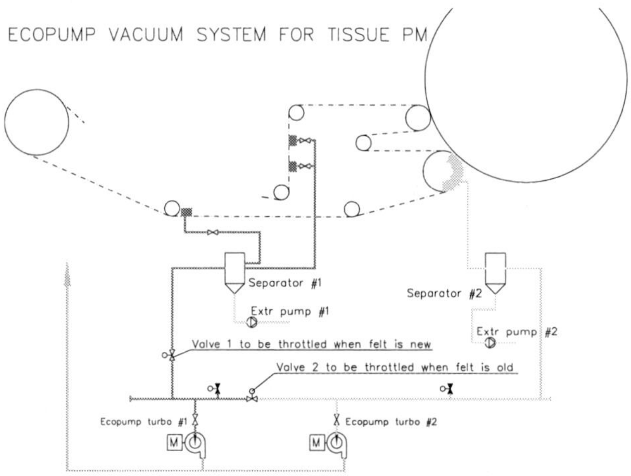

- Focus capacity on SPR (suction press roll) when felt is new.
- When felt ages, shift more capacity to Uhle boxes

More Profitability - Less Energy www.runtech.fi

Runtech Systems

SUCTION PRESS ROLL DEMAND

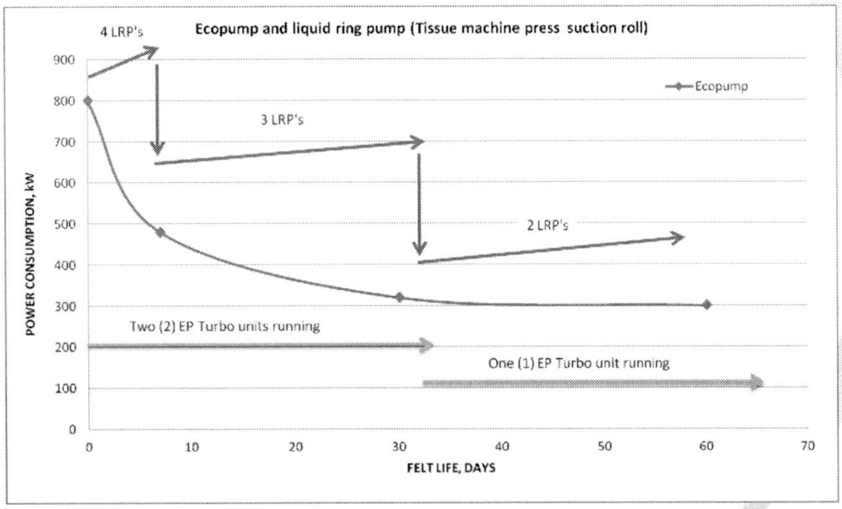

More Profitability - Less Energy www.runtech.fi

Runtech Systems

SUCTION PRESS ROLL DEMAND

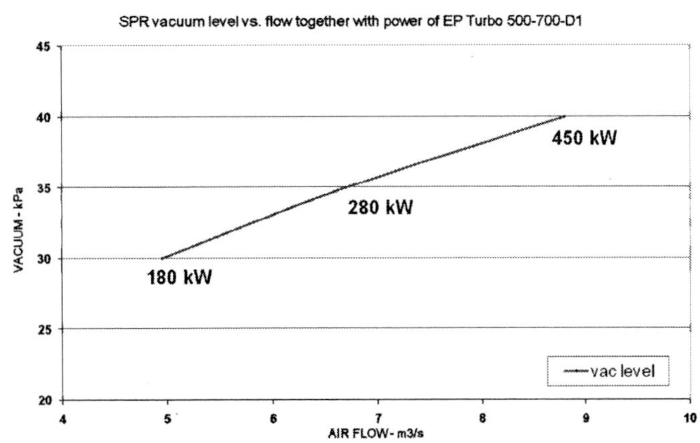

With speed controlled turbo blower, it's very easy to find optimum vacuum level. If vacuum level can be reduced, there is significant cost savings potential with EP Turbo.

More Profitability - Less Energy www.runtech.fi

Runtech Systems

BENCHMARKING – TISSUE MACHINES

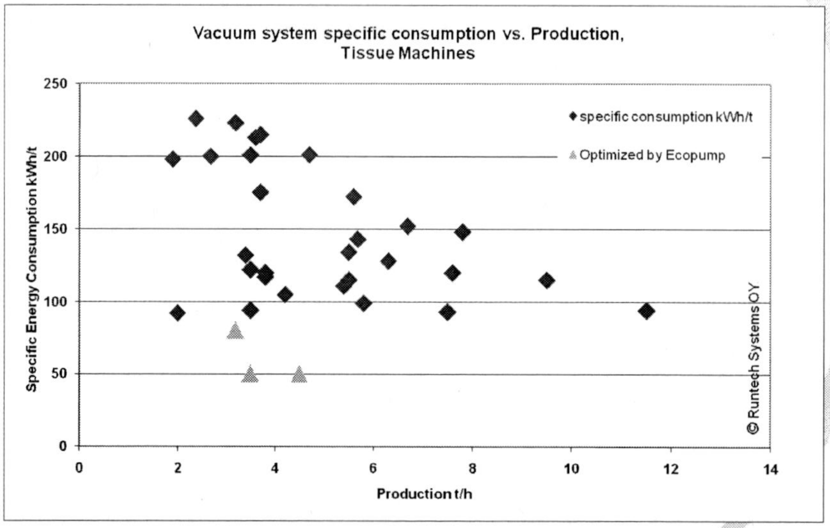

More Profitability - Less Energy www.runtech.fi

Runtech Systems

CASE STUDY – METSÄ TISSUE MÄNTTÄ PM1

Installation:

Three liquid ring pumps replaced with one EP400-700-D Turbo.

One old liquid ring pump was disconnected and removed during 10 hour shutdown. After that, machine operated for 1 month with backup vacuum capacity, and during that time turbo installation was finished.

During the next 12 hour shutdown, EP turbo blower was connected to process and tested. Total shutdown time for vacuum system rebuild was only 24 hours.

Pump	kW
LRP 1	220
LRP 2	264
LRP 3	264
LRP 4	188
TOTAL	936

Pump	kW
LRP 1	Stop
LRP 2	Stop
LRP 3	Stop
LRP 4	150
EP400	350
TOTAL	500

ENERGY SAVINGS: 400 kW (40% savings)

WATER SAVINGS: 200 000 m³ per year

More Profitability - Less Energy www.runtech.fi

Runtech Systems

CASE STUDY – CONFIDENTIAL MILL

Installation – PM1:

Two liquid ring pumps and two fixed speed blowers replaced with one EP500-700-S Turbo.

One old liquid ring pump is serving a felt Uhle box.

Pump	kW
1	160
2	180
3	0
4	106
5	156
6	250
TOTAL	852

Pump	kW
1	Stop
2	Stop
EP500	270
4	Stop
5	Stop
6	90
TOTAL	360

ENERGY SAVINGS: 500 kW (60% savings)
WATER SAVINGS: 250 000 m³ per year

Installation – PM2:

Units installed. Start-up is planned for February 2012.
Guaranteed power savings 500 kW.

More Profitability - Less Energy www.runtech.fi

Runtech Systems

REFERENCES FOR TISSUE

- Georgia-Pacific Nokia PM7, Finland
- Georgia-Pacific Cuijk PM4, Netherlands
- Georgia-Pacific Cuijk Converting Plant, Netherlands
- Metsä Tissue Mänttä PM1, Finland
- Metsä Tissue Mänttä PM7, Finland
- Unipak Tissue, Lebanon
- Confidential Tissue PMx, France
- Confidential Tissue PMx, South Africa
- Confidential Tissue PMx, South Africa
- Confidential Tissue PMx, France
- SCA Lucca PM4, Italy

Totally over 70 units installed, 25 units on order - as of January, 2012

More Profitability - Less Energy www.runtech.fi

NOVIMPIANTI
DRYING TECHNOLOGY

H.P. steam total production for a Yankee Dryer by Steam Generator & Post Burner on hood exhaust.

Tissue World Americas Miami, March 22nd, 2012

COMPANY

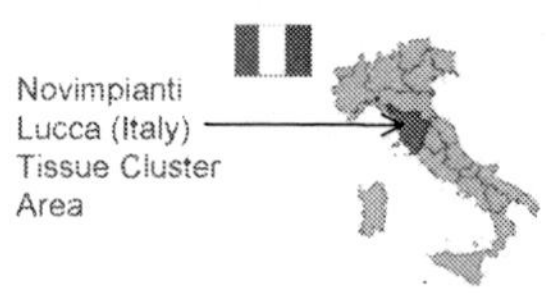

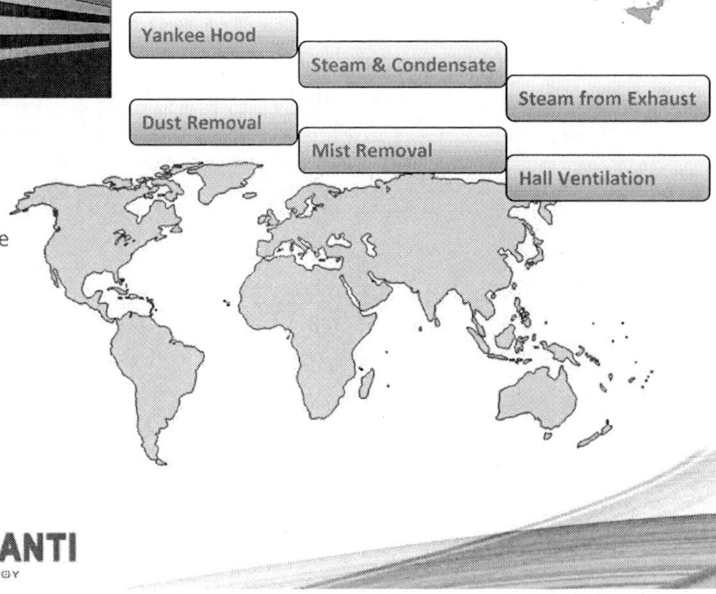

Company
- Started: 1980
- Q. Cert. ISO 9001
- Deliveries Worldwide

Processes:
- Drying
- Ventilation
- Environment

Areas:
- Tissue
- Paper and Board
- Corrugator

NOVIMPIANTI
DRYING TECHNOLOGY

YH/YD DRYING SPLIT CONCEPT

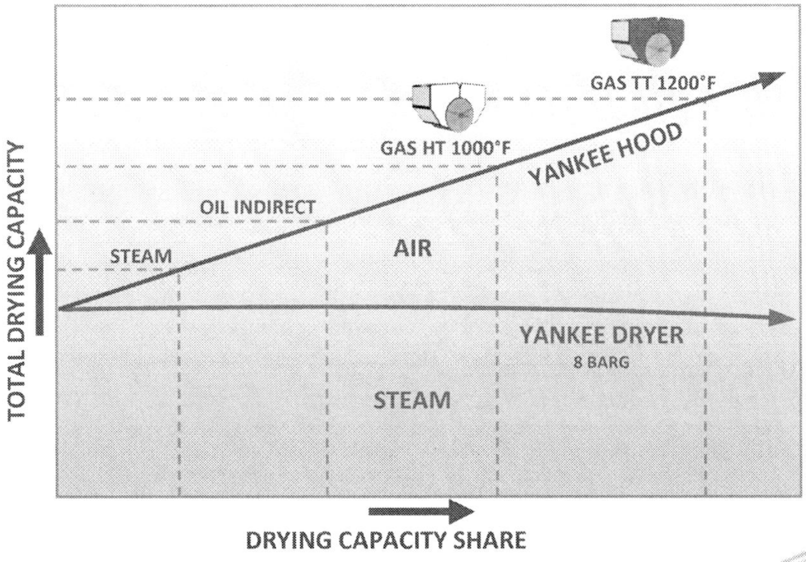

NOVIMPIANTI
DRYING TECHNOLOGY

YANKEE HOOD/DRYER ECOSTEAM GENERATOR

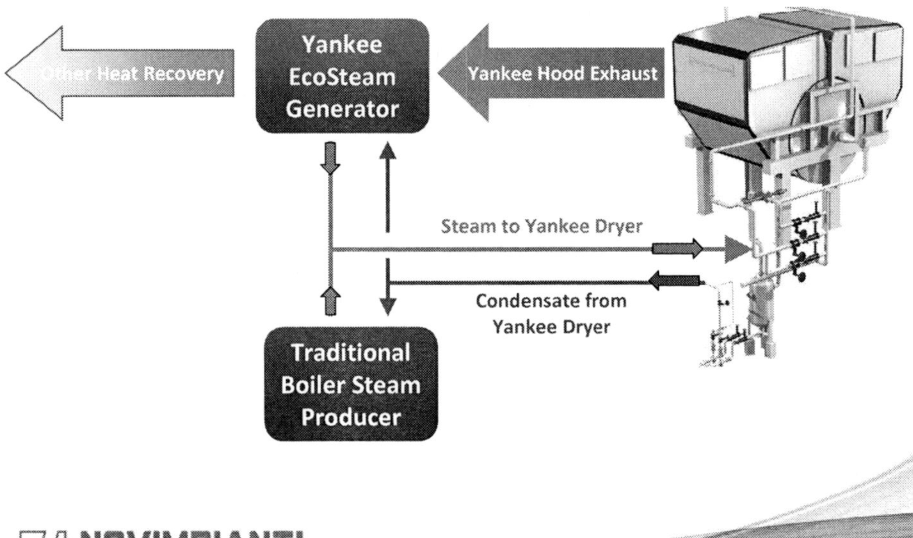

NOVIMPIANTI
DRYING TECHNOLOGY

YANKEE ECOSTEAM GENERATOR

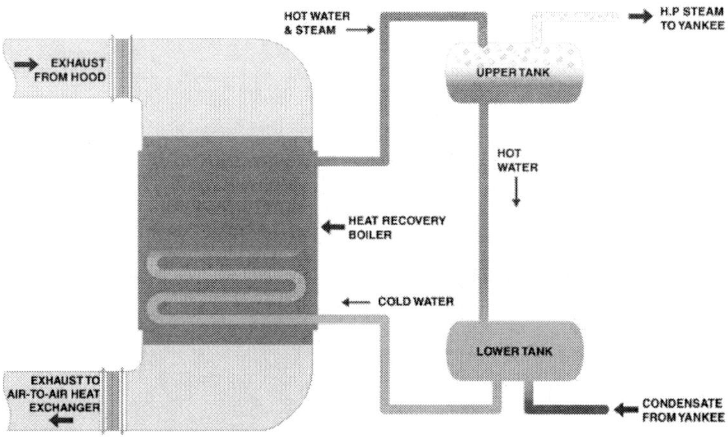

NOVIMPIANTI
DRYING TECHNOLOGY

STANDARD HT HOOD

Yankee dryer is completely fed by steam from traditional boiler, Yankee Hood and traditional boiler are both gas fired

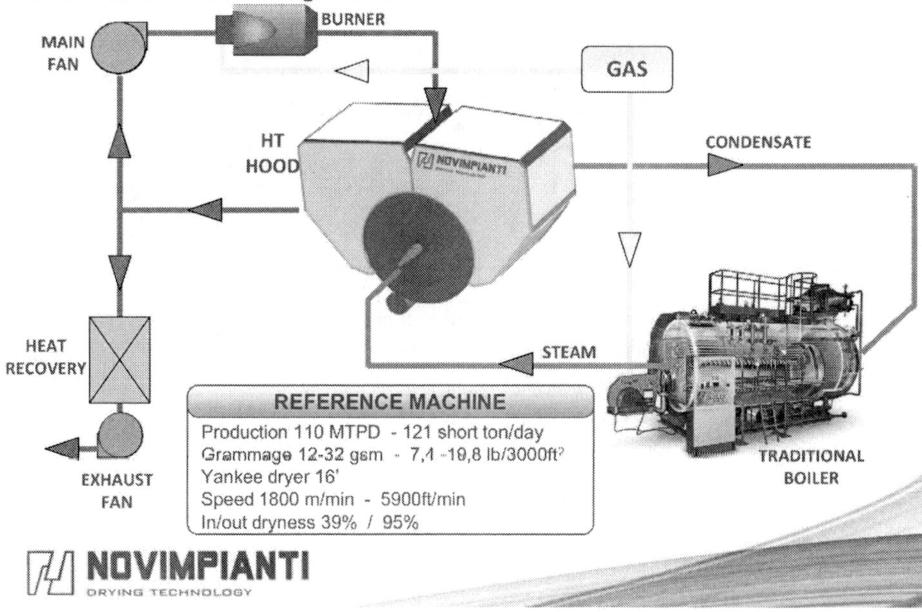

REFERENCE MACHINE
Production 110 MTPD - 121 short ton/day
Grammage 12-32 gsm - 7,4 -19,8 lb/3000ft²
Yankee dryer 16'
Speed 1800 m/min - 5900ft/min
In/out dryness 39% / 95%

NOVIMPIANTI
DRYING TECHNOLOGY

STANDARD HT HOOD

Yankee dryer is completely fed by steam from traditional boiler, Yankee Hood and traditional boiler are both gas fired

YANKEE HOOD		YANKEE DRYER	
Evaporated water	**3730kg/h - 8200lb/h**	Evaporated water	**2850kg/h – 6300lb/h**
Drying ratio	**57%**	Drying ratio	**43%**
Blowing temperature	**500°C – 932°F**	Cast Iron Dryer	
Blowing speed	**120m/s – 23600fpm**	Pressure	**7barg – 1000psi**
Exhaust humidity	**0,500**	Total Steam cons.	**4200kg/h – 9240lb/h**
Thermal (burner) 1098kWh/ton$_{paper}$ 3,75MMBTU/ton$_{paper}$		Thermal (boiler) 676kWh/ton$_{paper}$ 2,3MMBTU/ton$_{paper}$	
Total Specific Consumption 1774 kWh/ton$_{paper}$ **6,05 MMBTU/ton$_{paper}$**			

NOVIMPIANTI
DRYING TECHNOLOGY

STANDARD HT HOOD + YES - SBOX

Yankee dryer is fed by steam from traditional boiler, Yankee Hood and traditional boiler are both gas fired. Steam generator is used to feed a steam box.

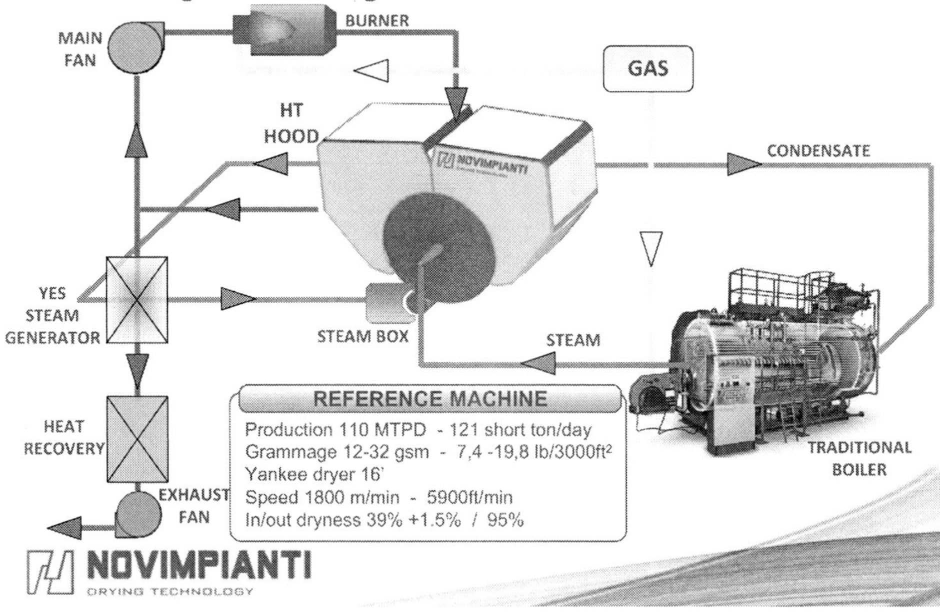

NOVIMPIANTI
DRYING TECHNOLOGY

STANDARD HT HOOD + YES - SBOX

Yankee dryer is fed by steam partially from YES steam generator and partially from traditional boiler, Yankee Hood and traditional boiler are both gas fired

YANKEE HOOD		YANKEE DRYER	
Evaporated water	**3320kg/h - 7300lb/h**	Evaporated water	**2850kg/h – 6300lb/h**
Drying ratio	**54%**	Drying ratio	**46%**
Blowing temperature	**500°C – 932°F**	Cast Iron Dryer	
Blowing speed	**108m/s – 21260fpm**	Pressure	**7barg – 1000psi**
Exhaust humidity	**0,500**	Total Steam cons.	**4200kg/h – 9240lb/h**
Thermal (burner) 1000kWh/ton_{paper} 3,41MMBTU/ton_{paper}		Thermal (boiler) 676kWh/ton_{paper} 2,3MMBTU/ton_{paper}	
Total Specific Consumption 1676 kWh/ton_{paper} **5,72 MMBTU/ton_{paper}**			

NOVIMPIANTI DRYING TECHNOLOGY

STANDARD HT HOOD + YES-M

Yankee dryer is fed by steam partially from YES steam generator and partially from traditional boiler, Yankee Hood and traditional boiler are both gas fired

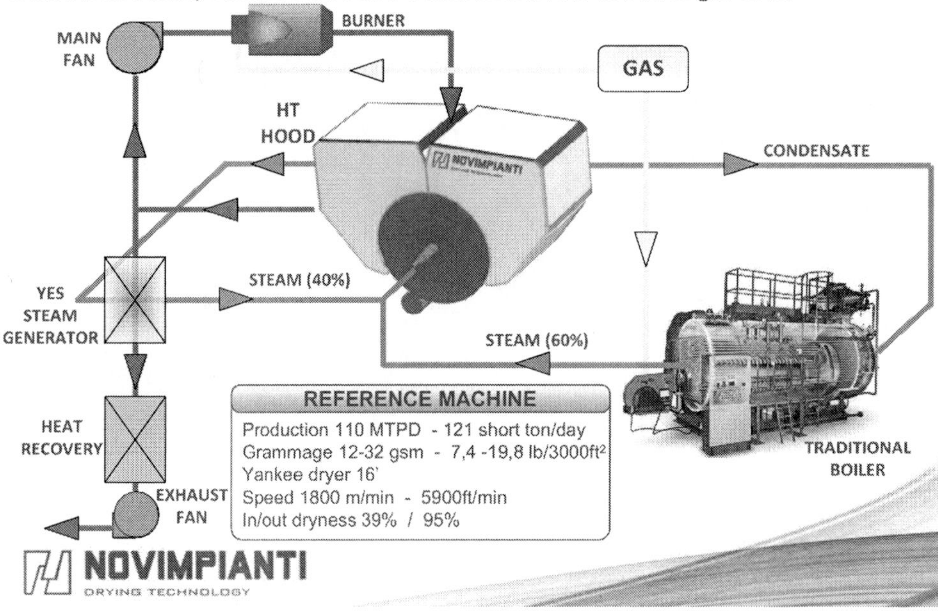

REFERENCE MACHINE
Production 110 MTPD - 121 short ton/day
Grammage 12-32 gsm - 7,4 -19,8 lb/3000ft²
Yankee dryer 16'
Speed 1800 m/min - 5900ft/min
In/out dryness 39% / 95%

NOVIMPIANTI DRYING TECHNOLOGY

STANDARD HT HOOD + YES-M

Yankee dryer is fed by steam partially from YES steam generator and partially from traditional boiler, Yankee Hood and traditional boiler are both gas fired

YANKEE HOOD		YANKEE DRYER	
Evaporated water	3730kg/h - 82400lb/h	Evaporated water	2850kg/h 6300lb/h
Drying ratio	57%	Drying ratio	43%
Blowing temperature	500°C – 932°F	Cast Iron Dryer	
Blowing speed	120m/s – 23600fpm	Pressure	7barg – 1000psi
Exhaust humidity	0,500	Total Steam cons	4200kg/h - 9240lb/h
Steam from YES	1550kg/h – 3400lb/h	Steam from Boiler	2650kg/h – 5830lb/h
Thermal (burner) 1098kWh/ton$_{paper}$ 3,75MMBTU/ton$_{paper}$		Thermal (boiler) 425kWh/ton$_{paper}$ 1,45MMBTU/ton$_{paper}$	
Total Specific Consumption 1523 kWh/ton$_{paper}$ **5,20 MMBTU/ton$_{paper}$**			

NOVIMPIANTI
DRYING TECHNOLOGY

TOP TEMPERATURE TT HOOD + YES-L

Yankee dryer is fed by steam partially from YES steam generator and partially from traditional boiler, Yankee Hood and traditional boiler are both gas fired

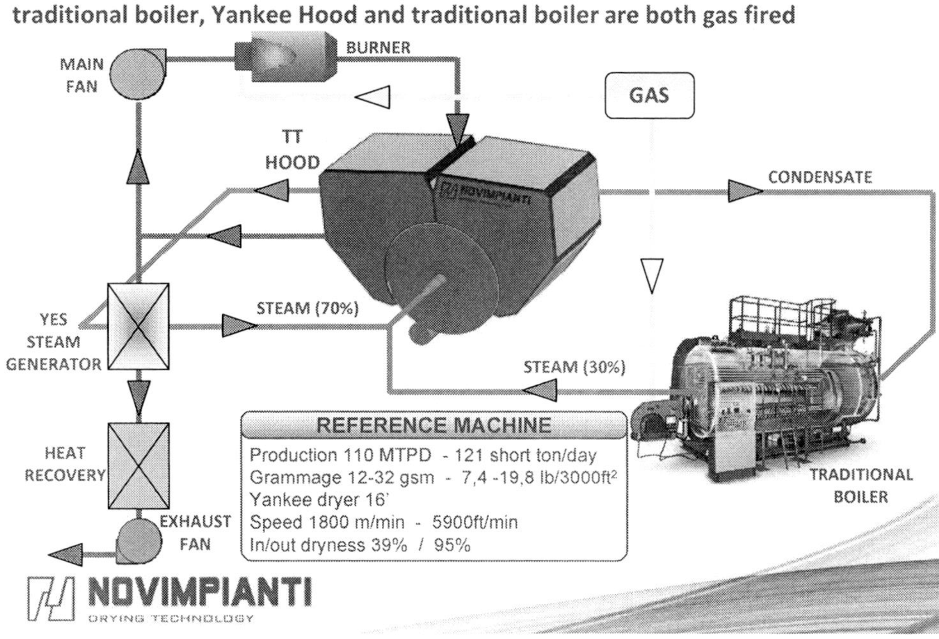

NOVIMPIANTI
DRYING TECHNOLOGY

TOP TEMPERATURE TT HOOD + YES-L

Yankee dryer is fed by steam partially from YES steam generator and partially from traditional boiler, Yankee Hood and traditional boiler are both gas fired

YANKEE HOOD		YANKEE DRYER	
Evaporated water	**3730kg/h - 82400lb/h**	Evaporated water	**2850kg/h - 6300lb/h**
Drying ratio	**57%**	Drying ratio	**43%**
Blowing temp.	**650°C – 1200°F**	Cast Iron Dryer	
Blowing speed	**95m/s – 18700fpm**	Pressure	**7barg – 1000psi**
Exhaust humidity	**0,500**	Total Steam cons.	**4200kg/h - 9240lb/h**
Steam from YES	**3000kg/h – 6600lb/h**	Steam from Boiler	**1200kg/h –2640lb/h**
Thermal (burner) 1132kWh/ton$_{paper}$ 3,86MMBTU/ton$_{paper}$		Thermal (boiler) 258kWh/ton$_{paper}$ 0,88MMBTU/ton$_{paper}$	
Total Specific Consumption 1391 kWh/ton$_{paper}$ **4,75 MMBTU/ton$_{paper}$**			

NOVIMPIANTI
DRYING TECHNOLOGY

TOP TEMPERATURE TT HOOD + YES-XL

Yankee dryer is fed by steam completely from YES steam generator and traditional boiler is only used as back-up.

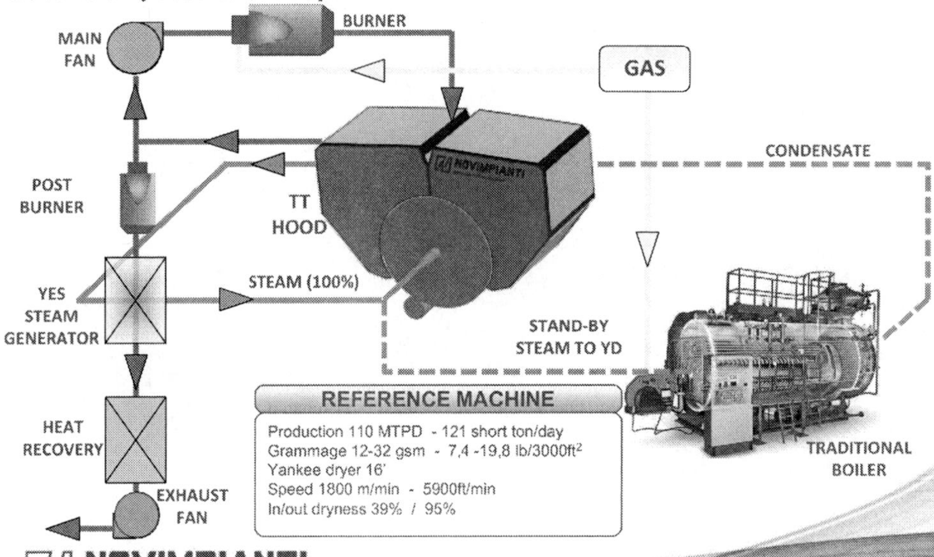

REFERENCE MACHINE
Production 110 MTPD - 121 short ton/day
Grammage 12-32 gsm - 7,4 -19,8 lb/3000ft^2
Yankee dryer 16'
Speed 1800 m/min - 5900ft/min
In/out dryness 39% / 95%

NOVIMPIANTI
DRYING TECHNOLOGY

TOP TEMPERATURE TT HOOD + YES-XL

Yankee dryer is fed by steam partially from YES steam generator and partially from traditional boiler, Yankee Hood and traditional boiler are both gas fired

YANKEE HOOD		YANKEE DRYER	
Evaporated water	3730kg/h - 82400lb/h	Evaporated water	2850kg/h - 6300lb/h
Drying ratio	57%	Drying ratio	43%
Blowing temp.	650°C – 1200°F	Cast Iron Dryer	
Blowing speed	95m/s – 18700fpm	Pressure	7barg – 1000psi
Exhaust humidity	0,500	Total Steam cons.	4200kg/h - 9240lb/h
Steam from YES	4200kg/h – 9240lb/h	Steam from Boiler	0
Thermal (burner) 1132kWh/ton_{paper} 3,86MMBTU/ton_{paper}	Thermal (post-burner) 229kWh/ton_{paper} 0,78MMBTU/ton_{paper}	Thermal (boiler) 0	
Total Specific Consumption 1362 kWh/ton_{paper} **4,65 MMBTU/ton_{paper}**			

NOVIMPIANTI DRYING TECHNOLOGY

SUMMARY OF TOTAL SPECIFIC ENERGY COMPARISON

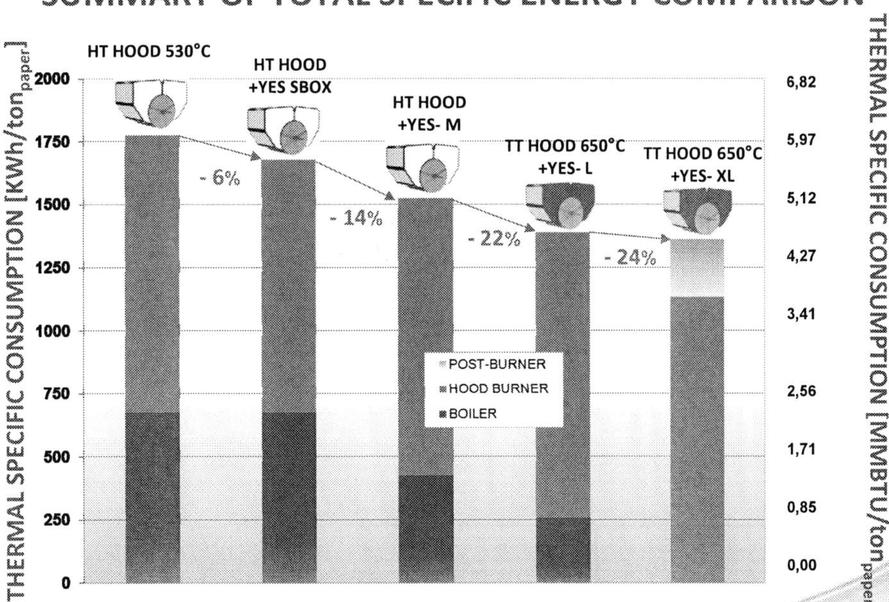

NOVIMPIANTI DRYING TECHNOLOGY

CASE HISORY: CONFIDENTIAL CUSTOMER (GERMANY) HIGH PRESSURE STEAM GENERATION SYSTEM BY NOVIMPIANTI

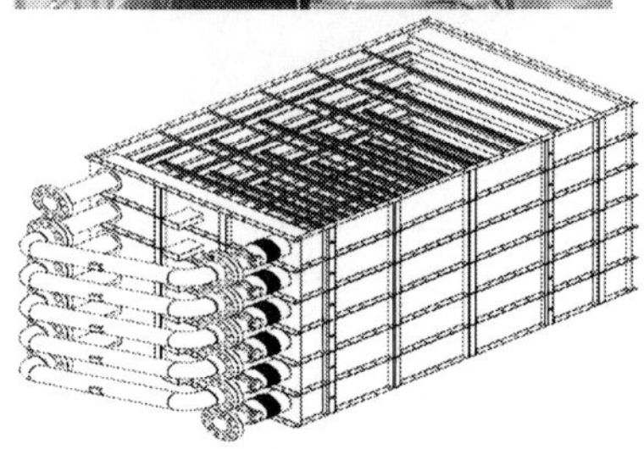

Tissue Machine Data

Production	194 MTD - 214 short ton/day	
Op. Yankee speed	1690 m/min - 5540 ft/min	
Op. steam pressure		6 bar – 87psi

New Steam Generation by Novimpianti

Start Up	2011	
Generation type	YES	
Ex temp.in	420 C°	790°F
Ex temp. out	240 C°	460°F
Recovered power	2,5 MW	8,54 MMBTU/h
Steam production	4,3 t/h	9,500 lb/h

NOVIMPIANTI
DRYING TECHNOLOGY

HIGH PRESSURE STEAM GENERATOR SCOPE OF SUPPLY CONFIDENTIAL CUSTOMERS

INCLUDED:

- •HEAT EXCHANGER
- •CIRCULATION PUMP
- •MAIN SEPARATOR
- •CONTROL VALVES
- •SAFETY VALVES
- •MEASURING INSTRUMENTS
- •ACCESSORIES (ball valves, check valves, filters...)
- •DUCTWORK (complete with dampers)
- •SUPERVISION (erection, commissioning and start-up)

NOT INCLUDED:

- •PIPINGS
- •INSULATIONS
- •MANPOWER FOR ERECTION

NOVIMPIANTI
DRYING TECHNOLOGY

HIGH PRESSURE STEAM GENERATOR – DCS PAGE

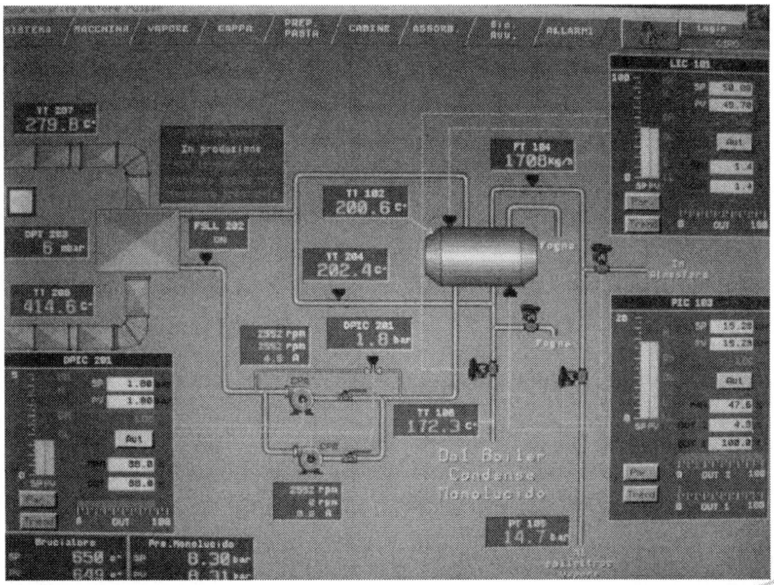

NOVIMPIANTI
DRYING TECHNOLOGY

BENEFITS SUMMARY FOR YES STEAM GENERATOR FOR EXISTING AND NEW TM

- Maximize Energy Saving and short payback period

- System certified for local rules (Boiler Type)

- Green Tech.> Reduces CO_2 emissions

- Easy to install with reduced machine shut down

- User and Maintenance friendly

NOVIMPIANTI
DRYING TECHNOLOGY

CONCLUSIONS
TODAY > TOMORROW > YANKEE ECO STEAM™

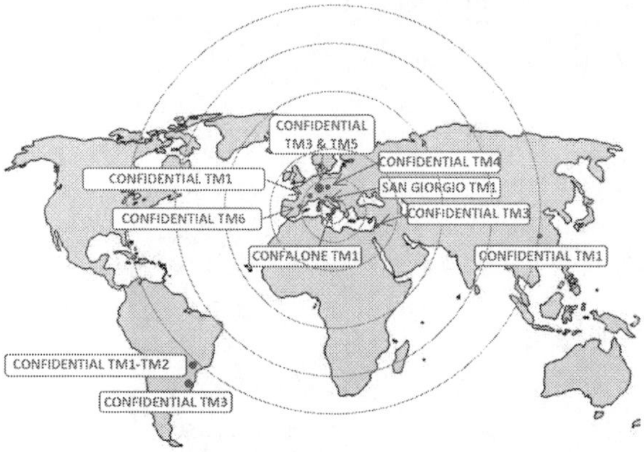

Yankee Eco Steam Generator is today a proven faster energy saving technology

Since the first installation in just 3 year we have supplied

12 YES Installations

4 are Top Corporate repeated orders and have under proposal and final negotiations ab.

15 Customer projects

Customized solutions and higher competitiveness are the key success factors of Novimpianti products Today and Tomorrow

NOVIMPIANTI DRYING TECHNOLOGY

Novimpianti thanks You for the attention !

For more info: www.novimpianti.com

Or contact: s.pecchia@novimpianti.com

NOVIMPIANTI DRYING TECHNOLOGY

Biogas for tissue production

Rip Penkert, Frank Kalany, and Bob Sewell, AMEC

Fredrik Öhman and Tom Karlson, ÅF

amec ÅF

Why biogas in tissue production?

- The forest and tissue industries are very green/sustainable today
  - Many operations have the potential to become even more sustainable
- Economy
  - Case-by-case basis
  - [illegible]tion using LPG in Europe
  - Tissue production using natural gas in USA
- Consumer demand for green prod[illegible] strong political advocacy are driv[illegible]
  - Incentives, carbon credits etc. may be [illegible]

amec ÅF

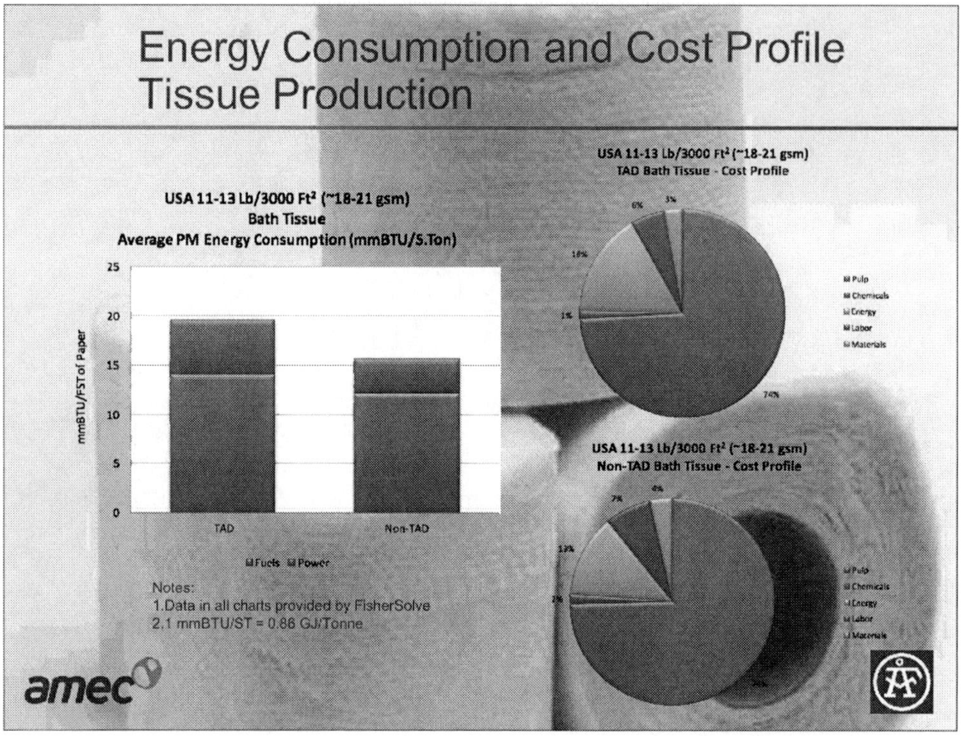
Energy Consumption and Cost Profile
Tissue Production
USA 11-13 Lb/3000 Ft² (~18-21 gsm)
Bath Tissue
Average PM Energy Consumption (mmBTU/S.Ton)
mmBTU/FST of Paper
25
20
15
10
5
0
TAD
Non-TAD
Fuels Power
USA 11-13 Lb/3000 Ft² (~18-21 gsm)
TAD Bath Tissue - Cost Profile
6%
3%
16%
1%
74%
Pulp
Chemicals
Energy
Labor
Materials
USA 11-13 Lb/3000 Ft² (~18-21 gsm)
Non-TAD Bath Tissue - Cost Profile
7%
4%
13%
Pulp
Chemicals
Energy
Labor
Materials
Notes:
1.Data in all charts provided by FisherSolve
2.1 mmBTU/ST = 0.86 GJ/Tonne
amec

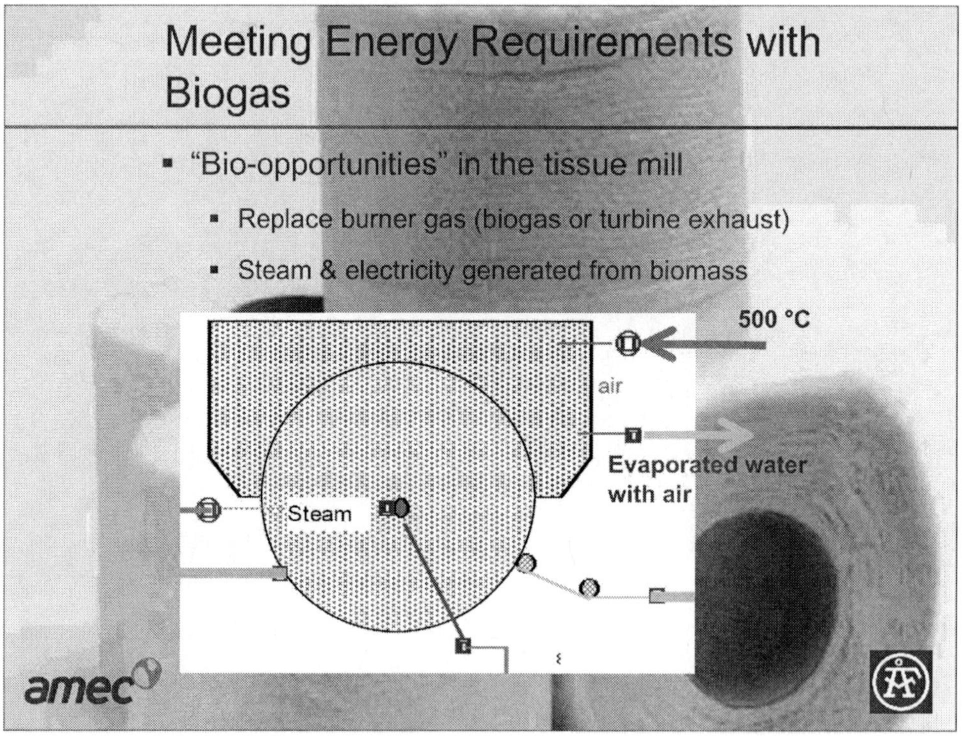
Meeting Energy Requirements with Biogas
"Bio-opportunities" in the tissue mill
Replace burner gas (biogas or turbine exhaust)
Steam & electricity generated from biomass
500 °C
air
Evaporated water with air
Steam
amec

Opportunities for bio-based gas supply

- Biogas produced internally in the mill
  - Anaerobic digestion of biological sludge
  - Anaerobic treatment of effluents
- Externally produced biogas
  - Sewage treatment plants
  - Landfill gas
- Biomass gasification
- Please note that some of the options discussed are still under development!

amec

Biogas produced internally in the mill

Example: Effluent treatment 700 000 adt/y pulp mill.
Tissue mills normally much smaller – integrated mills?

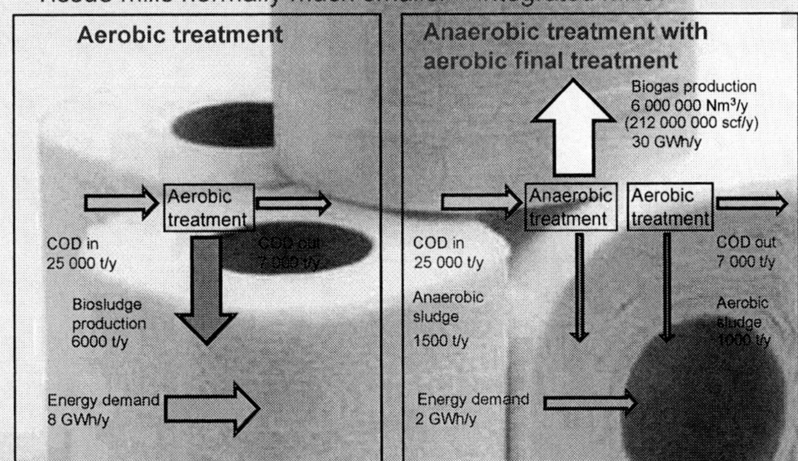

amec

Externally produced biogas

- What
  - Collected, cleaned and compressed at landfill or sewage treatment plants
  - Transported to mill via dedicated pipeline
  - Used at line pressure for direct firing
  - C[redacted]sed in mill if feeding to a gas turbine
- Considerations
  - As collected, the heating value of this gas is usually 40%[redacted]s heating value
  - Re[redacted]its may be available
  - Proven technology – For example, BMW in Spartanburg County South Carolina generates 11 MW by burning landfill gas in gas turbines generators

amec

Biomass gasification

- Technology under development – close to commercialization
- Cost-effective gas cleaning?
- Examples of air-blown gasifier technologies
- On-si[redacted]grated complexes

5 MW Bioneer gasifier, Jalasjärvi, Finland

MEVA INNOVATION AB

amec

nexterra

Comparison of heat values

Type of gas	Heat value (MJ/Nm3)	Heat value (btu/scf)
Natural gas	40	1070
Upgraded biogas	35	940
Raw biogas	18-28	480-750
Syngas (oxygen blown)	10-20	270-540
Syngas (air blown)	3-15	80-400

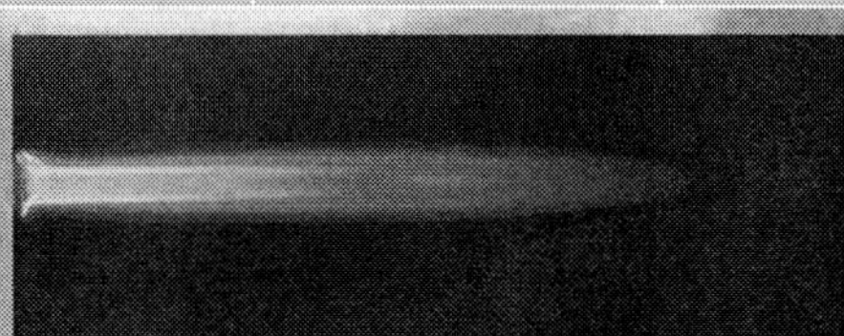

amec

Biogas - Conclusions

- Biogas is being used today to make Tissue mills and other manufacturing facilities more green and sustainable
- Consumer demand for green products and strong po[illegible]dvocacy are drivers
  - In some cases also economy
- Potential [illegible]cations [illegible] minimal risk, well pro[illegible]ended biogas) to unproven or infant technologies
- Biogas supply
  - integration with surrounding community
  - on-site generation
- Case-by-case evaluation necessary. The [illegible] consultants can help with these evaluations.

amec

Tissue World Americas 2012 Conference

Laminating Glues for Design Embossing

Colored Adhesives or Sticky Inks ?

Knut Möller, KAPP-CHEMIE GmbH & Co. KG, D-Miehlen

Introduction

Printing and Color Embossing on Bathroom Tissue

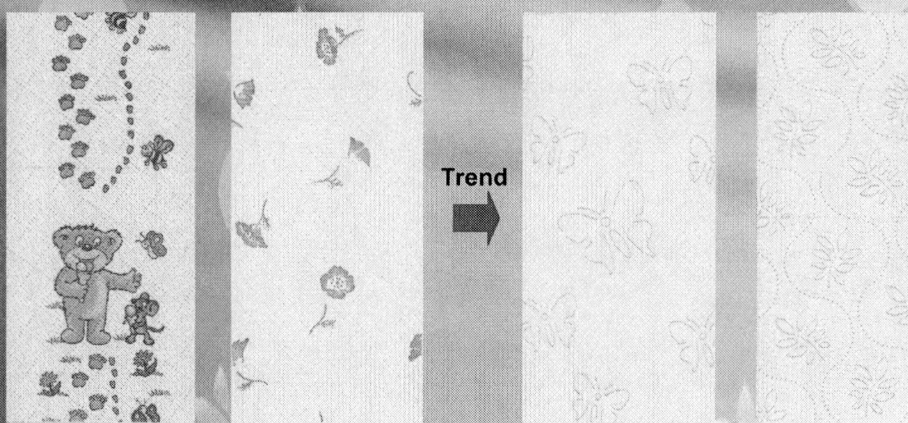

All examples taken from commercially available products

Introduction
Printing on Kitchen Towel
Trend
All examples taken from commercially available products

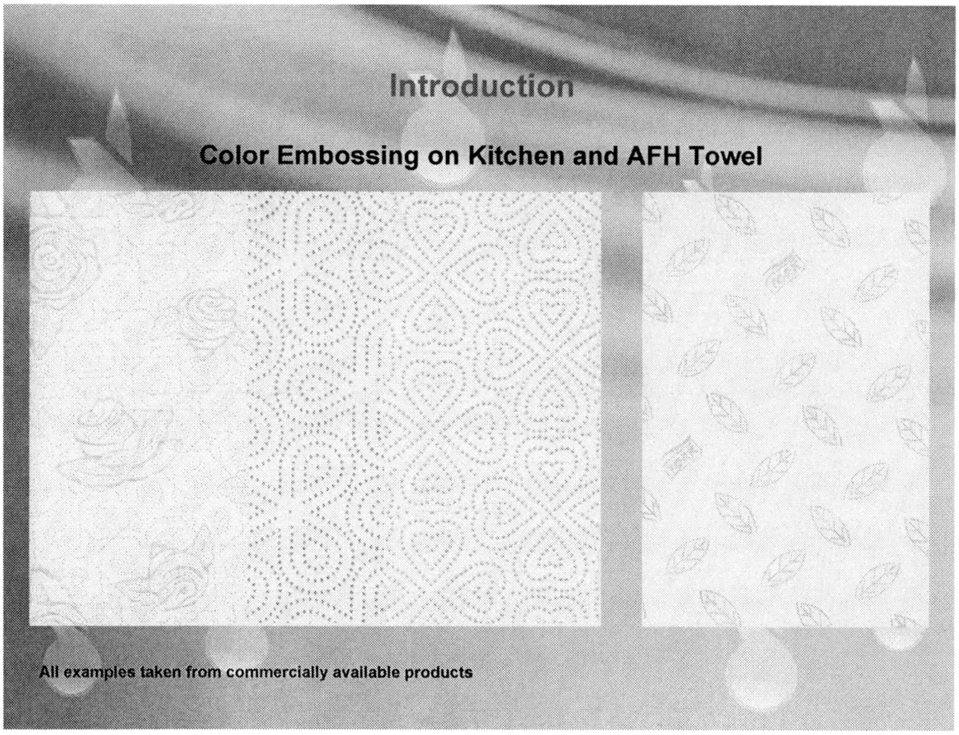
Introduction
Color Embossing on Kitchen and AFH Towel
All examples taken from commercially available products

Printing Inks

Requirements, Especially for Tissue

- ✓ **Ideal shades**
  - Customized following e.g. Pantone scales
  - CMYK for four color process printing
- ✓ **Superior runnability**
- ✓ **Color Intensity**
  - Good ink transfer, surface wetting
  - Ink holdout, only little penetration
- ✓ **Excellent bleeding resistance against**
  - standard liquids (EN 646:2006)
  - aggressive household cleaners
  - high temperatures
- ✓ **Rub-off resistance**
- ✓ **All ingredients approved for food contact**

Color Fastness by Film Formation

Pigments Embedded in Binder Film

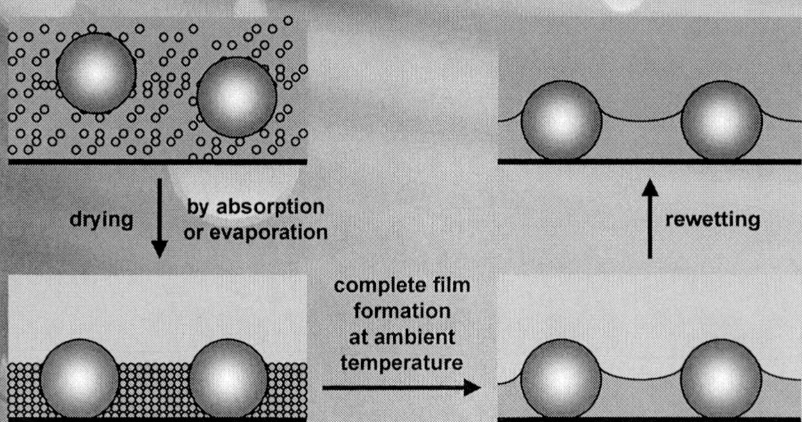

Laminating Glues

Requirements, Especially for Tissue

- ✓ **Good adhesive strength**
  - Affinity to cellulose fibers
  - Molecular weight
- ✓ **Excellent runnability**
  - No deposits
  - No splashing nor web formation
- ✓ Full water solubility
  - Dried film resolvable with water
- ✓ Controlled penetration through plies
  - Viscosity
  - Solubility
- ✓ All ingredients approved for food contact

Top-to-Top Embossing (schematic)

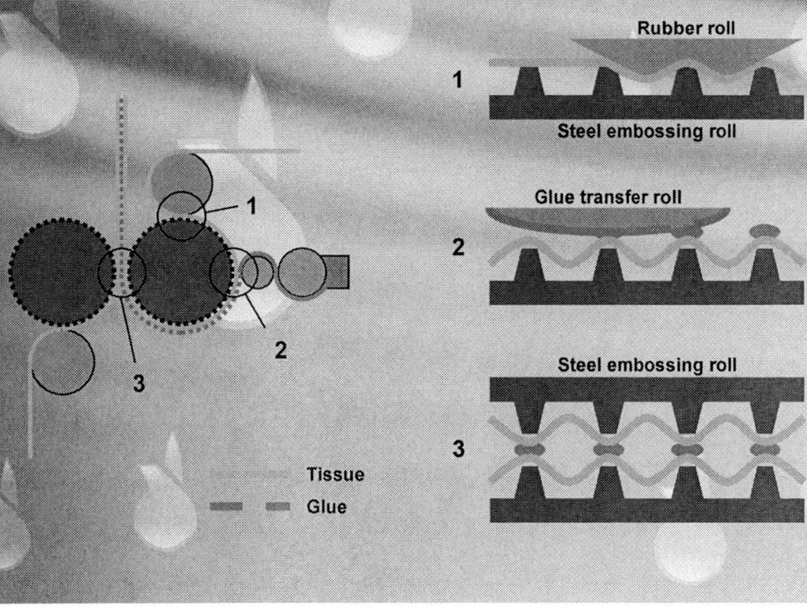

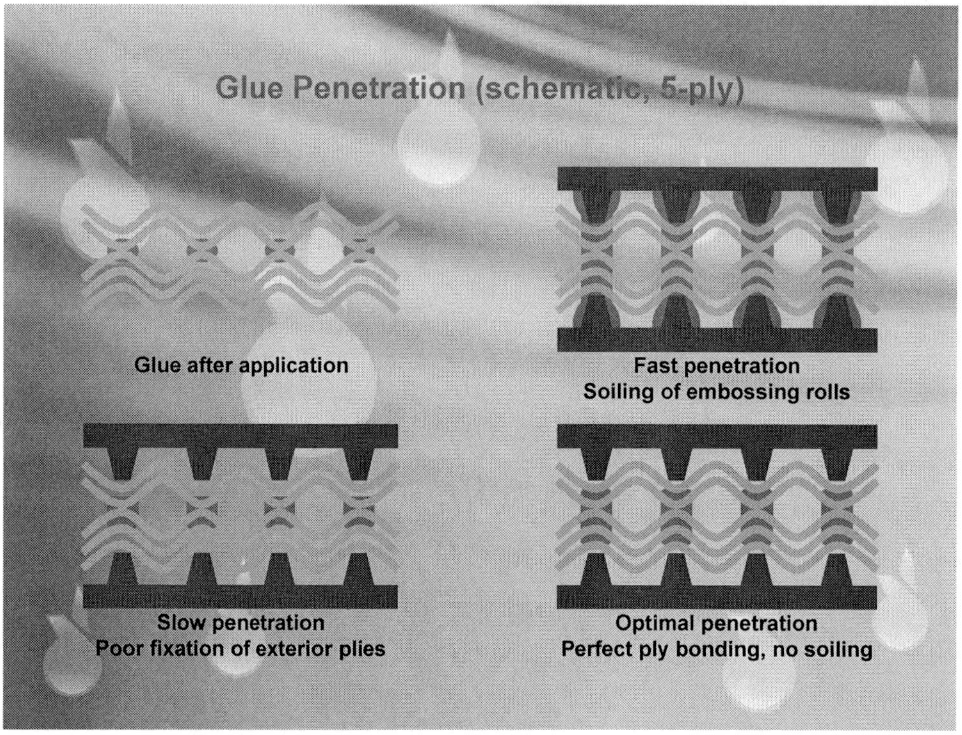

Laminating Glues

Factors Influencing the Glue Penetration

- ✓ **Glue viscosity**
  - Molecular weight
  - Concentration
  - Solubility
  - Glue polymer type
  - Functional additives
- ✓ **Dwell time**
  - Distance between application and embossing
  - Machine speed
- ✓ **Temperature**
- ✓ **Tissue quality**
- ✓ **Nip pressure**
- ✓ **Design and material of embossing rolls**

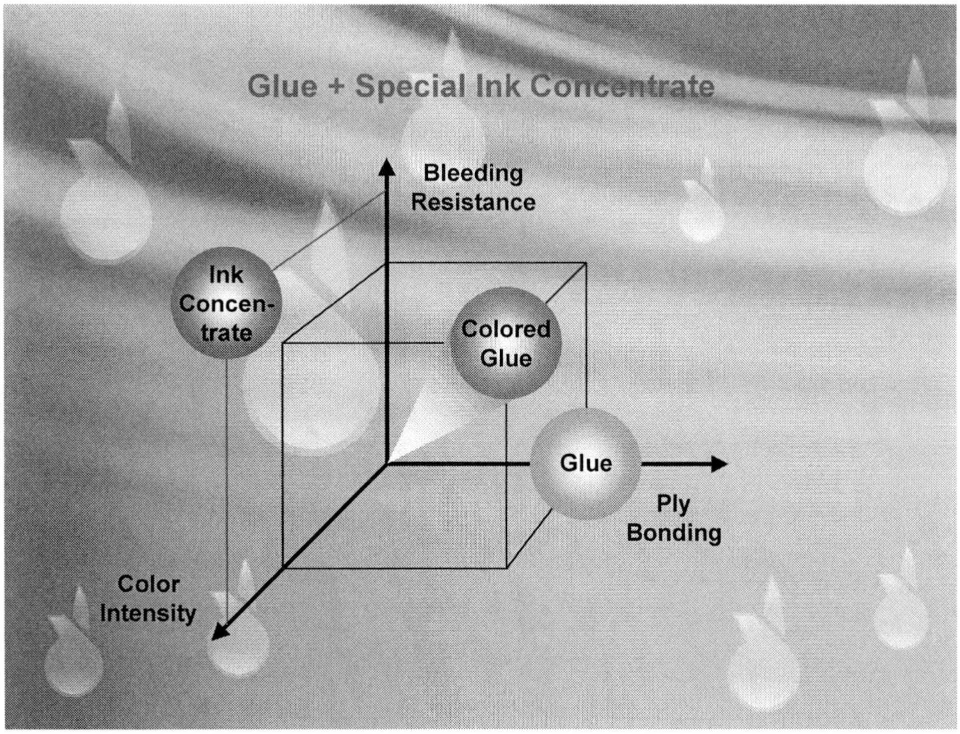
Glue + Special Ink Concentrate
Bleeding Resistance
Ink Concen-trate
Colored Glue
Glue
Ply Bonding
Color Intensity

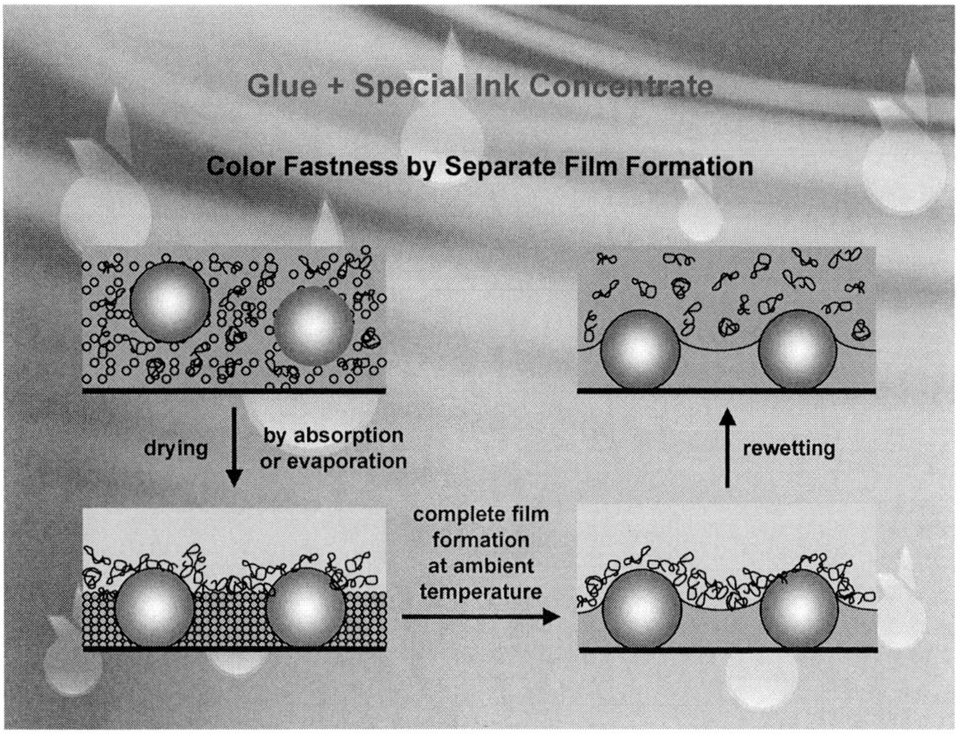
Glue + Special Ink Concentrate
Color Fastness by Separate Film Formation
drying
by absorption or evaporation
complete film formation at ambient temperature
rewetting

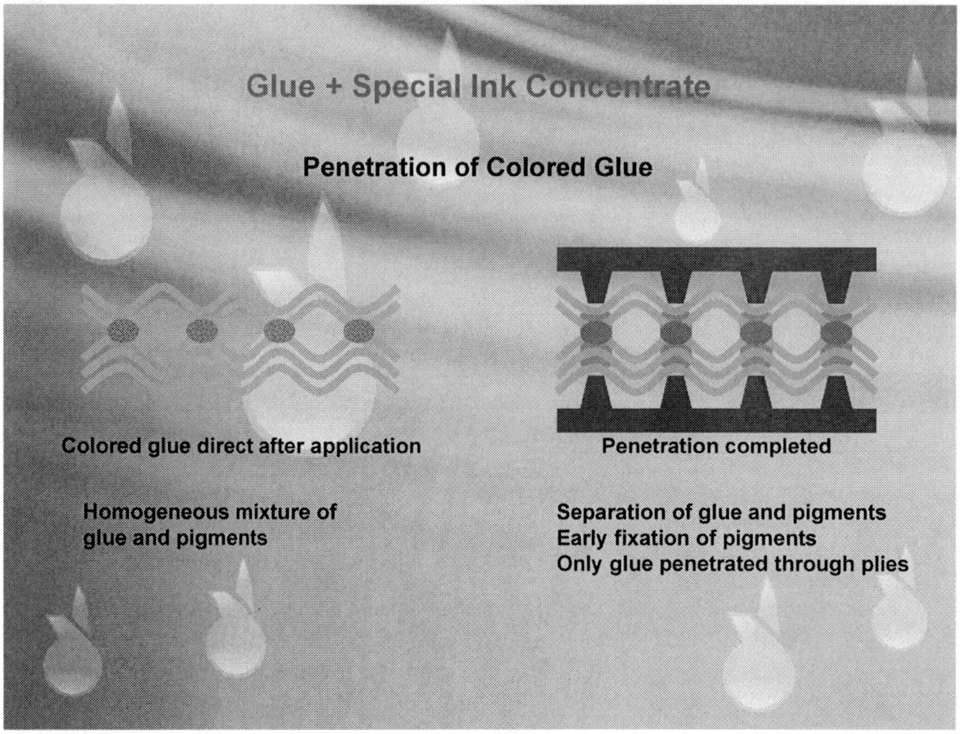
Glue + Special Ink Concentrate
Penetration of Colored Glue
Colored glue direct after application
Penetration completed
Homogeneous mixture of glue and pigments
Separation of glue and pigments
Early fixation of pigments
Only glue penetrated through plies

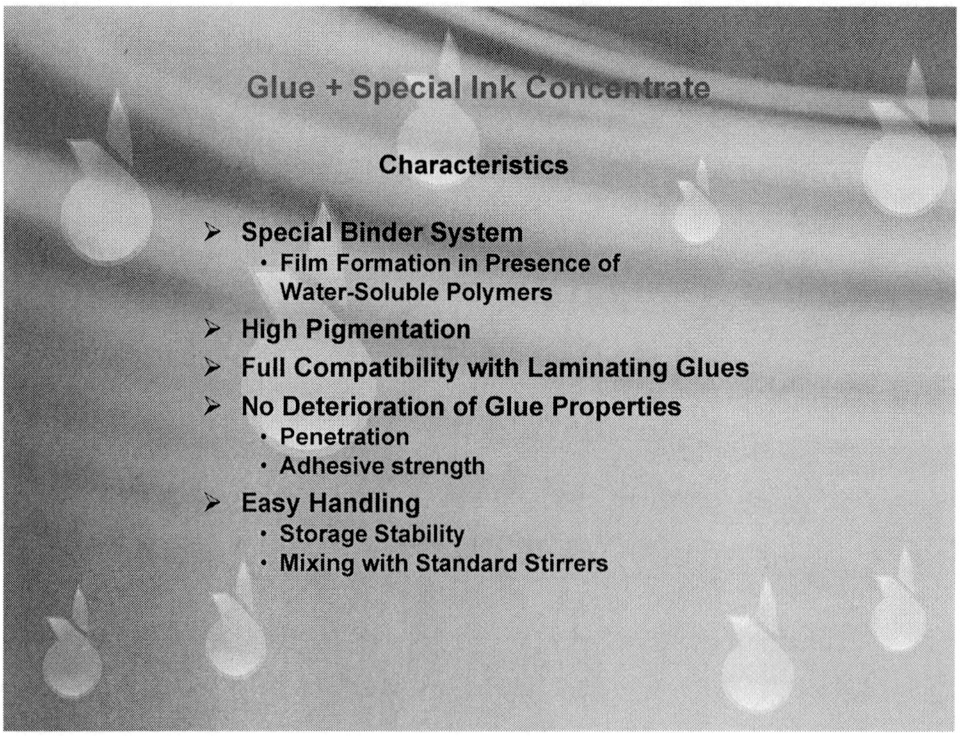
Glue + Special Ink Concentrate
Characteristics
Special Binder System
Film Formation in Presence of Water-Soluble Polymers
High Pigmentation
Full Compatibility with Laminating Glues
No Deterioration of Glue Properties
Penetration
Adhesive strength
Easy Handling
Storage Stability
Mixing with Standard Stirrers

Glue + Special Ink Concentrate

Make-Up Techniques Used in Practice

- **Separate Systems**
  - Mixing on-site, preferably using an automatic dispensing and mixing unit
  - Quick and easy change between colorless glue and different color shades
  - Ink concentrates must be fully compatible with all glue qualities used
  - Quality insurance on-site
- **Ready for Use Systems**
  - Favored for extended production runs of products having one single design
  - Quality guaranteed by supplier
- **Concentrated Colored Glues**
  - Economic version of ready for use systems

Defect Prevention

Color Lines

All examples taken from commercially available products

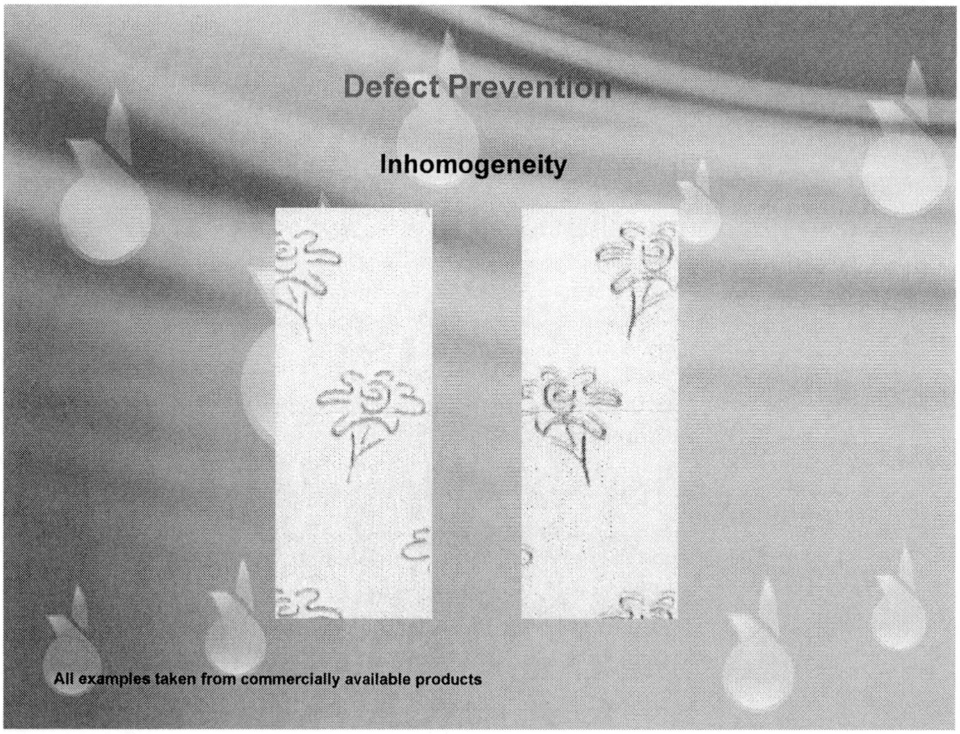
Defect Prevention
Inhomogeneity
All examples taken from commercially available products

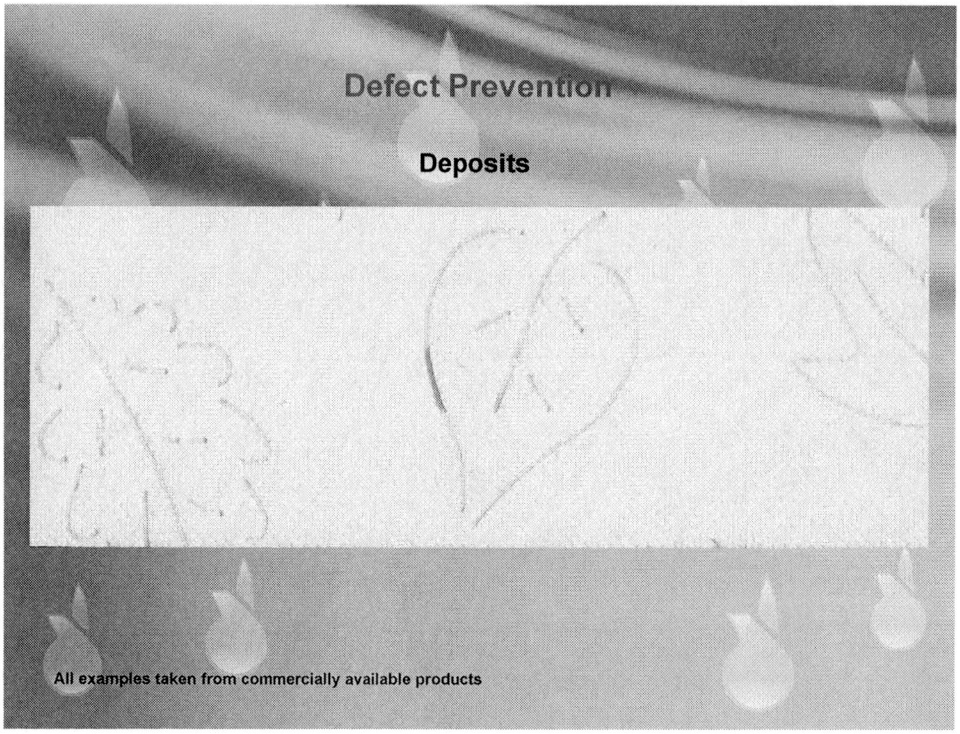
Defect Prevention
Deposits
All examples taken from commercially available products

Combination with Printing Inks and Colored Tissue

2-ply Bathroom Tissue | **3-ply Bathroom Tissue**

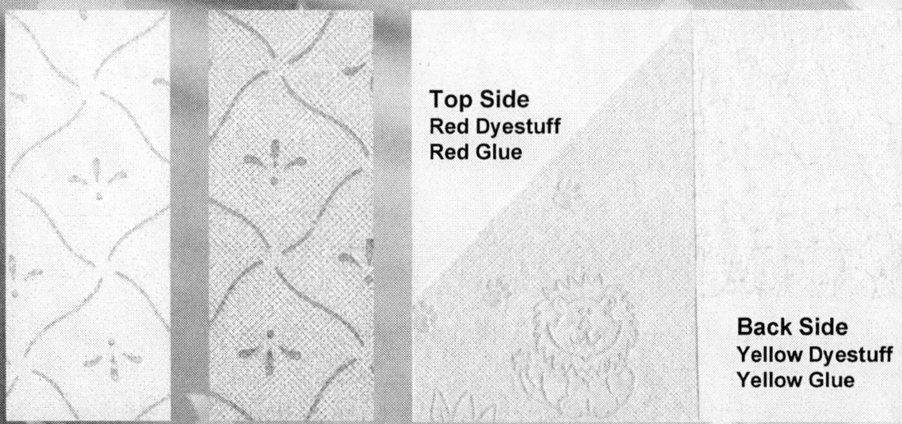

All examples taken from commercially available products

Conclusion

- Color embossing may supersede flexographic printing at least for bathroom tissue
- Only special ink concentrates for laminating glues can guarantee full color fastness at high color intensities (no pigment bleeding)
- Only special binder systems may ensures complete pigment embedding and early pigment particle fixation in multi-ply tissue products
- Full compatibility of ink and glue systems is a fundamental requirement
- The glue penetration has to be adjusted to avoid colored deposits
- Color embossing can be combined with dyed or printed tissue; color combinations have to be chosen carefully to avoid dull mixed colors

Tissue World Americas 2012

***Converting and Printing** session*

TCO spells informed investment

By Scott Johnson and Justen Smith, Futura North America Inc

There are many cautionary tales across all cultures which carry the message that buying cheap is not necessarily a good idea. Remember the three pigs and their houses of straw, timber and brick? No prizes for remembering who was the most satisfied with his choice of construction material.

We would like to take the argument further by suggesting that lower purchase costs in any field can actually turn out to be the more expensive option in the long run.

This is because of an important concept which we refer to as Total Cost of Ownership, or TCO.

Our analysis using TCO suggests that if you buy cheaper, you likely pay a higher price in the end.

Of course you can argue that a higher system price doesn't guarantee more reliability. But TCO is not as simple as that. The TCO approach to purchase decisions, rather than looking merely at the upfront investment cost, is the more logical figure against which to measure the value of an investment.

One might assume that the project cost would remain the most significant figure in the equation. The truth is, however, that the project cost is far lower than the accumulated operating costs over the lifetime of the whole converting line. Spending slightly more on equipment in order to achieve lower operating costs tends to be worthwhile.

Definition

We define Total Cost of Ownership according to its principal components: Preparation, Project and Operation. Once these phases are put in the context of the investment lifespan, then you have a realistic Total Cost of Ownership, or TCO. The aim is to minimize the TCO, but you need to measure it properly in the first place, to make a valid comparison.

The three phases of TCO

The focus on balancing capital cost with operating cost is what is often lacking in the preparation phase of TCO analysis, and this can make all the difference to the eventual profitability of an investment. So we need to make accurate assessments of comparative downtime for different technologies, reliability, ease of operation, start-up progression and levels of waste.

Establishing project cost might seem simple compared with operating costs, but provided you know what you expect to produce, and have accurate data concerning the technology you have in mind to produce these products and capture market share, then putting a figure on operating cost should be straightforward.

Here are typical examples of the technical information which a converting operation will need in order to establish operating cost:

- **Average attained speed and average efficiency of the operation**
- **Average waste the equipment design produces**
- **Amount of downtime to make a product changeover**
- **Complexity of production (number of changeovers)**
- **Amount of downtime to clean the equipment**
- **Impact of equipment design on product quality**

Then we need to put numbers on some of the other variables, as follows:

- **Product specifications**
- **Output rates for SKU for each unit in the line**
- **Cost of and usage of raw materials**
- **Operating Equipment Efficiency**
- **Labor requirement and cost**
- **Training needs**
- **Parts replacement and service frequency**
- **Energy consumption per unit produced**

Such calculations are all part of the preparation phase. It's the lowest-cost phase of the whole project, but it has the most impact on its overall success.

Case studies

Of course you would like to see some numbers to back up the theory. In the example shown today, the purchase of a rewinder illustrates how the higher output from the preferred option one reduces the unit cost, which could increase profit by substantial percentages. Over the lifetime of a converting line, we are talking about numbers you really can't ignore.

We can also show today examples of investments in cutting systems for kitchen towel and toilet tissue. By increasing production and reducing rejects, the profit yielded from these elements of the line could also reach a significant sum.

Assuming a converting line lasts for an average 20 years, a smart technology choice can typically save millions of dollars over the lifetime of the equipment. In almost any scenario the user benefits to be gained from choosing the more efficient and effective equipment are substantial.

These figures are based on actual product specifications and extrapolated. But actually, the more challenging the conditions, the greater the benefits. If a paper is bulkier and more absorbent or just super soft, then while these features are great from the end-user's perspective, the product can be tougher to convert. But this is where a well-informed choice of converting solution has even more bearing on TCO.

Complete line or upgrade

So whether you are considering the purchase of a complete converting line, or upgrading one or more elements, bear in mind that the cheaper system in terms of project cost could prove more costly to own. It possibly won't perform as well either.

TCO is an objective process, and that is something which any supplier should welcome. The real beneficiary is the user, of course, who ends up with converting solutions which have been specified on the basis of all the relevant information, not just some of it.

TISSUE WORLD AMERICAS

Gambini

Miami, March 22nd 2012

cmggroup.it

TISSUE WORLD AMERICAS

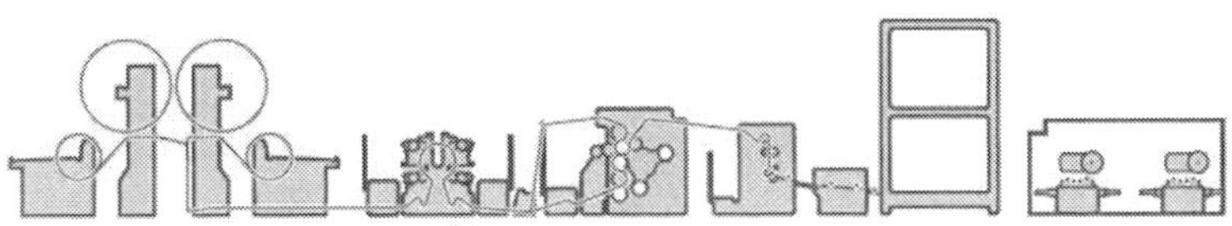

INNOVATIVE TISSUE MARKET MACHINERY

Miami, March 22nd 2012

Gambini

cmggroup.it

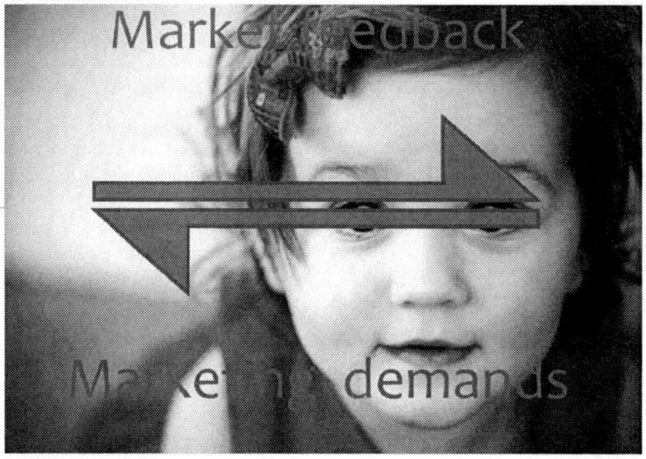

Miami, March 22nd 2012

Gambini

cmggroup.it

Market requests

Miami, March 22nd 2012

Gambini

cmggroup.it

TISSUE WORLD AMERICAS

TOUCHMAX

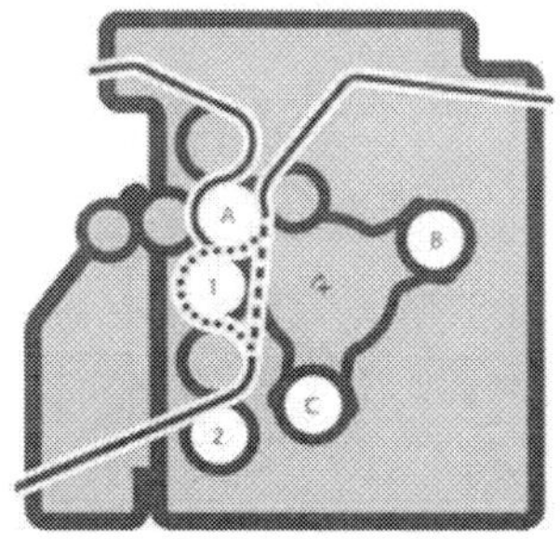

The safest + most versatile embosser

Miami, March 22nd 2012

Gambini

cmggroup.it

TISSUE WORLD AMERICAS

TOUCHMAX

Miami, March 22nd 2012

Gambini

cmggroup.it

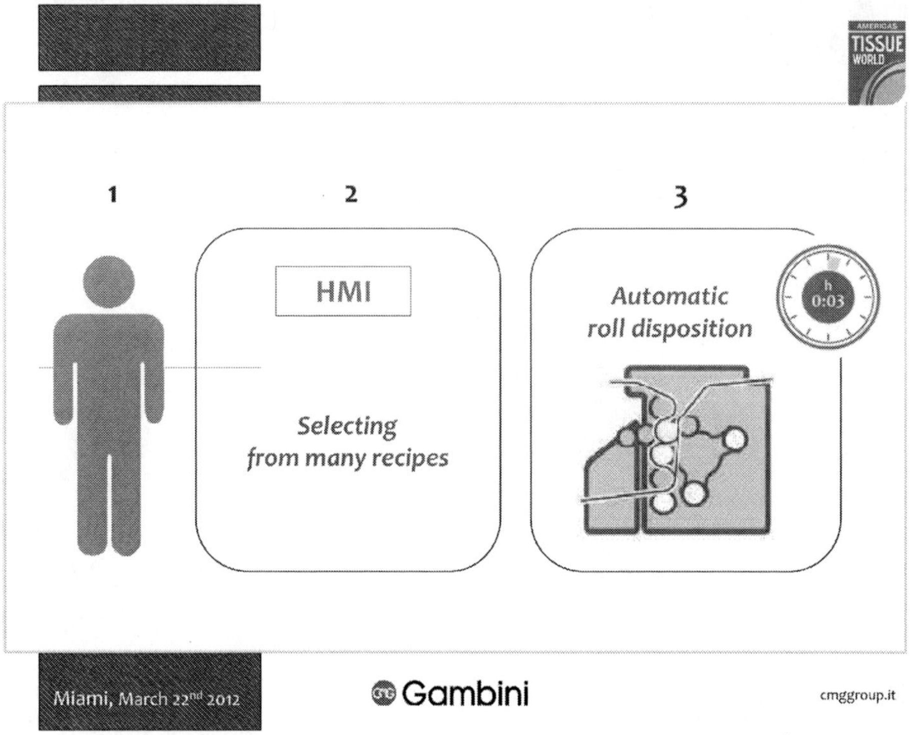
[redacted]
TISSUE WORLD
1
2
3
HMI
Selecting
from many recipes
Automatic
roll disposition
h
0:03
Miami, March 22nd 2012
Gambini
cmggroup.it

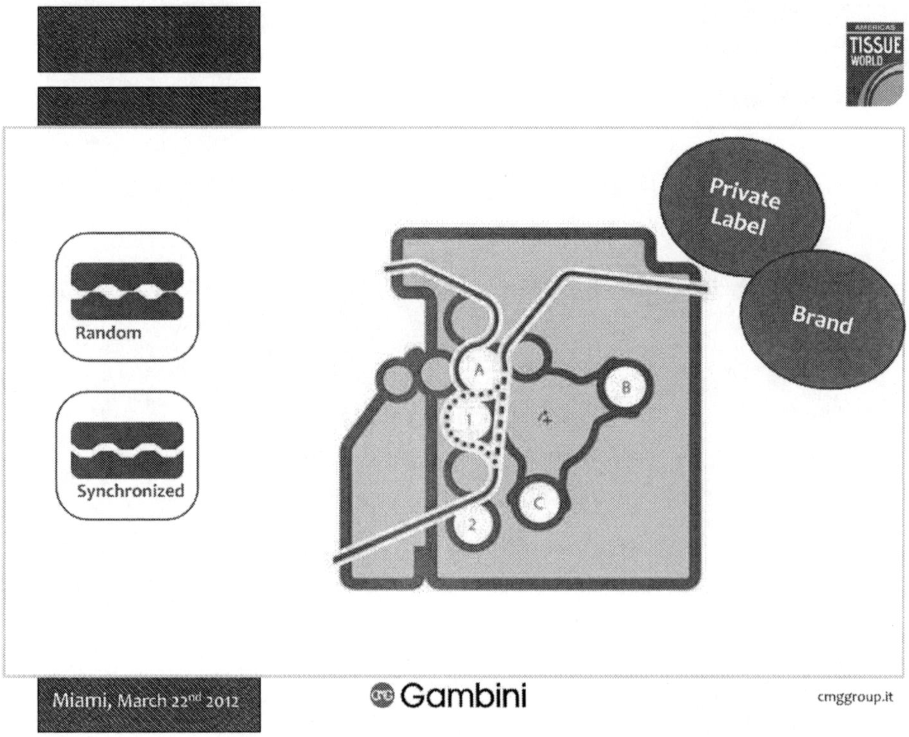
[redacted]
TISSUE WORLD
Random
Synchronized
Private
Label
Brand
A
B
C
1
2
4
Miami, March 22nd 2012
Gambini
cmggroup.it

Market trend

Private label shares and brand shares are destined to even out

European Market

Comparable markets

North America & Latin America

Miami, March 22nd 2012 — Gambini — cmggroup.it

TOUCHMAX

FLEXIBILITY

SPEED

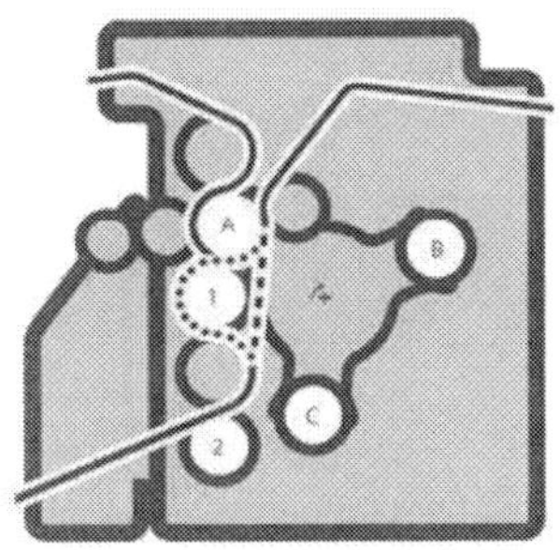

SAFETY AT WORK

SIMPLICITY

Competitive Prices

Miami, March 22nd 2012 — Gambini — cmggroup.it

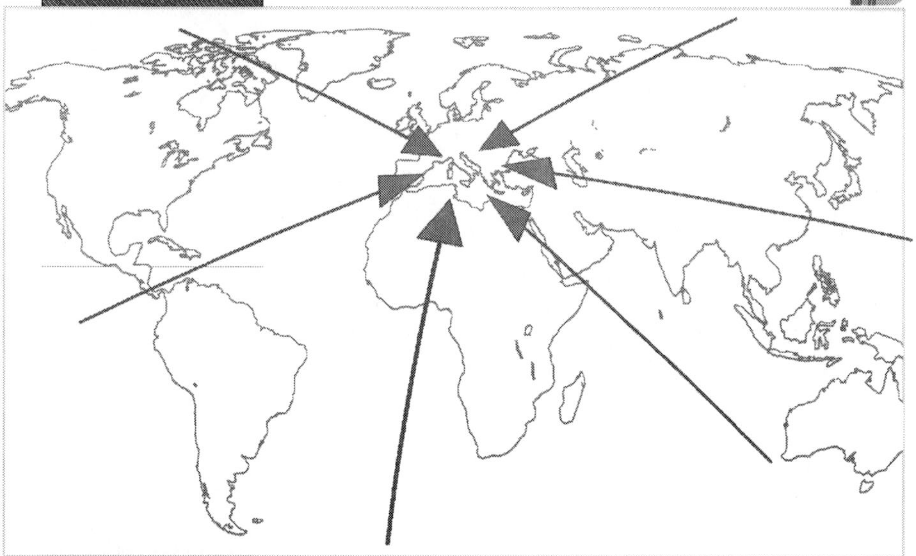

Miami, March 22nd 2012

Gambini

cmggroup.it

Thank you for your kind attention!

Miami, March 22nd 2012

Gambini

cmggroup.it

TMC
TISSUE MACHINERY COMPANY
Integrate to Innovate:
The Next Steps in Packaging Efficency
Presented by:
Davide Pasqualotto
TMC – Global Sales Manager

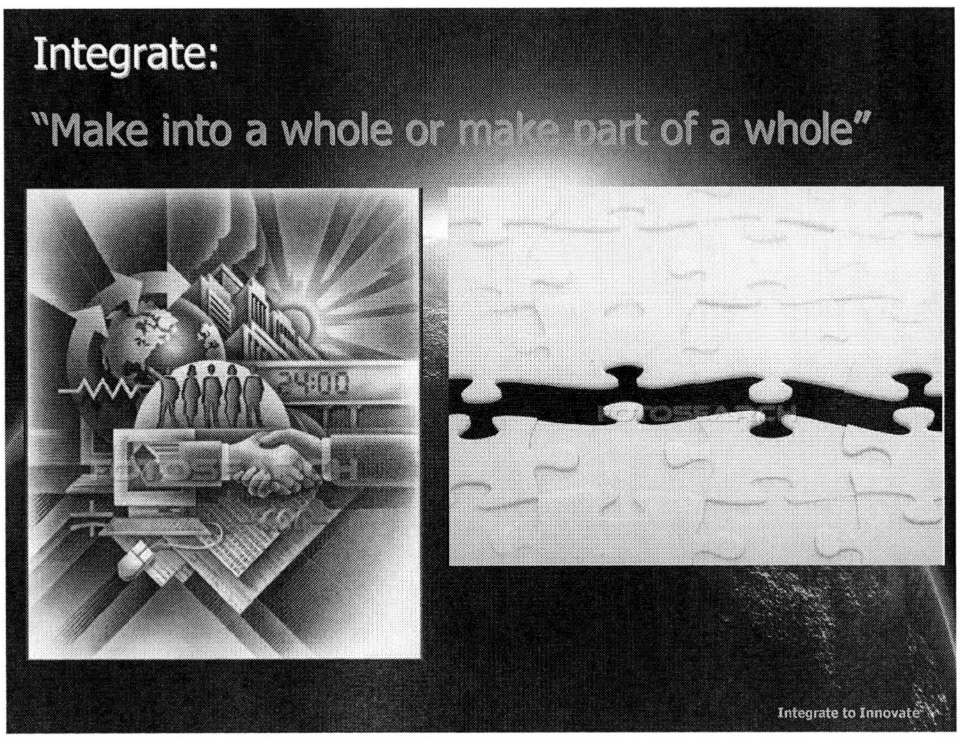
Integrate:
"Make into a whole or make part of a whole"
Integrate to Innovate

Integrate: Step One
"Supply the packaging machines we need and link them all together"
Integrate to Innovate

Integrate: Step Two
Step One:
"Supply the packaging machines we need and link them all together"
Step Two:
"And Integrate Equipment from third supplier"
Integrate to Innovate

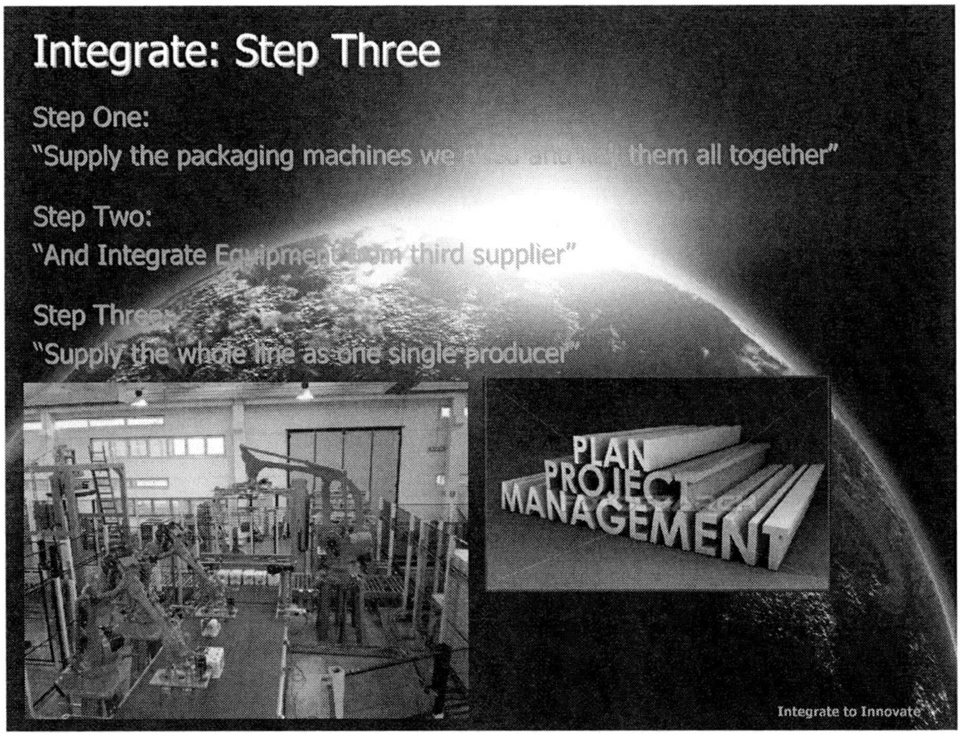
Integrate: Step Three
Step One:
"Supply the packaging machines we
them all together"
Step Two:
"And Integrate
third supplier"
Step Thr
"Supply the whole line as one single producer"
PLAN
PROJECT
MANAGEMENT
Integrate to Innovate

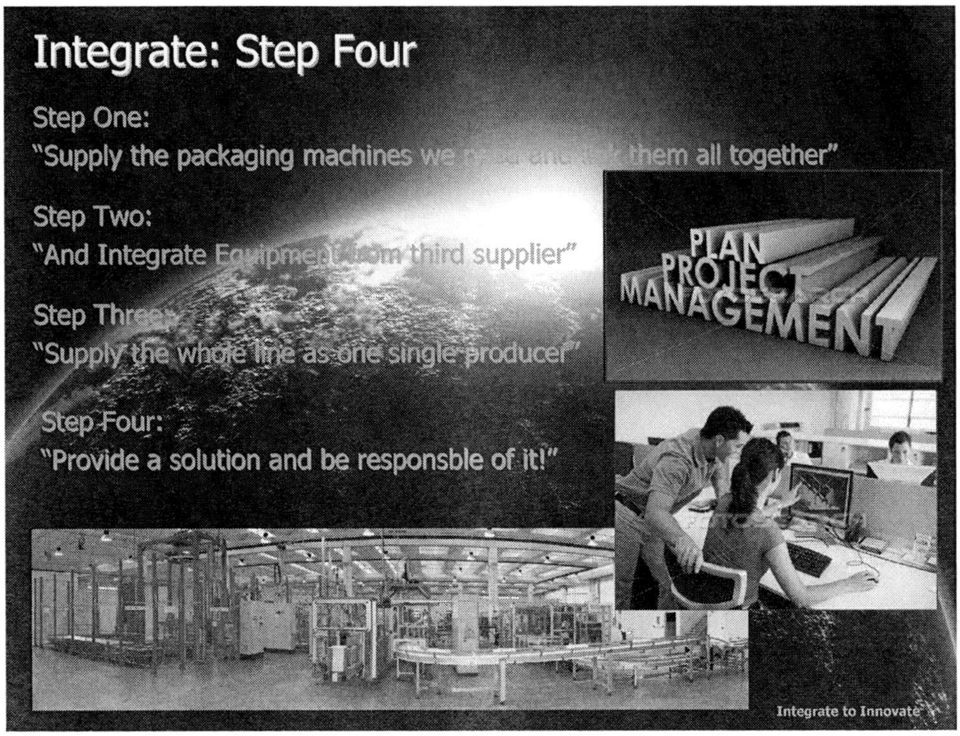
Integrate: Step Four
Step One:
"Supply the packaging machines we
them all together"
Step Two:
"And Integrate
third supplier"
Step Thr
"Supply the whole line as one single producer"
Step Four:
"Provide a solution and be responsble of it!"
PLAN
PROJECT
MANAGEMENT
Integrate to Innovate

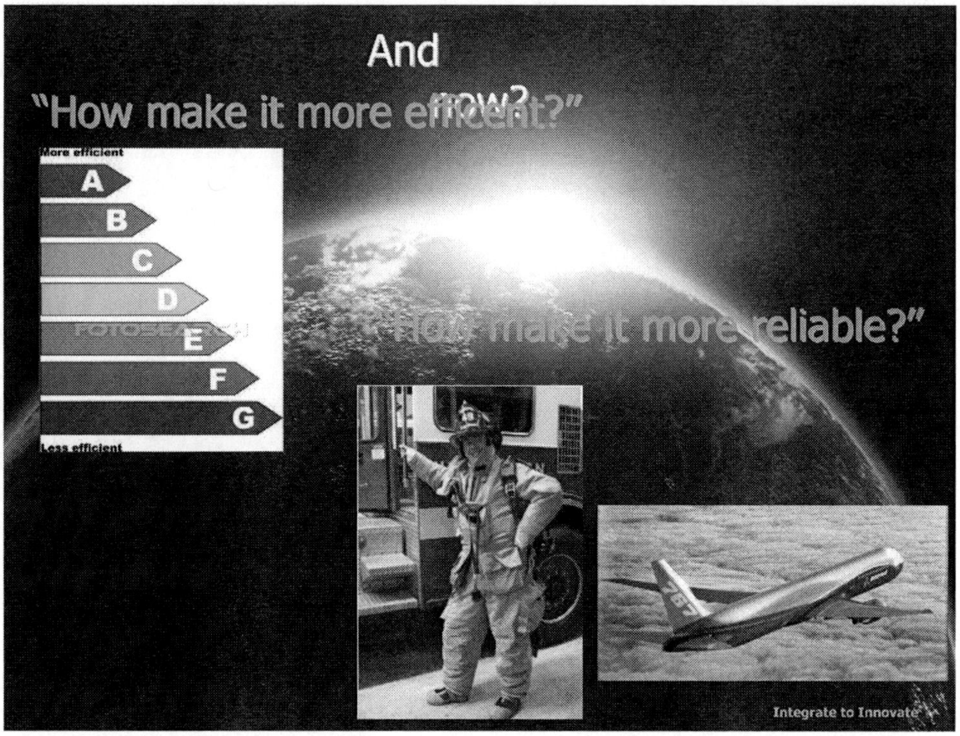
And now?
"How make it more efficient?"
More efficient
A
B
C
D
E
F
G
Less efficient
"How make it more reliable?"
Integrate to Innovate

Extending the integration into an innovative machine
TMC
Integrate to Innovate

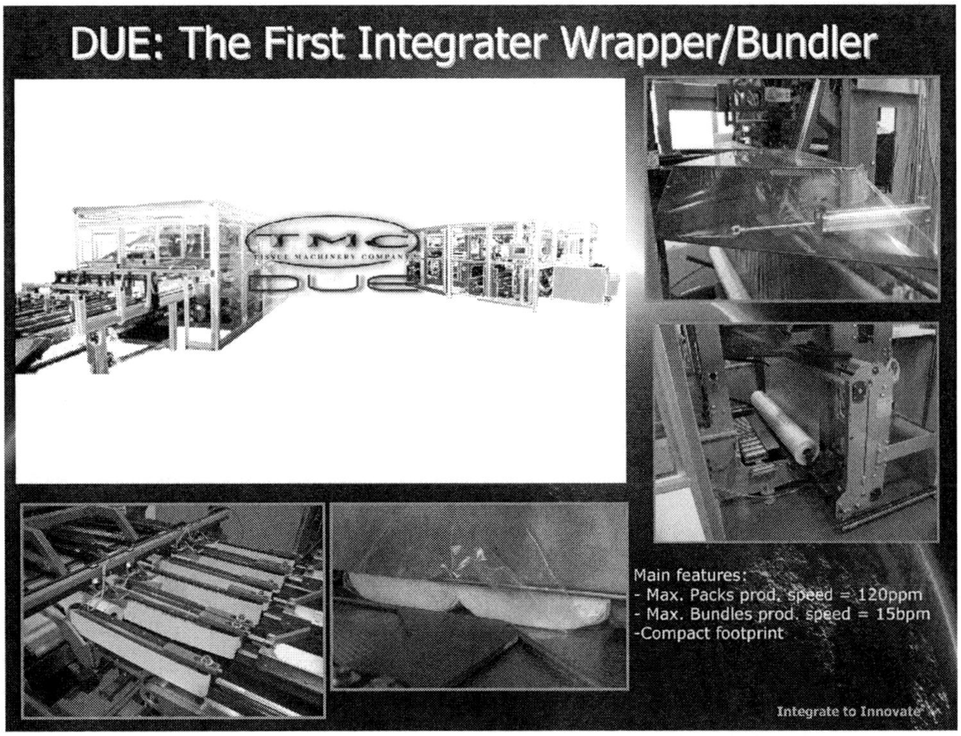
DUE: The First Integrater Wrapper/Bundler
Main features:
- Max. Packs prod. speed = 120ppm
- Max. Bundles prod. speed = 15bpm
-Compact footprint
Integrate to Innovate

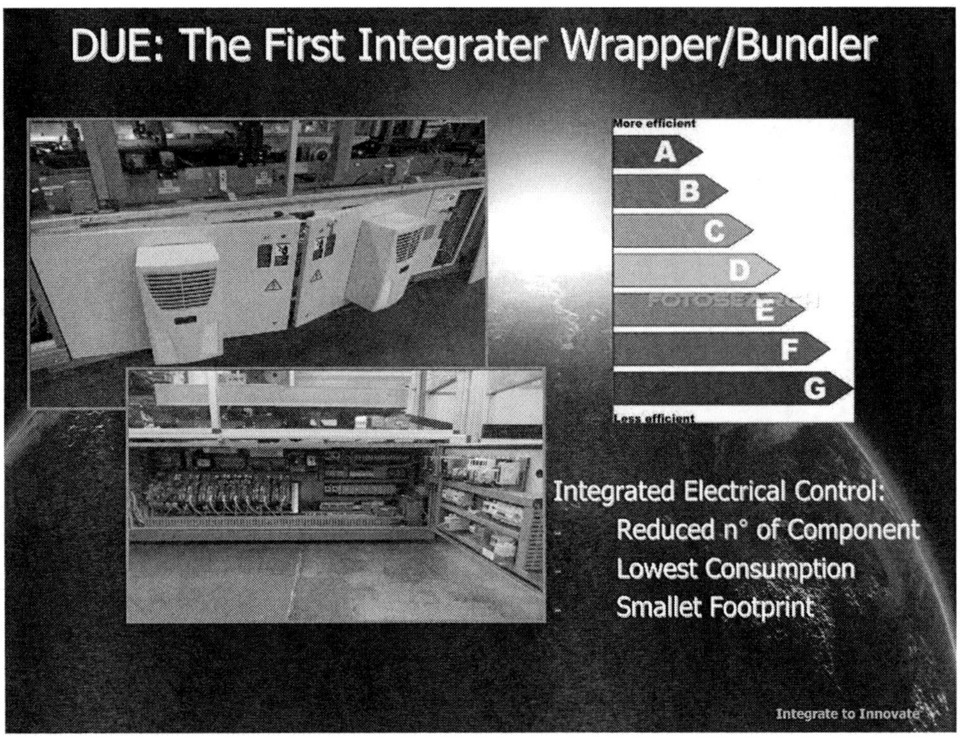
DUE: The First Integrater Wrapper/Bundler
More efficient
A
B
C
D
E
F
G
Less efficient
Integrated Electrical Control:
- Reduced n° of Component
- Lowest Consumption
- Smallet Footprint
Integrate to Innovate

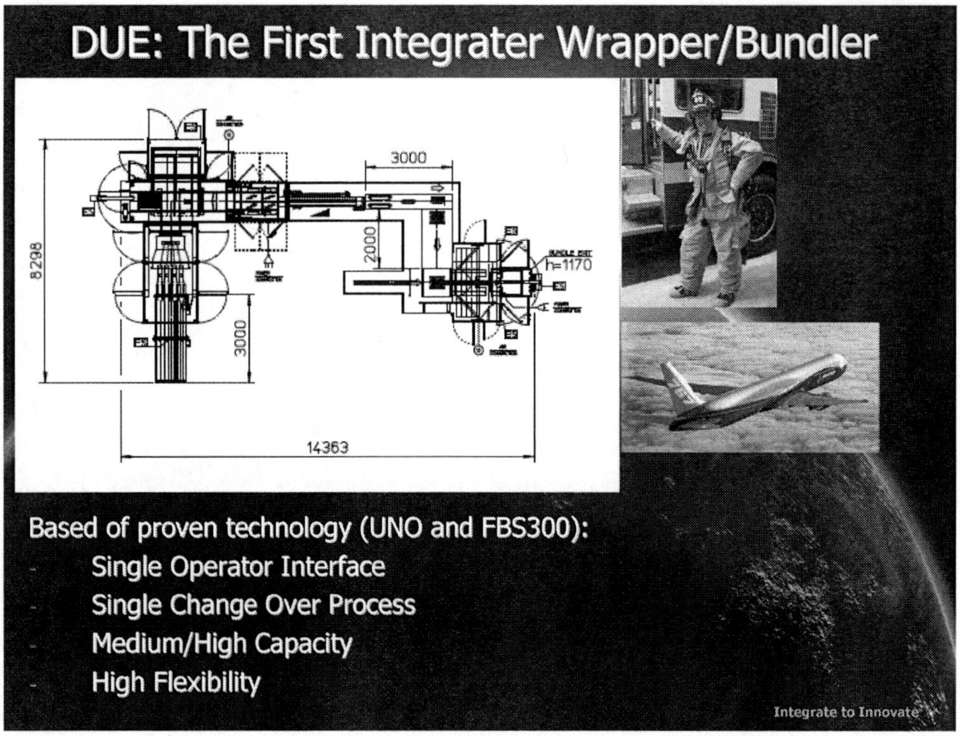
DUE: The First Integrater Wrapper/Bundler
3000
2000
8298
3000
14363
h=1170
Based of proven technology (UNO and FBS300):
- Single Operator Interface
- Single Change Over Process
- Medium/High Capacity
- High Flexibility
Integrate to Innovate

Integrate to Innovate:
- Our Customer requires integrated solution
- We've been challenged to be Sytem Integrator
- We're challenged to be Solution Supplier
- We're now extending the integrated requirement into a new machine desing
- Reliable relationship is in our commitment
- Being innovative looking for customer needs
Solution across the lineSolution across the line

Integrate to Innovate:
The Next Steps in Packaging Efficency
Presented by:
Davide Pasqualotto
TMC – Global Sales Manager

Compact Packaging of Tissue Paper

OPTIMA
EXCELLENCE IN PACKAGING

The Concept

OPTIMA

Use the opportunity of cost savings and packaging improvements for sustainability by
film packaging vs. cardboard packaging

- ► **Reduce your environmental footprint – "go green"**
- ► **Reduce consumption of raw materials**

= SUSTAINABILITY
= COST SAVINGS

How can we help?

OPTIMA

- **Differentiate from the competition**
- **Volume reduction in finished packs up to 33%**
- **Cost Savings through**
  - less raw materials
  - less warehousing space and costs
  - less transport costs
- **Save the environment through**
  - reduced trucks on the road
- **Meet the sustainable requirements for retail outlets**

COST SAVINGS through:

OPTIMA

Reduced volume of finished packs

- in warehouse at manufacturing site
- transportation by manufacturer
- less handling by distribution channels
- and final customer

Example for SPACE CONSUMPTION

OPTIMA
EXCELLENCE IN PACKAGING

Incoming packaging raw materials at your facility

Required Space per year

OPTIMA
EXCELLENCE IN PACKAGING

(4,800 units/shift = 14,400 units/day x 350 days)

	RAW MATERIALS	FINISHED PRODUCTS
BOXES	881 truckloads per year	5,950 truckloads per year
FILM BAGS	14 truckloads per year	5,250 truckloads per year

© OPTIMA packaging group GmbH

Less trucking cost (raw materials)

Boxes:

220 flat boxes / pallet x 26 pallets / truck = 5,720 boxes / truck
14,400 units/day ÷ 5,720 = 2.5 trucks / day **= 881 trucks / year**

Bags:

6,000 bags / box = 12,000 bags / pallet x 26 pallets / truck = 312,000 bags / truck
14,400 units/day ÷ 312,000 = 0.04 trucks / day **= 14 trucks / year**

Savings:

881 - 14 = 867 transports at an average of $500 each : **$ 433,500 / year**

Less trucking cost ( )

Boxes:

32 (4x8) filled boxes/pallet x 26 pallets/truck	=	832 filled boxes/truck
(512 products per pallet)		
14,400 filled boxes / day ÷ 832	=	17 trucks / day
x 350 days	=	**5,950 trucks / year**

Bags:

36 (4x9) filled bags/pallet x 26 pallets/truck	=	936 filled bags/truck
(720 products per pallet)		
14,400 filled bags / day ÷ 936	=	15 trucks / day
x 350 days	=	**5,250 trucks / year**

Savings:

700 trucks / year x $500 / transport = **$350,000 /year**

Total Savings

OPTIMA
EXCELLENCE IN PACKAGING

Cost reduction for packaging material (bag vs. box)
Cost reduction through less trucking costs

Packaging material (savings between bags and boxes)	**$**	**2,016,000 / year**
Trucking cost (packaging raw material)	**$**	**433,500 / year**
Trucking cost (final packaging)	**$**	**350,000 / year**

$2,799,500 Savings

Examples of Filling Process

OPTIMA
EXCELLENCE IN PACKAGING

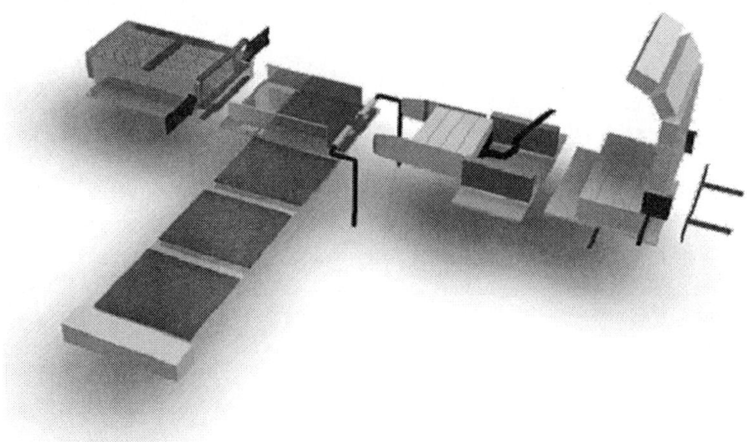

© OPTIMA packaging group GmbH

Examples

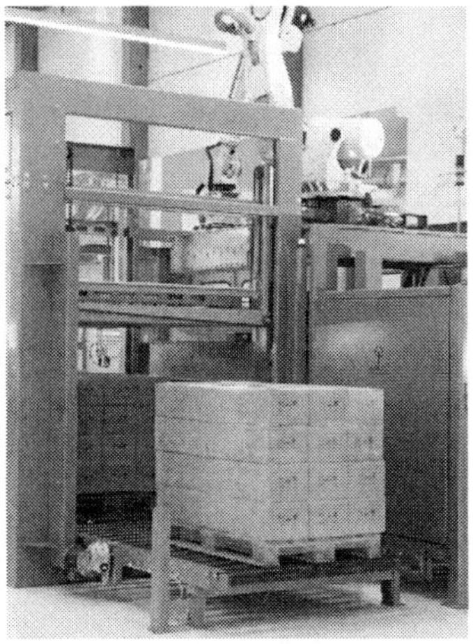

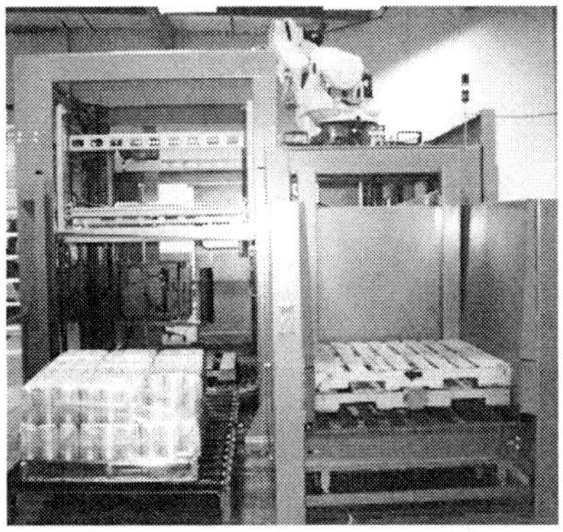

Schneider
Packaging Equipment Co., Inc.

INNOVATIVE END-OF-LINE SOLUTIONS

"Overall Equipment Effectiveness: How Next Generation Automation Gives Real Improvements"

Terry Zarnowski – Dir. Sales & Marketing

Schneider Packaging Equipment Company, Inc.

www.schneiderequip.com

Schneider
Packaging Equipment Co., Inc.
INNOVATIVE END-OF-LINE SOLUTIONS

Why OEE

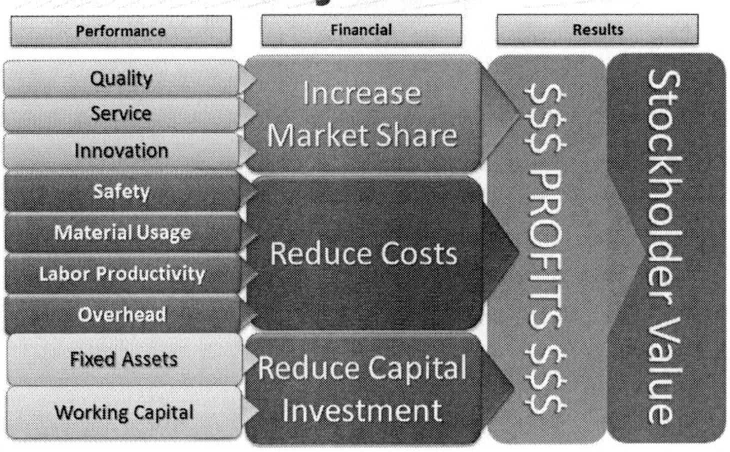

www.schneiderequip.com

Schneider Packaging Equipment Co., Inc. INNOVATIVE END-OF-LINE SOLUTIONS

How Does OEE Work?

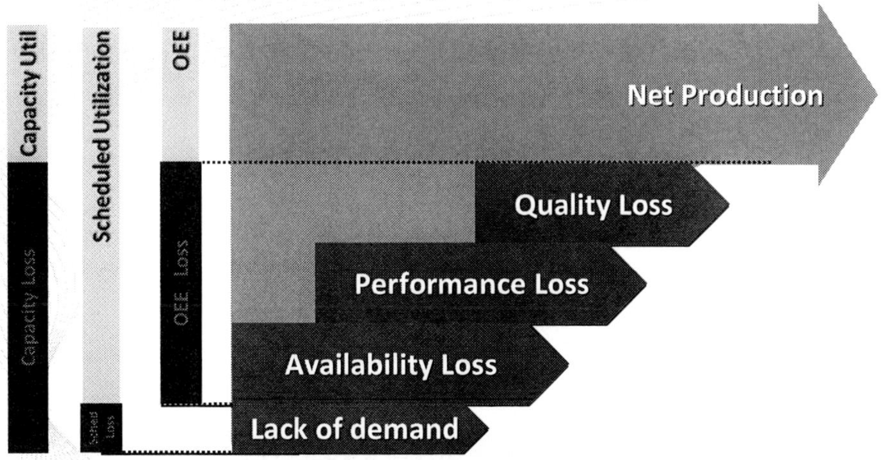

www.schneiderequip.com

Schneider Packaging Equipment Co., Inc. INNOVATIVE END-OF-LINE SOLUTIONS

OEE – The Equation

- The ‘power’ of a simple Equation
  - Availability – Uptime as a % of scheduled time
  - Performance Rate – % Actual units produced per minute to target units per minute
  - Quality Rate - % of units produced that are salable

OEE (%) therefore equals:

Availability X Performance Rate X Quality Rate

www.schneiderequip.com

Schneider Packaging Equipment Co., Inc. INNOVATIVE END-OF-LINE SOLUTIONS

The Process of OEE – Case Packer Example

- The Process - simplified
  - Define possible loss production events
  - Track those events
  - Understand the data
  - Act on the data
- Process can be manual for older equipment and automated on new equipment
  - Level of automation depends on the ROI!

www.schneiderequip.com

Schneider Packaging Equipment Co., Inc. INNOVATIVE END-OF-LINE SOLUTIONS

Data Collection & Analysis

Manual OEE Tracking form

Opeator:	117
Date:	1/19/2012
Shift:	2
Machine S/N:	2947-0110

Production Counter @ Start of Shift:	47,568
Production @ End of Shift:	55,757
# Cases scrapped over shift:	5

Max. rate of machine:	20.0	cpm	Availability:	79.17%	Min. theoretical cost per case:	$	4.25
Average rate over shift:	17.1	cpm	Performance:	85.30%	Actual cost per case:	$	6.30
Actual cases produced:	8,189		Quality:	99.94%	Max. theoretical cases producible:		9600
Scheduled shift minutes:	480		OEE:	67.49%	*Potential + Revenue if 100% OEE*	$	*22,761.83*
Unplanned Downtime:	40				Max theretical profit per case:	$	1.28
Planned Downtime:	60				**Actual profit per case:**	$	(0.77)
Log of downtime events:					*Potential + Profit @ 100% OEE*	$	*18,564.10*

Time	Product Select	# minutes Lost	Breakdown	Case Pick-off Case Jam	Erect Case Jam	Bad minor flap	Bad major flap fold	Bad Seal	Accumulator jam	laner jam	Infeed product jam	Change over	Meetings/Breaks/ Lunch	Training	Maintenance		Quality Inspection	Comment - Describe downtime event in detail
09:17	16	2		x														
09:22	16	2		x														
09:24	16	2		x														Looks like cup #3 on erect needs replacement
09:35	16	8	x												x			cup #3 replaced
14:21	16	2			x													kdf glued together
14:23	16	2			x													kdf glued together - bad lot?
14:37	16	15															x	QC kdf lot rejected, loaded new lot
15:30	36	7										x						Change to product #36, love that auto - adjust!

www.schneiderequip.com

Schneider Packaging Equipment Co., Inc. INNOVATIVE END-OF-LINE SOLUTIONS

Automation and OEE

- Most equipment produced today has the ability to collect loss data – specifically capture downtime events, the reasons that caused them, and the duration the event occurred. That same equipment can, if so equipped, track scheduled time, ID of the operator(s), and keep track of product in vs. completed products out, and track it's own production data. All the essentials in calculating OEE and providing the data essential in making change.
- Along with collecting loss data, the machine can also be enhanced to collect key data and trend that data with correlation to loss production events. Actually alert the maintenance or quality dept. in some circumstances.

www.schneiderequip.com

Schneider Packaging Equipment Co., Inc. INNOVATIVE END-OF-LINE SOLUTIONS

Automation and OEE

Downtime Accumulator

Total Runtime	Total Downtime
540.0	488.6
Not in Auto (%)	**Machine Faults (%)**
0.4 %	0.7 %
Blocked or Jam (%)	**Safety System (%)**
0.7 %	0.3 %

OEE Data

OEE Factor	Shift 1	Shift 2
Availability	92.44	90.48
Performance	98.04	97.49
Quality	99.97	99.94
OEE	90.60	88.15

Of course if the machine captured and calculated the data…..

www.schneiderequip.com

Schneider Packaging Equipment Co., Inc. INNOVATIVE END-OF-LINE SOLUTIONS

Machine Enhancements for Improved OEE

- Auto-adjust
  - Reduce setup, changeover time and setup errors
- Online interactive documentation
  - Points operator to exactly where on the machine the problem has occurred and instructions
  - If maintenance logs on to the HMI, goes to the appropriate assembly drawing and schematics
  - Maintenance reminders based on Failure Analysis
- Additional operational status monitoring
  - Motor currents
  - Air consumption
- Auto KDF loading with dunnage removal
  - Uninterrupted production during reload
- Easier access for improved operation and maintenance

www.schneiderequip.com

Schneider Packaging Equipment Co., Inc. INNOVATIVE END-OF-LINE SOLUTIONS

To Learn More

Contact me: tzarnowski@schneiderequip.com

The Alliance for Innovation and Operational Effectiveness: www.pmmi.org/AIOE

Schneider offers complete end of line 'Tissue Tough' solutions for all of your rolled and folded products. **See us at booth 960**.

Case packing, palletizing, and fully integrated lines.

www.schneiderequip.com

Tissue converting engineering **TISSUE WORLD 2012**

AMERICAS TISSUE WORLD

Pulsar PRESENTATION
MIAMI - 2012

Pulsar

Tissue converting engineering **TISSUE WORLD 2012** Pulsar

From case histories to practical solutions:

NEW TRENDS IN CONVEYOR LINES SETUP AND MANAGEMENT

ENERGY SAVING

MAINTENANCE

SAFETY

TISSUE WORLD 2012

Along the years we have witnessed an evolution in the converting area.

Conveyors were usually considered simple connection devices between different machines.

This generated what we call THE STANDARD APPROACH

Switch on the conveyors and let us work!

TISSUE WORLD 2012

Today conveyor lines transfer not only paper rolls but logic and information too. We have lived with this new experience in the field together with our customers and we have collected a lot of data and know-how.

The new requirements have generated a NEW APPROACH

We need a system to manage the speed of the machines without involving the operator ...

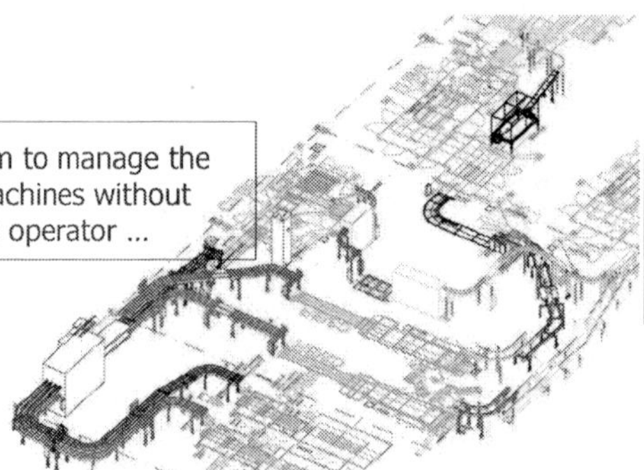

We have just what you need and we can tailor it on your specific needs

ENERGY SAVING

On standard conveyor lines electric motors and gearboxes generate the major energy consumption

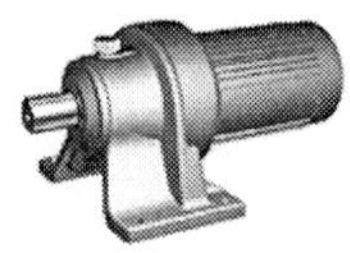

A standard conveyor system with

a mid number (24)

of mid-sized motors (0.75 kW)

working 6 day/week for 2 shifts

has a power consumption bill of about 20 times the cost of the motors!

=

Better to check the energy bill rather than the pure cost of the component!

Reducing energy consumption

No magic formulas but accurate engineering!

Reduction of installed power	Careful layout analysis
High efficiency components	Less energy used = increased efficiency
Speed management	Correct speed regulated by format and flow rate
Correct maintenance	Higher efficiency over the years and reduced downtime

REDUCING INSTALLED POWER BY ACCURATE LAYOUT ANALYSIS ...

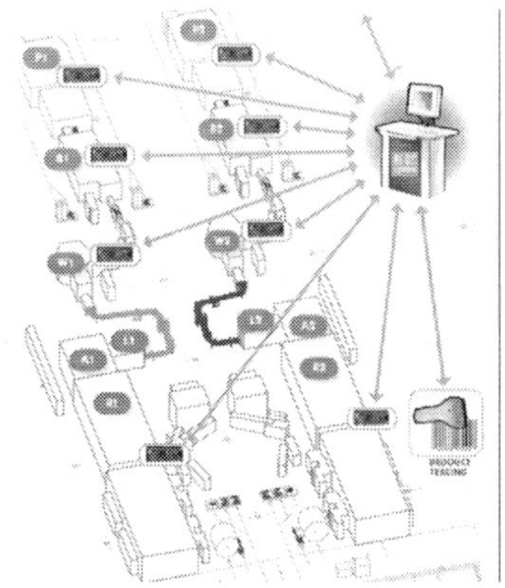

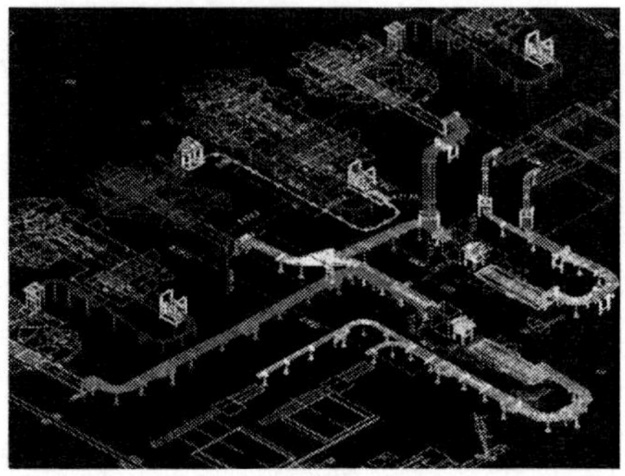

... BUT WITHOUT AFFECTING THE CORRECT CONTROL OF "BACKLOG" AND "BACK-PRESSURE"

RULE OF THUMB: AT LEAST 3 MOTORS FOR EVERY MACHINE-MACHINE CONNECTION:

1) SECTION CONNECTED TO UPSTREAM MACHINE (I.E. LOGSAW)
2) SECTION DEDICATED TO BACKLOG FUNCTION
3) SECTION CONNECTED TO DOWNSTREAM MACHINE (I.E. WRAPPER)

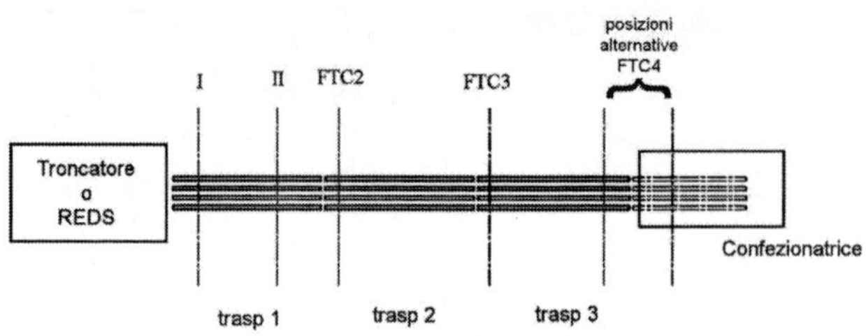

NEW KIND OF CONVEYOR COMPONENTS WARRANT LESS ENERGY CONSUMPTION PER KG OF PRODUCT

VS

High energy consumption + 30%

Low energy consumption - 30%

CONVEYOR SPEED MUST TAKE CARE OF:

1) PRODUCT DIMENSIONS AND SHAPE
2) FLOW RATE (PIECES/MINUTE)
3) PRESSURE MANAGEMENT (SHORTER OR LONGER PRODUCT ACCUMULATION)
4) GAP REQUIRED BETWEEN PRODUCTS

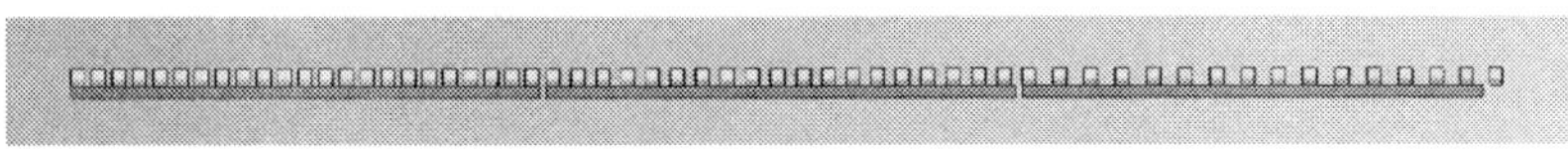

SPEEDS MUCH HIGHER THAN NECESSARY BRING:

1) EXCESSIVE ENERGY
2) POSSIBLE DAMAGE IN THE ACCUMULATED PRODUCT
3) PRODUCT INSTABILITY ALONG THE CONVEYORS (GAPS, CURVES)
4) EXCESSIVE NOISE

HIGHER CONVEYOR SPEEDS DON'T PRODUCE MORE PAPER!

Tissue converting engineering
TISSUE WORLD 2012
Pulsar GROUP

CORRECT MAINTENANCE ENSURES LOW FRICTION, SMOOTHER RUNNING OF CHAIN CONVEYORS AND NO DOWNTIME

PULSAR'S FIELD EXPERIENCE HAS BEEN TRANSFORMED INTO MATHS AND FORMULAS EMBEDDED IN THE SOFTWARE. THE OPERATOR PANEL GIVES INDICATIONS AND WARNING FOR MAINTENANCE AND COMPONENT REPLACEMENT

Life of chain in converting area (roll conveyor)

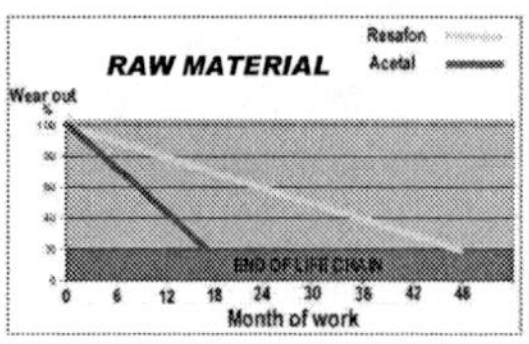

Life of slide rail in converting area (roll conveyor)

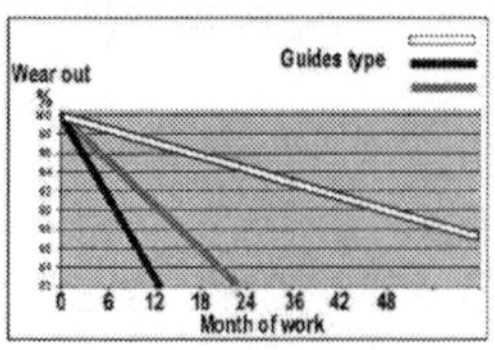

TISSUE WORLD 2012

CONVEYOR SPEED MUST BE:

1) CHANGED EASILY AND FROM A REMOTE CONNECTION
2) SPEED PARAMETERS MUST BE SAVED IN RECIPES FOR SPECIFIC WORKING CONDITIONS AND RECIPES MUST BE LOADED WHEN REQUIRED
3) EXTENDED ON A WIDE RANGE TO MEET THE PRODUCTION NEEDS FOR EVERY FORMAT OF ROLL / PACK / BAG

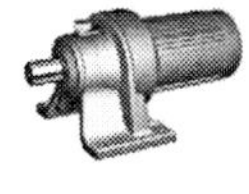

GEARBOX MOTOR

INVERTER FIELD-BUS CONNECTED

PANEL

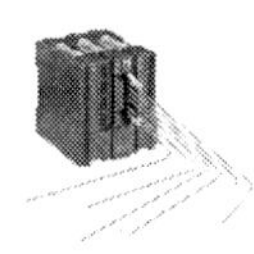

PLC

© 2012 Pulsar - www.pulsarbgroup.com
TISSUE WORLD 2012
Page 13

TISSUE WORLD 2012

OUR CONVEYOR LINES INCORPORATE THE "BASE HARMONIZER" SOFTWARE: CONVEYORS ' SPEED FOLLOW THE ACTUAL SPEED OF THE MACHINES CONNECTED.

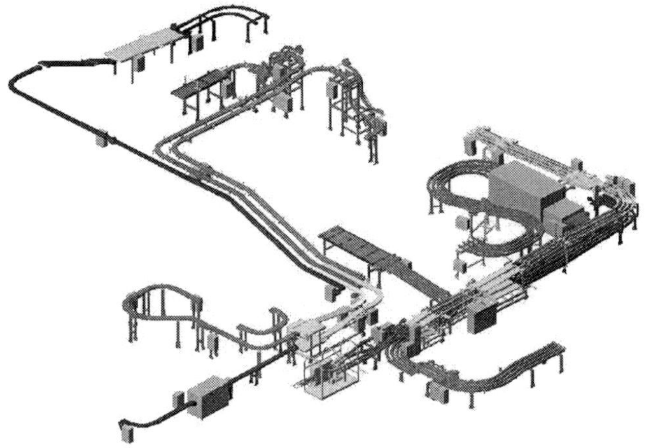

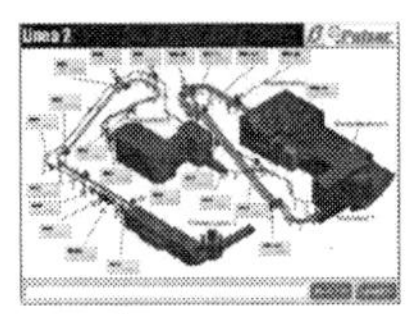

© 2012 Pulsar - www.pulsarbgroup.com
TISSUE WORLD 2012
Page 14

THE NEXT STEP "HARMONIZER PLUS" REGULATES ALSO THE PRODUCTION SPEEDS OF THE MACHINES TO MAXIMISE THE PRODUCTION EFFICIENCY.

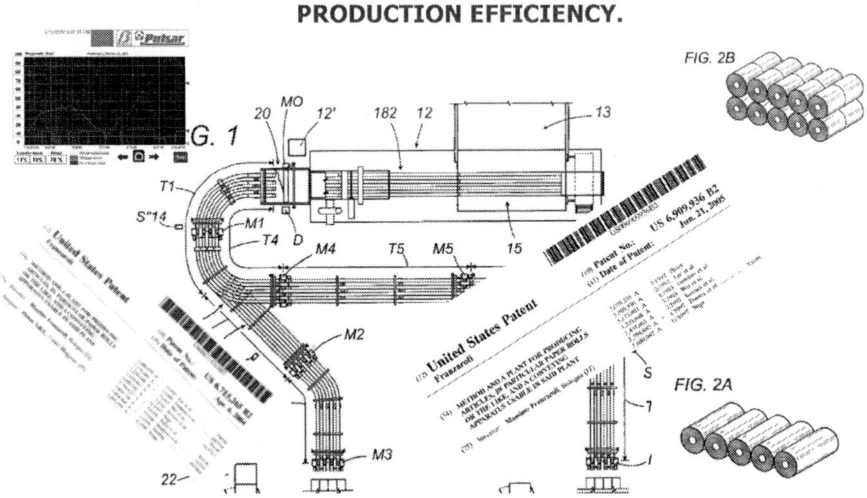

Tissue converting engineering TISSUE WORLD 2012 β GROUP Pulsar

HARMONIZER PLUS

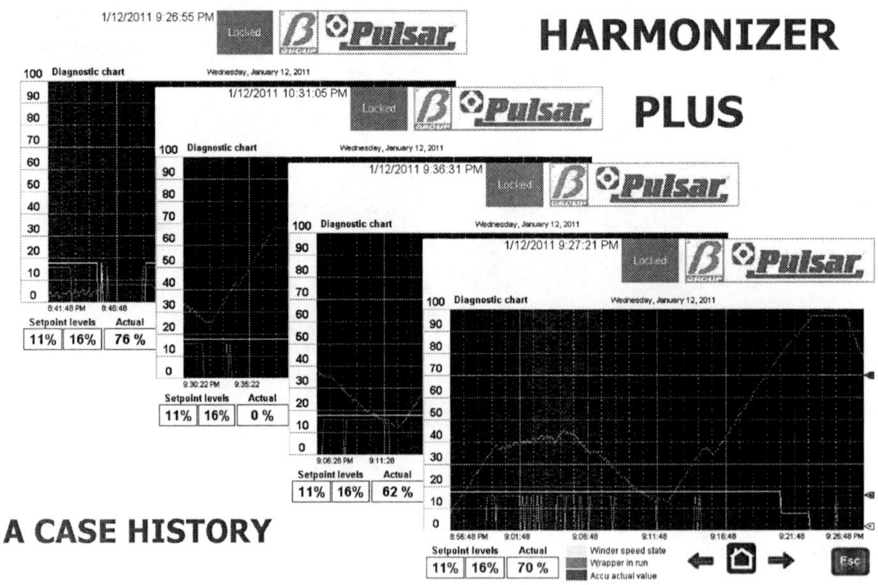

A CASE HISTORY

W+D LANGHAMMER

KÖRBER PAPERLINK

Tissue World Americas 2012 Conference

Stephan Schlottoff

PART OF YOUR SUCCESS

W+D-Langhammer GmbH

40 Years of TISSUE Experience

KÖRBER PAPERLINK

1971 Eisenberg

1991 Freiberg

2007 Eisenberg

2011 Freiberg

Founded in 1971 by Heinrich Langhammer
Tissue Innovations since 1974
Part of Körber since 2008

KÖRBER

W+D-Langhammer GmbH

Retail Tissue Market North America

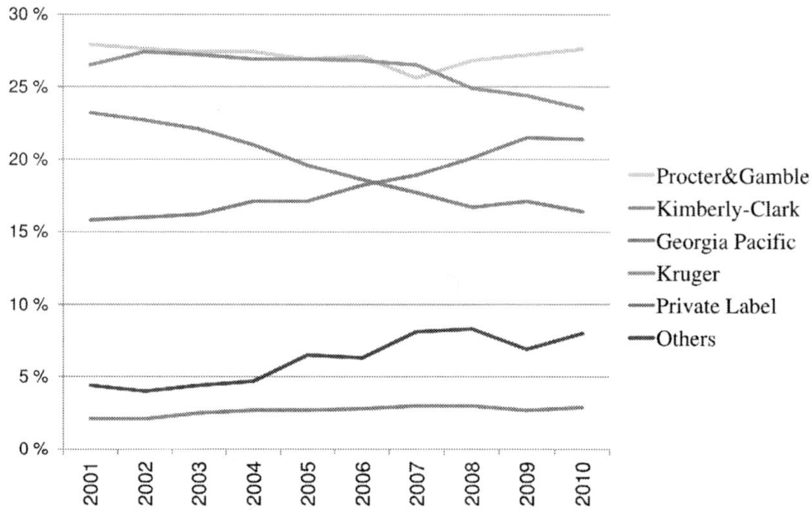

Source: Euromonitor International

W+D-Langhammer GmbH

Challenges

CHALLENGES 2020

- Changing Market
- Flexible Operation
- Logistic Costs
- Eco-Friendly

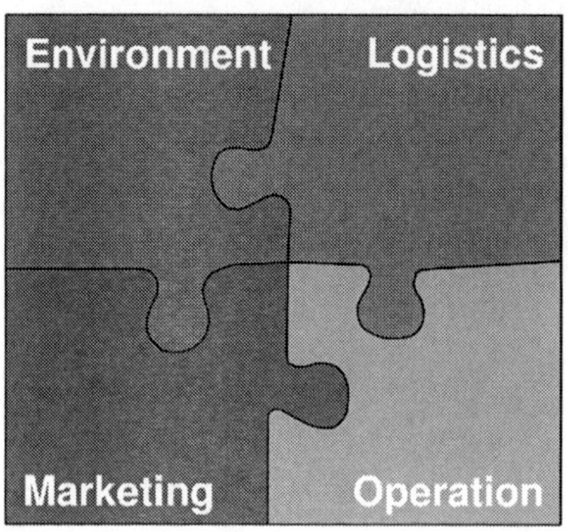

W+D-Langhammer GmbH

CMB 202 - LA01

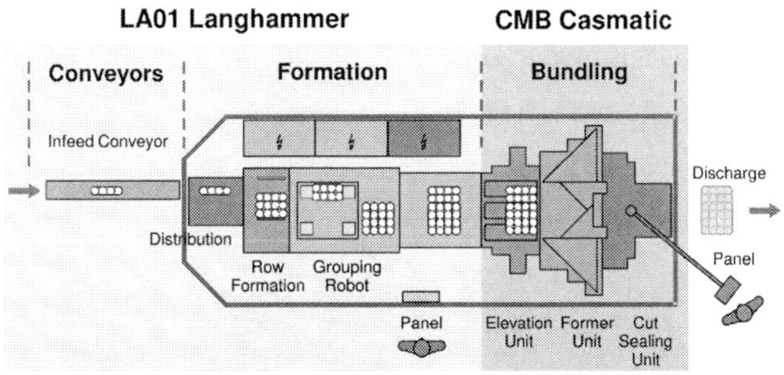

Distribution | Row Formation | Grouping Robot | Elevation Unit | Former Unit | Discharge

W+D-Langhammer GmbH

CMB202-LA02

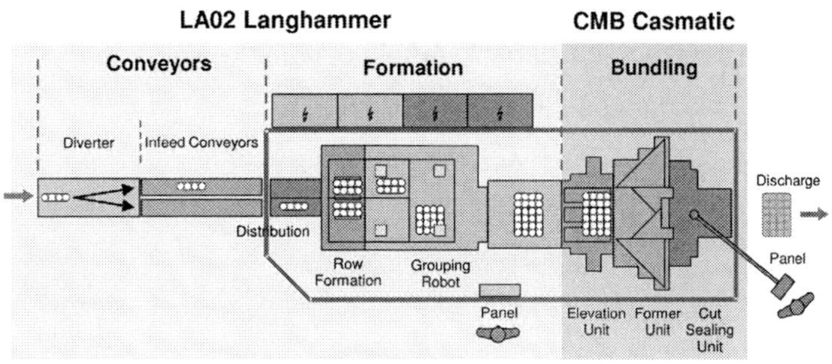

Options

- **Upender**
- **Stacker Unit**
- **Combi Stacker & Upender**

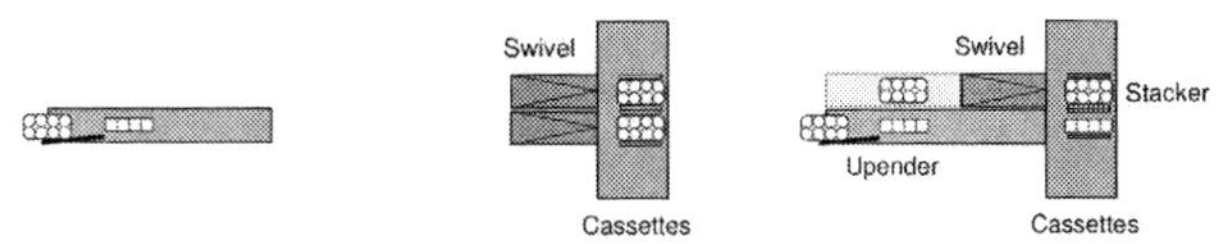

W+D Langhammer GmbH

Customer Benefits

OPERATION

- Easy creation of new formats at HMI.
- Small footprint, straight product flow.
- Reduced changeover time.
- No format parts needed.

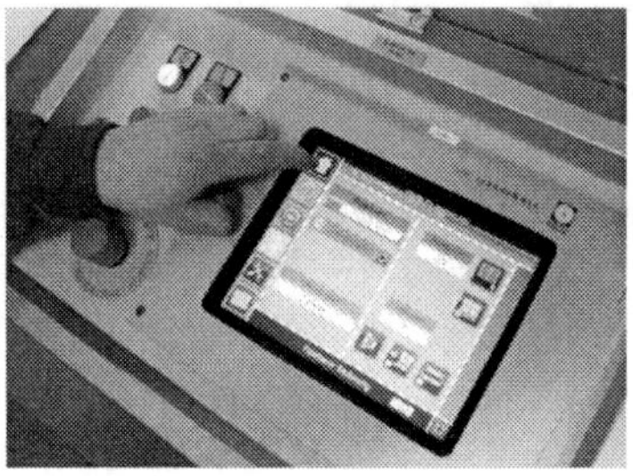

W+D Langhammer GmbH

Customer Benefits

LOGISTICS

- Flexible configuration of packs within bundles.
- Flatpacks & Upended Packs in one Bundle.
- Over 94% pallet volume occupied.
- Enhanced stability of the pallet.

W+D Langhammer GmbH

Best Practices

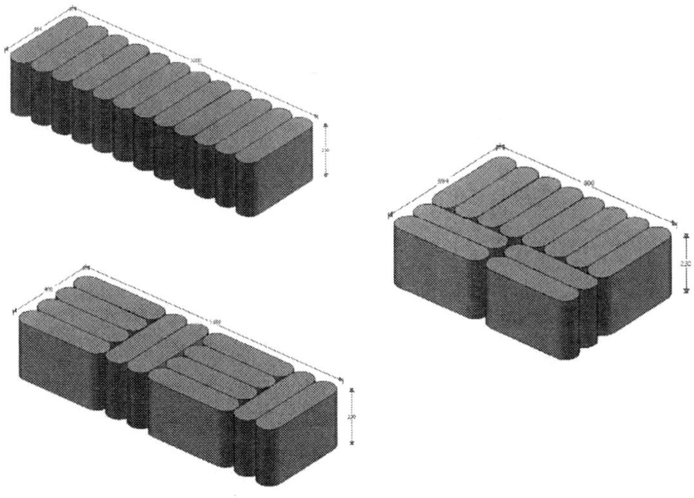

W+D Langhammer GmbH

Best Practices

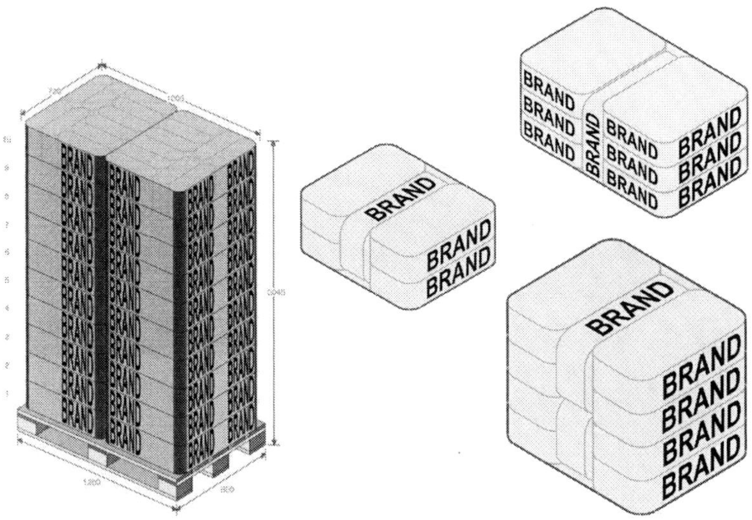

W+D Langhammer GmbH

Summary

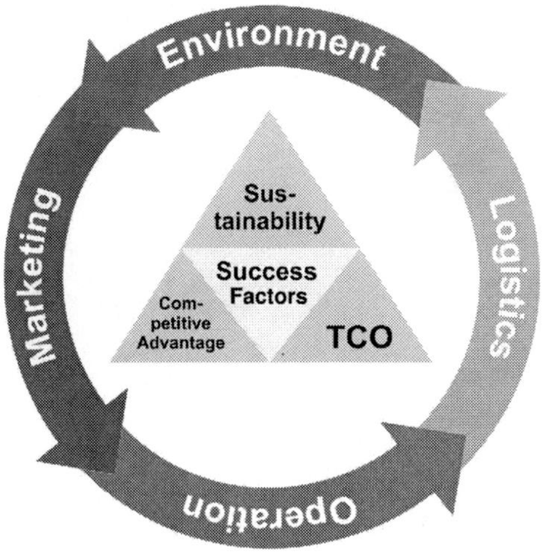

W+D-Langhammer GmbH

Bundling & Palletizing Technology

VIDEO

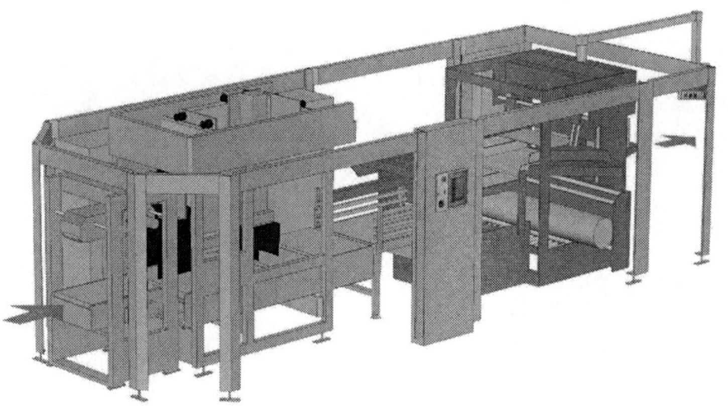

W+D LANGHAMMER

KÖRBER PAPERLINK

Thank you very much for your attention!

Please visit our booth for further information.

PART OF YOUR SUCCESS

Reducing Costs and Improving Operations with Automatic Guided Vehicles (AGVs)

- What is an AGV and how does it work?
- What are the features and benefits of AGVs?
- Where have AGVs be used successfully in tissue mills?
- Questions

JBT Corporation

1

What is an AGV?

- Automatic Guided Vehicle (AGV)
- Mobile Robot/Transport Unit
- Unmanned / Fully Computer Controlled & Real Time Monitored
- Battery Powered
- Intended to replace a person on a fork, clamp truck, or tow vehicle

2

Equipment Communication

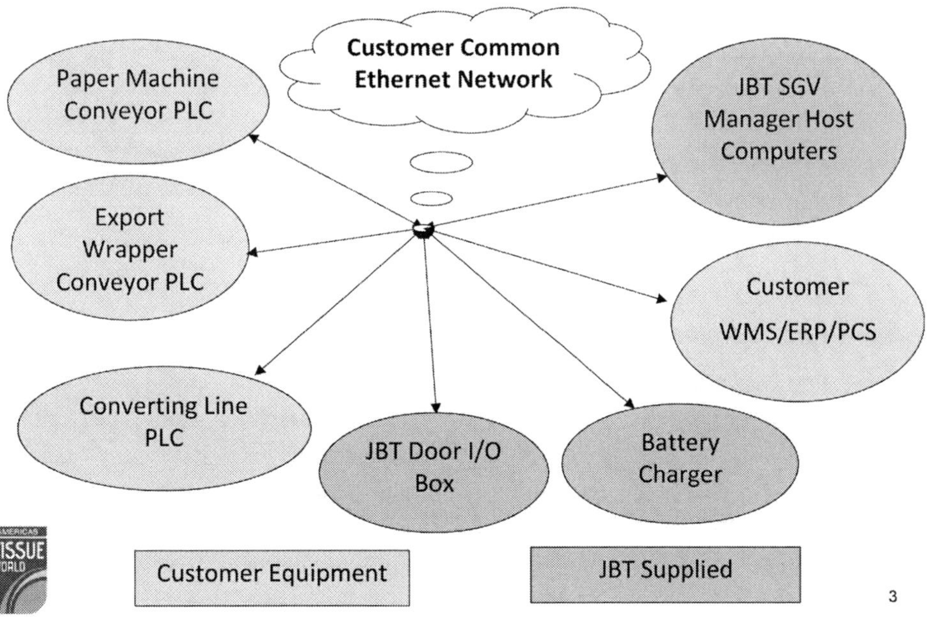

3

Features of AGVs

- Highly flexible and scalable
- Not a permanent obstacle
- Durable/long service life
- Requires minimal infrastructure
  - Ideal for automating existing facilities
- Expect high utilization/uptime
- Very Safe: 360 degrees obstacle detection
- Typically Good ROIs for 16/5 operations
- Typically Great ROIs for 24/7 operations

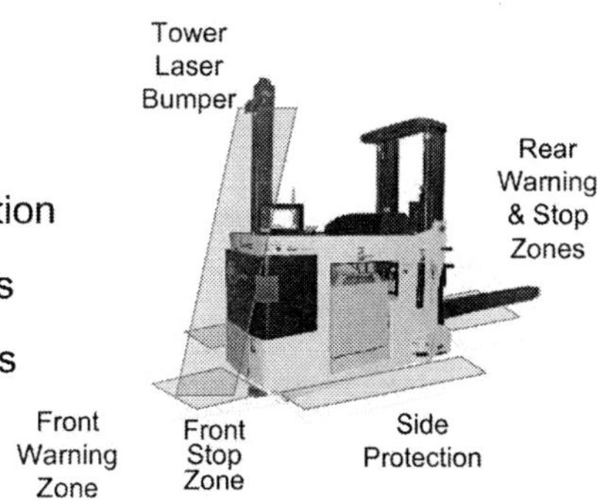

4

Benefits of AGVs

- Reduce labor costs
- Reduced product and facility damage
- Reduce/Eliminate errors
  - Right product to the right location at the right time!
- Built in redundancy – not a single point of failure
- Increase productivity
- Improve safety
- Tracks material – reduces losses

AMERICAS TISSUE WORLD

5

Tissue Industry AGV Applications

Parent Rolls:

- Pick-up at the paper machines
- Storage/Retrieval in warehouse
- Delivery to converting lines
- Pole, Outrigger or Roll Clamp vehicles

AMERICAS TISSUE WORLD

Tissue Industry AGV Applications

Packaging Materials

Deliver....

- Empty pallets to palletizer
- Poly wrap to packaging machine
- Boxes (KDF) to case packing machine
- Unload trailers

Tissue Industry AGV Applications

Finished Goods

- Movement from palletizer to stretch wrapper
- Storage/Retrieval in the warehouse
- Automatic trailer loading

Tissue Industry AGV Applications

Miscellaneous Material Movement

- Move full waste (broke) containers and replace with empty. Weigh containers if required.
- Move full and empty core containers

AMERICAS TISSUE WORLD

JBT Corporation

Questions?

Mike Thornton
Paper Industry Manager
772-569-6463
mike.thornton@jbtc.com

John Bean Technologies (JBT) Corp.
400 Highpoint Drive
Chalfont, PA 18914 USA
Phone: 215-822-4600
Fax: 215-822-4553
sgv.sales@jbtc.com

AMERICAS TISSUE WORLD

10

New Insights into the Application of High Yield Pulp (HYP) in Tissue and Towel

Yajun Zhou[1], Zhirun Yuan[2], Jimmy Jong[2], Stephan Lariviere[2], Francis Fournier[3], Vincent Landry[3]

1. Tembec
2. FPInnovations
3. Kruger Products

Tembec FPInnovations Kruger Products

Typical Fiber Mix in Tissue/Towel in North America

- Tissue
  - 60-70% HW Kraft
  - 30-40% SW Kraft

- Towel
  - 100% SW Kraft
  - 10-20% HYP in some grades

HYP Replacing HW Kraft or SW Kraft in Tissue/Towel

	SW Kraft	HW Kraft	HW HYP
CSF, mL	684 (486)	539 (448)	383 (310)
ISO brightness, %	89	89	85
FQA Fines, %	3 (9)	3 (9)	17 (21)
Bulk, cm^3/g	2.1 (1.7)	2.2 (2.1)	2.8 (2.7)
Tensile, N.m/g	37 (59)	22 (28)	25 (26)
Absorbency, g water/g fiber)	1.0	1.0	1.8

Data in brackets are post refining

HYP has higher bulk and absorbency

Topics to Discuss

- HYP vs. Kraft and DIP in Sustainability

- Pilot tissue/towel trials with HYP at FPInnovations

HYP Vs. Other Pulps

-Sustainability

- Fiber usage
- Effluent discharge
  - Treated effluent flow
  - BOD5
  - COD
  - TSS
  - AOX
- Solid waste
- Air emissions
  - NOx
  - SOx
  - Particulate
  - CO2
- Energy usage
  - Electricity
  - Fossil fuels
  - Biofuels

Process Comparison

Process	HYP	Kraft	DIP
Raw materials	Wood chip	Wood chip	Mixed office waste made with Kraft pulp originally
Fiber Separation	Mechanical refining	Chemical cooking	Re-pulping and chemical deinking
Washing/cleaning	Simple screening/cleaning	Multi-stage brown stock washing/cleaning	Multi-stage washing/cleaning
Bleaching	Two-stage alkaline peroxide	Multi-stage O_2 or Caustic extractions, and ClO2 & peroxide bleaching	Reductive bleaching and alkaline peroxide
Waste	Fiber and sludge (could be burned for generating heat and electricity)	Fiber and sludge (could be burned for generating heat and electricity)	Fiber, ink & fillers and sludge (can not be burned)

HYP Vs. DIP Vs. Kraft Pulps

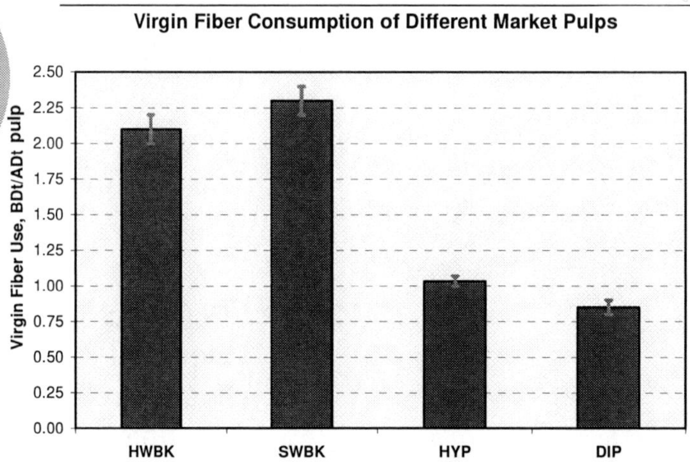

HYP almost has the twice yield of Kraft pulps

Effluent Discharge

	Unit	SW Kraft	HW Kraft	HYP	DIP
Flow	m^3/ADt pulp	55.3	47.9	12.8	22.3
BOD_5	kg/ADt pulp	1.5	1.8	0.6	1.8
COD	kg/ADt pulp	17.8	15.1	13.2	/
TSS	kg/ADt pulp	2.3	2.1	1.9	1.5
AOX	kg/ADt pulp	0.13	0.13	0	0

HYP mill can be zero-effluent mill, which has no effluent discharge into the environment.

HYP has low water consumption and environmental impact and is AOX-free

Solid Waste to Landfill

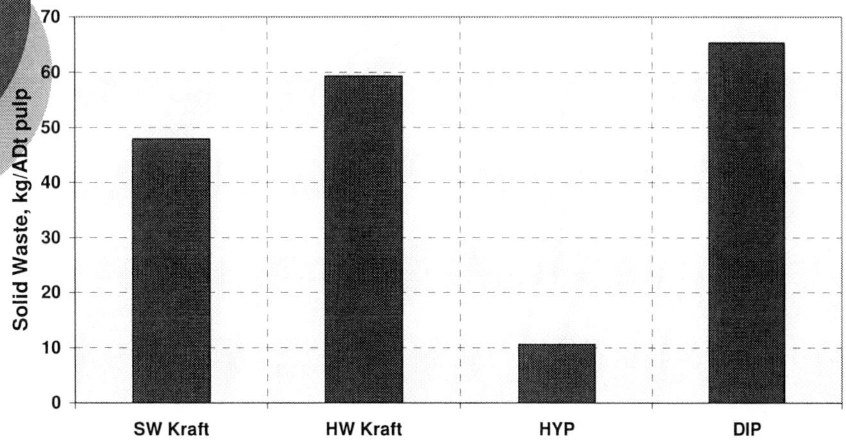

HYP has low solid waste to landfill

Air Emission

	Unit	SW Kraft	HW Kraft	HYP	DIP
No_x	kg/ADt pulp	1.68	1.56	0.23	0.23
Particulates	kg/ADt pulp	0.86	0.68	0.27	0.07
CO_2	kg/ADt pulp	209.2	219.2	100.5	/

No enough GHG data available for DIP

HYP has low air emission

Energy Consumption

Energy Consumption of Different Market Pulps (Based on EU's BAT)

Parameter	Unit	BK[2]	HYP	DIP[3]
Process heat consumption	GJ/ADt	10-14	1.2	6.0-7.0
Power generation	GJ/ADt	2-2.5	0	1.0-1.3
Total steam heat	GJ/ADt	12-16.5	1.2	7.0-8.3
Total biofuel input (LHV)	GJ/ADt	24-27	1.8	10.0-12.0
Total fossil fuel (LHV)[1]	GJ/ADt	1.5-1.8	0	0
Total power consumption	kWh/ADt	600-800	2000-3000	600-700
Total power generation	kWh/ADt	700-1000	0	275-360
Power sales	kWh/ADt	0-300	0	0
Total energy input	GJ/ADt	25-28	9.8-12.6	11.2-13.5

[1] Includes lime kiln and other direct process fuel usage
[2] BK mill without auxiliary boiler
[3] EKONO's estimate

EKONO
A JAAKKO PÖYRY Company

Summary of Environmental Footprint of HYP

- Maximize use of resources
  - Twice as much pulp yield per tree
  - Less water and fewer chemicals used in pulp production
- Lower environmental impact
  - AOX-free
  - Low green house gas (GHG) emission
  - It is close to or exceeds DIP in all measurable sustainability parameters
- Tembec HYP is FSC-certified

FSC

Pilot Tissue/Towel Trial/Converting

- Pilot PM trial at FPInnovations
  - Three weeks
  - No major PM runnability issues
- Converting in a pilot plant
  - No major issues were reported

Fiber Mix of Tissue Trials

Sample	SW kraft (%)	HW kraft (%)	HWHYP (%)
1	40	60	0
2	40	45	15
3	40	35	25
4	40	25	35

In the trial, Aspen HYP was used to replace HW kraft

Tissue Softness

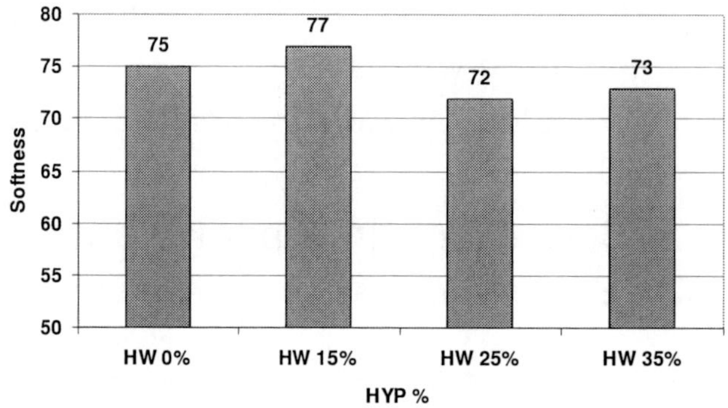

HYP does not significantly affect tissue softness

Tissue Strength

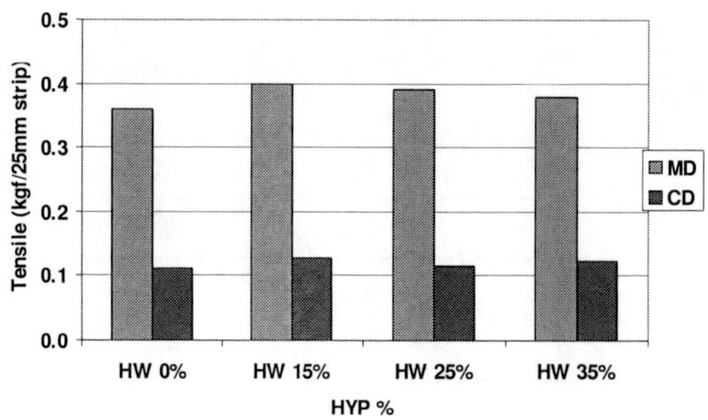

HYP does not significantly affect tissue strength

Tissue Bulk

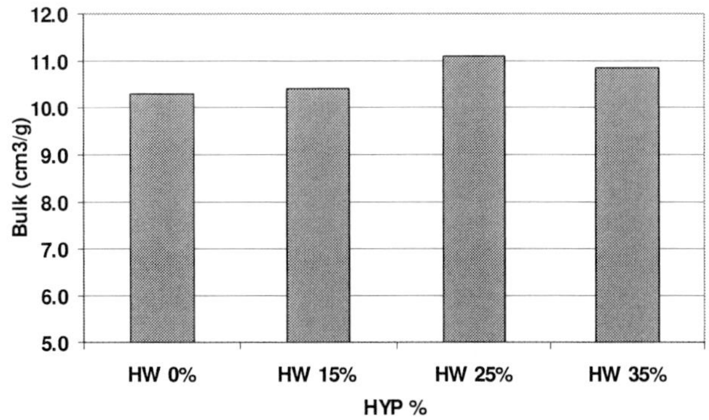

HYP improves tissue bulk

Tissue Brightness

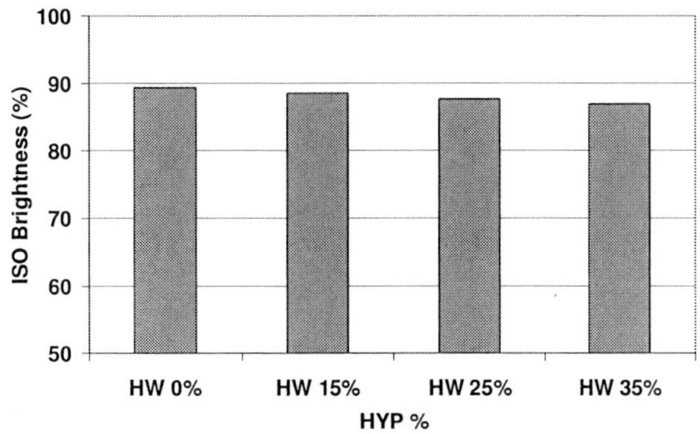

HYP does not significantly decrease tissue brightness

Towel Trial

Sample	SW kraft (%)	HW HYP (%)
1	100	
2	90	10
3	80	20

In the trial, Aspen HYP is to replace SW Kraft

Towel Water Absorbency

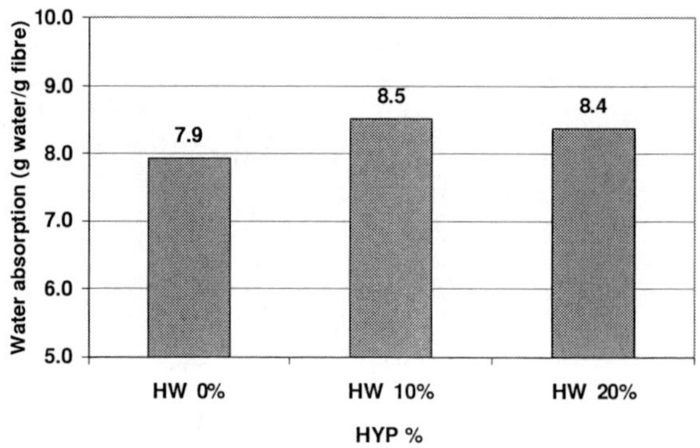

HYP improves the total water absorbency

Towel CD Wet Strength

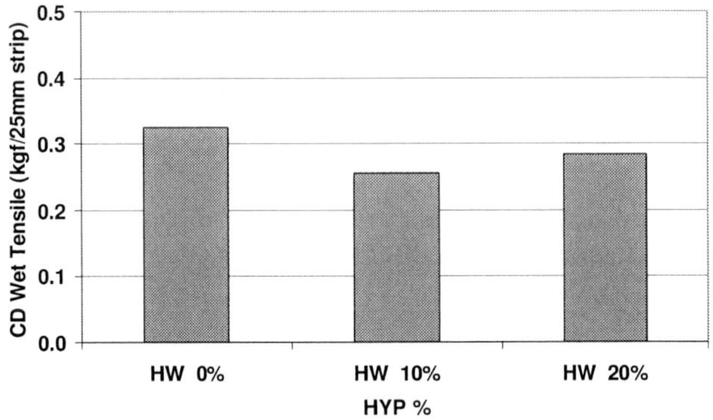

Wet strength is developed by wet strength agent, and can be improved by the process optimization

Towel Dry Strength

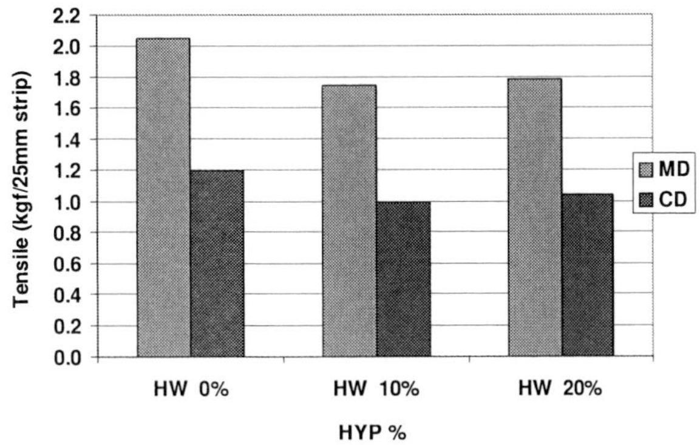

HYP slightly decreases the towel strength, but is still within the spec (> 1.6)

Towel Brightness

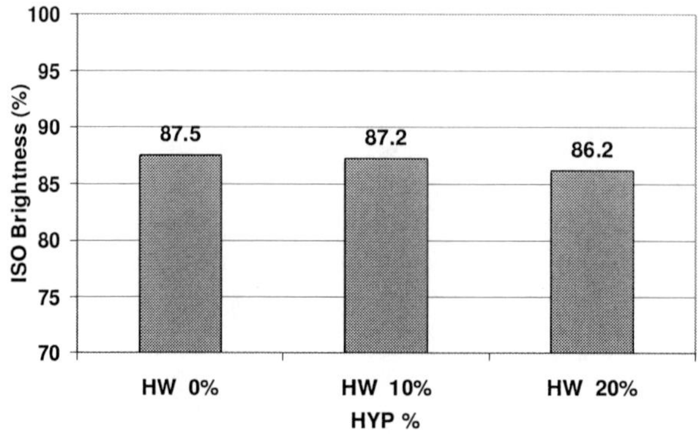

HYP does not significantly decrease towel brightness

Summary of Pilot Trials

- HYP does not significantly affect/decrease tissue softness and strength
- HYP improves towel absorbency and does not significantly impact the strength

HYP Advantages

- Cost reduction by replacing HW Kraft in tissue
- Cost reduction and improvement in absorbency by replacing SW Kraft in towel
- Green message of HYP
  - FSC and low environment impact
  - HYP is a good alternative to DIP in sustainable tissue/towel making

OptiFiner Pro

First mill experience with a new refiner concept

metso

OptiFiner Pro

Production data

	Pro 1	Pro 2	Pro 3	Pro 4	Pro 5
Production, max (tpd)	140	240	350	530	800
Power, max (kW)	630	1000	1500	2500	4000
Speed (rpm)	1500	1000	750	600	500
Idling power (kW)	47	60	95	135	165
Filling diameter (mm)	340	480	620	750	900

One OptiFiner Pro can replace **two** traditional refiners and still deliver **40% electrical energy savings**.

2 | © Metso

metso

First mill experiences with a new refiner concept

SAPPI Biberist, Switzerland

Sonoco, Germany

Customer	Size	Application	Delivery
Koehler Greiz, Germany	Pro 1	DIP trim refnining	2010
SAPPI Biberist, Switzerland	Pro 1	Eucalyptus	2010
Sonoco Deutschland, Germany	Pro 1	OCC	2011
Confidential, Slovenia	Pro 1	Eucalyptus	2011
Confidential, Austria	Pro 3	Eucalyptus	2011

metso

Sappi Biberist PM6: reduce energy consumption, improve stock quality

- 70,000 tpy of uncoated office grades
- 80 % Brazilian eucalyptus
- 6 conical refiners

Project target

- Reduce energy consumption
- Improve eucalyptus refining
- Reduce softwood usage

4 | © Metso

metso

Refining curves show the target is possible to achieve

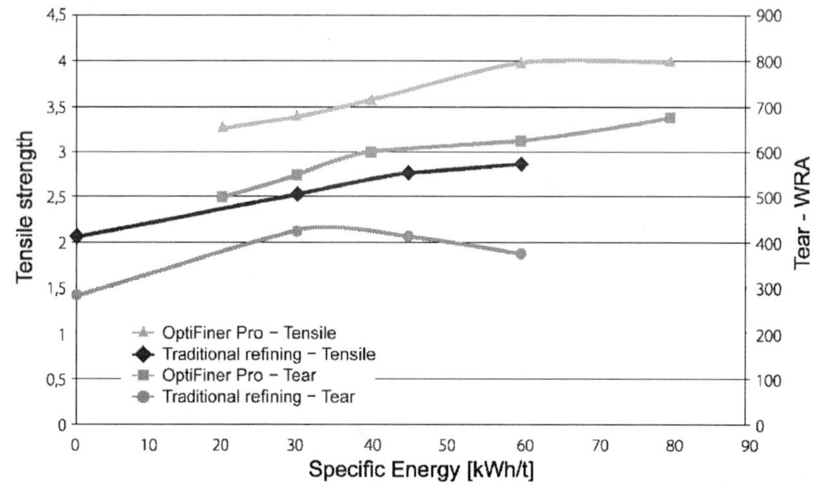

New refiner meets the operational guarantees

- Maximum capacity of 35 l/s at 5%
- 30% lower energy consumption when compared to 2 x RF-3 refiners at 30 SR level
- Same or better quality as with 2 x RF-3 refiners
- 60% energy saving at 23 SR level when compared to old refiners

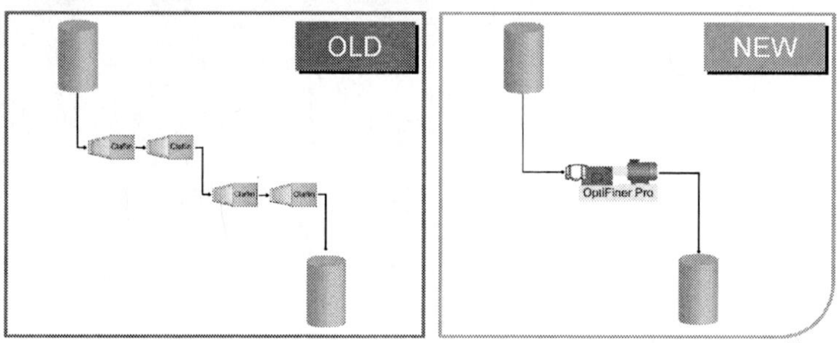

Values before and after OptiFiner Pro installation

	4 x Claflin 101	OptiFiner Pro	Change
Gross Power [kW]	465	305	-34 %
Flow [l/min]	1500	1350	
Cs [%]	4.2	5.3	
Production [t/h]	3.8	4.3	13 %
SRE [kWh/t] *)	60	60	0 %
°SR	20	24	20 %
Porosity Bendtsen	3300	2300	-30 %
Bulk [cm^3/g]	1.85	1.76	-5 %
Stretch [%]	1.66	1.11	-33 %
Breaking Length [km]	2.9	3.7	29 %
Tear [mNm] (80 gsm)	362	747	106 %

7 | © Metso

metso

Sonoco: reduce energy consumption

- 35,000 tpy of core board
- 100% recycled fiber
  - 30%-50% tobacco & fruit boxes
  - 50% high quality store waste
- 3 conical refiners
  - 2 x RF-2 400 kW, 750 rpm
  - RF-1 200 kW, 1000 rpm

8 | © Metso

metso

New refiner meets the operational guarantees

- Maximum capacity of 27 l/s at 3.5-4.4%
- 30% lower energy consumption when compared to 2 x RF-2 refiners at 42-45 SR level

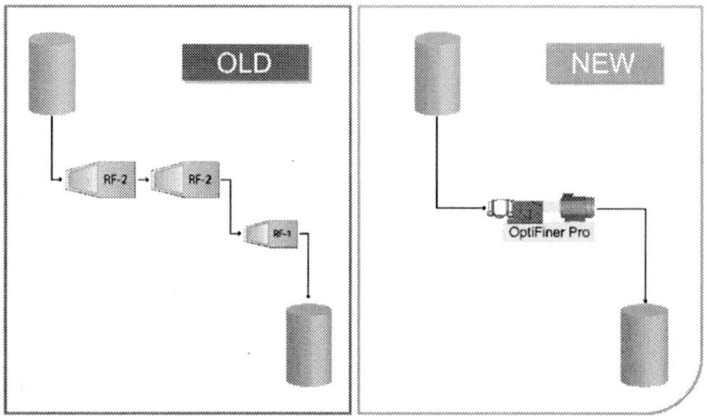

9 | © Metso

metso

THE GRAPHIC ON THIS PAGE WAS NOT AVAILABLE ON THE E-MEDIA VERSION OF THIS PAPER.

10 | © Metso

metso

Compression Refining

Flexible Fibres and low Fines Content,
Potential Benefits for Tissue producers

Miami Beach, March 23 2012: Jocco Dekker, Gulden Yilmaz

WAGENINGEN UR
For quality of life

Current Refining Technology

- Increase fibre bonding:
  Tissue Strength

WAGENINGEN UR
For quality of life

Current Refining Technology

- Rough fibre treatment
- High shear and friction
  - ➢ High specific energy consumption
  - ➢ Lot of fines: - Slow dewaterin
    - High drying energy

WAGENINGEN UR
For quality of life

Compression Refining

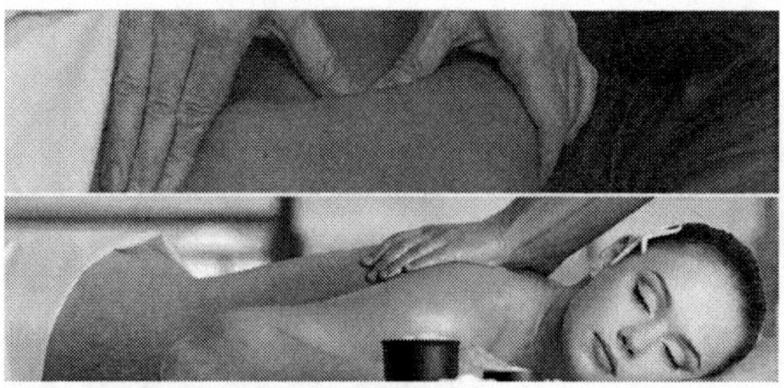

- Mild fibre treatment for virgin fibres
  - Homogeneous
  - Mainly compression forces
  - Increases fibre flexibility
  - Minimal fines generation

WAGENINGEN UR
For quality of life

Compression Refiner, Invented by WUR-FBR

- Small Pilot in Wageningen, The Netherlands
  - 10-50 kg/h continuous
  - 7.5 kW
  - 4-20% Consistency

WAGENINGEN UR
For quality of life

Trials with mill pulp

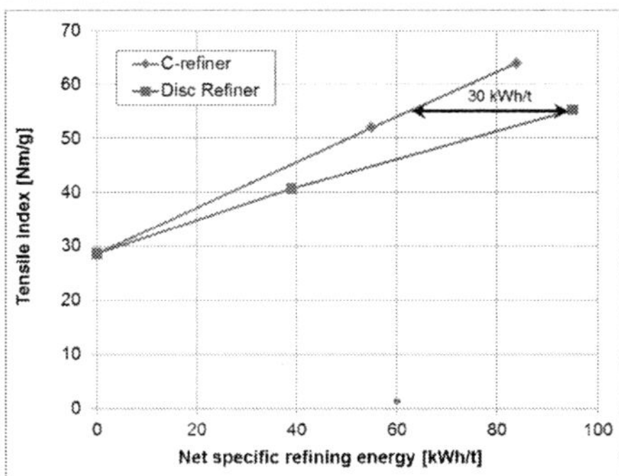

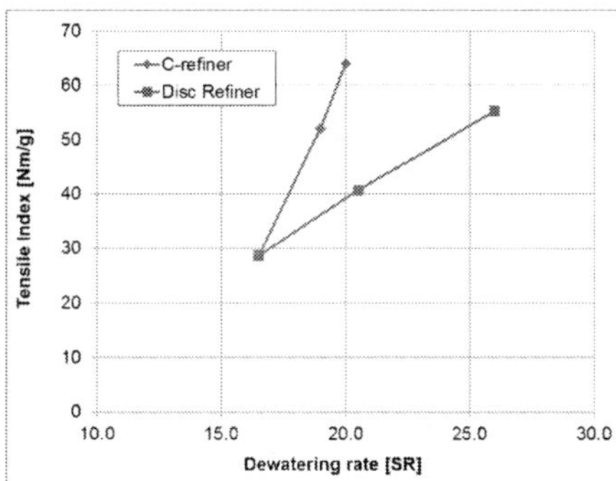

- At same strength:
  - Lower specific refining energy (27% saving)
  - Faster dewatering (lower SR)

WAGENINGEN UR
For quality of life

Trials with mill pulp

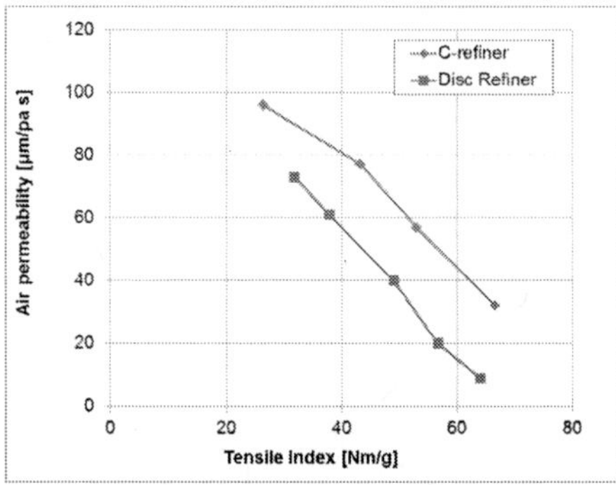

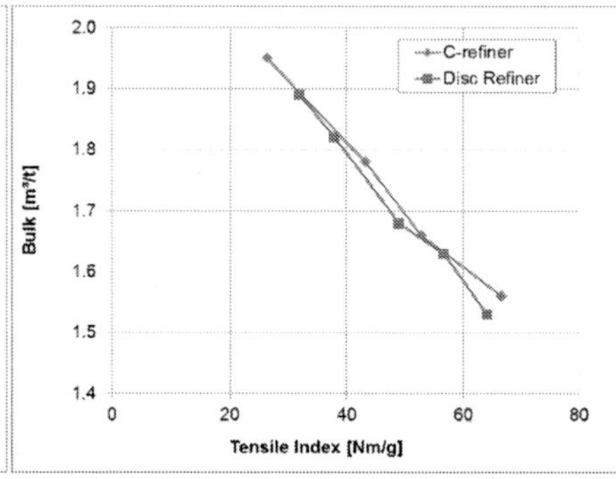

- At same strength:
  - Higher air permeability
  - Equal bulk

WAGENINGEN UR
For quality of life

Industrial compression refiner

Challenge: Industrial size compression refiner

- First attempt not successful
- Knowledge available for up scaling
- Partners for up scaling welcome

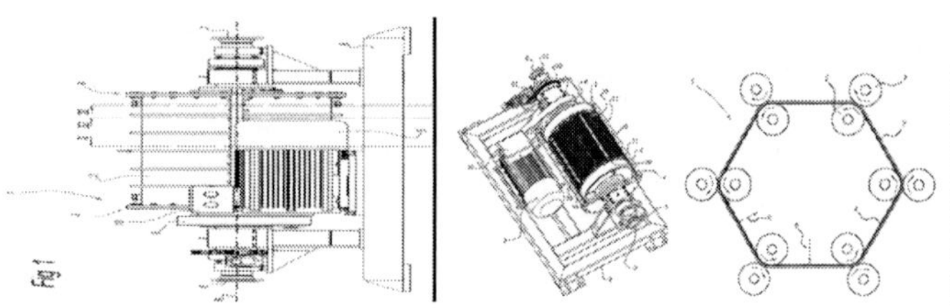

Patents:
WO2008056978
EP 2 278 065

WAGENINGEN UR
For quality of life

Potential benefits for Tissue producers

Compression refining can give:

- Energy savings:
  - Refining energy
  - Faster dewatering
  - Drying energy
- Better product performance
  - Softer products
  - Stronger products
  - Higher air permeability
- Higher profits

WAGENINGEN UR
For quality of life

Join the benefits of Compression Refining

- Trials with your pulp
- Join us to develop industrial compression refiner

THANK YOU

jocco.dekker@wur.nl

WAGENINGEN UR
For quality of life

Carbon Fiber Composite Materials for Tissue Rewinding Reel Spools

Mark Peretti, VP of Engineering, Double E Company

OVERVIEW

This subject explores the state of the art in the use of carbon fiber composite materials to manufacture tissue rewinding spools. Using the unique physical properties of carbon fiber, it is possible to produce a reel spool which has higher critical speeds, is lighter in weight, and costs less than conventional steel and aluminum versions. This allows the OEM to design more productivity into their new machines and allows end users to increase productivity on existing machines.

Several unique designs will be discussed along with relevant case studies.

INTRODUCTION

Tissue makers and converters are constantly trying to meet market demands by improving productivity and increasing the throughput of their equipment. Many times, these improvements have to be accomplished while simultaneously reducing costs by minimizing waste. The only ways to address all of these concerns are to upgrade existing equipment or to install new machines. Since the installed base of existing equipment is large, the opportunity for improvement tends to focus on this sector. Therefore, the goal of increasing productivity and throughput becomes limited by the physical constraints of the existing equipment when using conventional engineering materials to solve these problems.

Most machine components and converting equipment are constructed of aluminum and steel materials. For most situations involving the trade-off between strength and weight, these materials work well. They work well and are cost effective. However, when there are circumstances that require high strength, light weight, and low inertia, there is an alternative.

Carbon fiber composite materials have unique physical properties of high strength and high rigidity with a density four times less than steel. These properties lend themselves to solving application problems such as higher critical speeds, lower inertia, and lighter weight for reel spools.

COMPARISON OF SHAFT HOUSING MATERIALS

MATERIAL	TUBE MODULUS OF ELASTICITY (E) 10^6 PSI [GPa]	ULTIMATE STRENGTH TUBULAR FORM 10^3 PSI [MPa]	DENSITY lb/in³ [g/cm³]	ENDURANCE RATING	STIFFNESS TO DENSITY RATIO
FIBERGLASS	2 [13.7]	50 [344]	0.09 [2.5]	GOOD	23
ALUMINUM (6061-T4)	10 [68.95]	35 [241]	0.1 [2.76]	POOR	100
ALUMINUM (7075-T6)	10 [68.95]	83 [572]	0.1 [2.76]	POOR	100
TITANIUM (6 AL-4V)	16 [110.30]	140 [965]	0.16 [4.43]	GOOD	100
STEEL - 4130 (COLD WORKED)	30 [207]	110 [758]	0.28 [7.75]	GOOD	107
STEEL - 4140 (HARDENED & TEMPERED)	30 [207]	160 [1103]	0.28 [7.75	VERY GOOD	107
CARBON FIBER COMPOSITE (33 X 10^6 FIBER FILAMENT MODULUS)	17 [117]	250 [1723]	0.06 [1.66]	EXCELLENT	283
CARBON FIBER COMPOSITE (56 X 10^6 FIBER FILAMENT MODULUS)	30 [206]	263 [1813]	0.06 [1.66]	EXCELLENT	500
CARBON FIBER COMPOSITE (78 X 10^6 FIBER FILAMENT MODULUS)	34 [234]	205 [1413]	0.06 [1.66]	EXCELLENT	567

The following solutions were developed by using the unique properties of the carbon fiber composite materials.

HIGH SPEED / LIGHTWEIGHT REEL SPOOLS – TISSUE WINDING

The output of many tissue machines is sometimes limited by the critical speed or the rotating inertia of the reel spools. The physical constraints of the machine design and several laws of physics govern the maximum achievable speed or lowest possible inertia.

The following equations apply in these cases:

$$\textit{Critical Speed} = 3.13\sqrt{1/\Delta\text{ self loaded}}$$

$$\textit{Self-Loaded Deflection} = \Delta\text{ self loaded} = WL^3 / 48EI$$

And for rotating inertia:

$$\textit{Mass Moment of Inertia for Tube} = I_m = m/8\,(d_o^2 - d_i^2)$$

Where m=pi x L x density x $(d_o^2 - d_i^2)$ mass of the rotating tube and d_o and d_i are the outer and inner diameters of the tube.

By using the unique properties of carbon fiber composite materials, the relative critical speed of reel spools can be doubled when you combine the effects of reduced self-loaded deflection and rotating inertia. The following chart outlines the effects for various diameter tubes (assumes similar wall thickness and no internal weight effects).

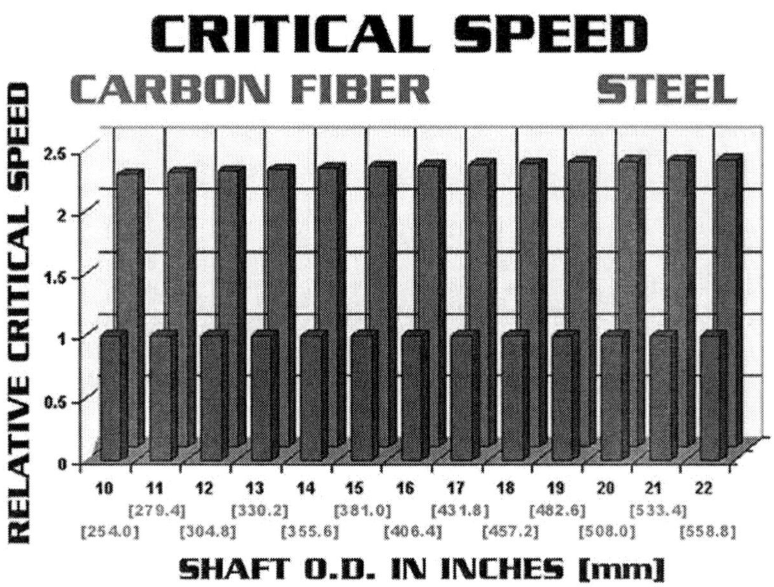

REEL SPOOL APPLICATION #1

In order to achieve maximum reel weight during winding and achieve the smallest possible core diameter for the given machine width or span, we needed to take advantage of the stiffness and low density of carbon fiber composite. The required operating speed was 3600 FPM with a 10.5" diameter core and a support span of 244". A low rotating inertia was required to allow for a core assist drive to be used to maximize the reel diameter at 84".

The image below is of the final reel spool configuration.

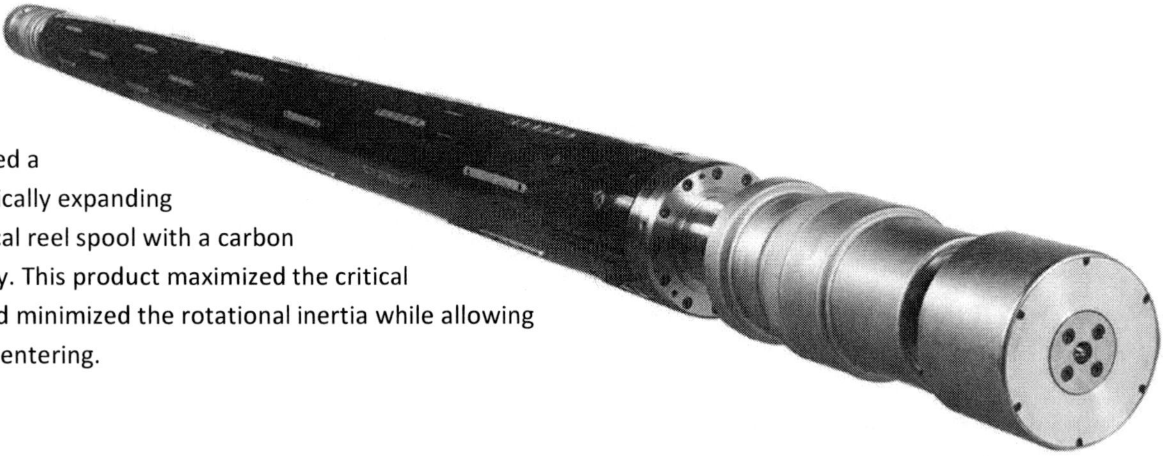

We applied a pneumatically expanding mechanical reel spool with a carbon fiber body. This product maximized the critical speed and minimized the rotational inertia while allowing for core centering.

REEL SPOOL APPLICATION #2

The goal was to replace a non-gripping reel spool with a gripping type while increasing the critical speed of the reel spool. This would allow for increased productivity by winding at faster speeds. The desired critical speed was 5500 FPM which could not be achieved with the current non-gripping, steel-body, dead shaft reel spool. Conventional gripping reel spool technology was available in steel, but the outside diameter of the reel spool would have to increase to 18 inches due to the span of 326 inches. The application required a 16 inch diameter because all of the post-wind converting equipment was sized for the current (16 inch) reel spools.

We designed and manufactured a unique hybrid carbon fiber composite reel spool to meet the application requirements. By using the low inertia properties of the carbon fiber, it was possible to use the tubing as a backbone for extrusions that contained the gripping elements. The unique blend of stiffness and low inertia met the critical speed requirement, and the spool was balanced above the 5500 FPM speed requirement.

The image below is of the final reel spool configuration.

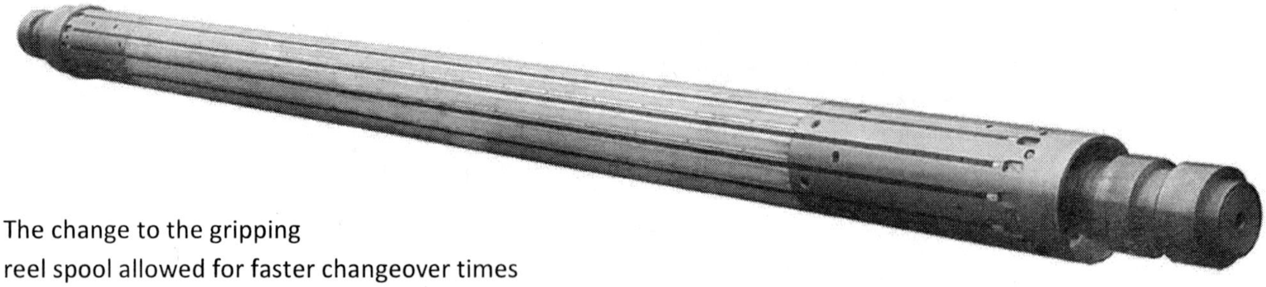

The change to the gripping reel spool allowed for faster changeover times between winds, while the diameter of the reel spool remained unchanged. No downstream converting equipment had to be altered.

REEL SPOOL APPLICATION #3

The most recent innovation in the use of carbon fiber composite materials to manufacture tissue rewinding spools is the HCRS-2000. The design incorporates the use of carbon fiber composite materials along with aluminum expanding strip elements to produce an extremely strong, lightweight and durable structure. This allows us to achieve the desired increase in performance at the same cost of current metal-only reel spools.

Load Capacity - similar to the capabilities of steel and better than aluminum –

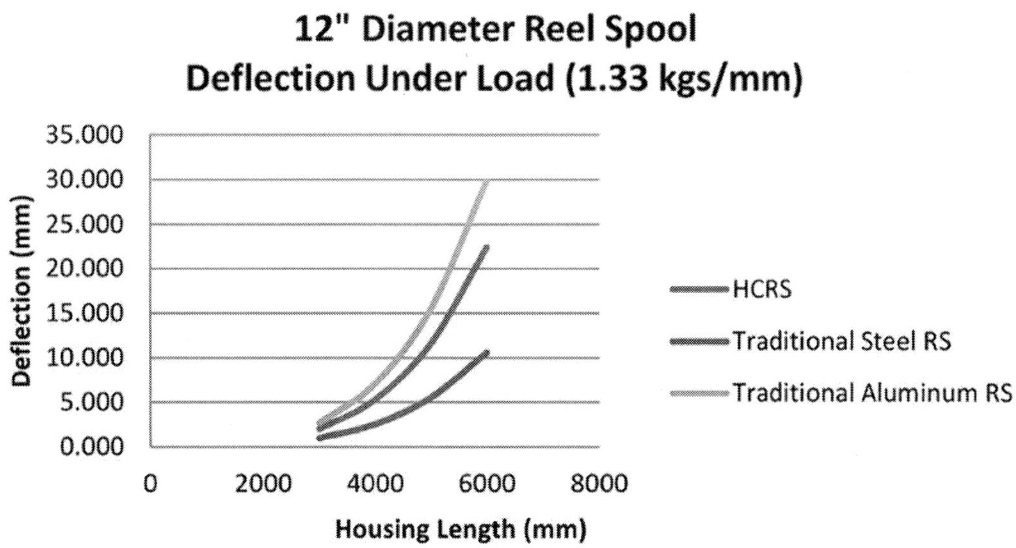

Higher Critical Speeds – faster throughput and increased productivity –

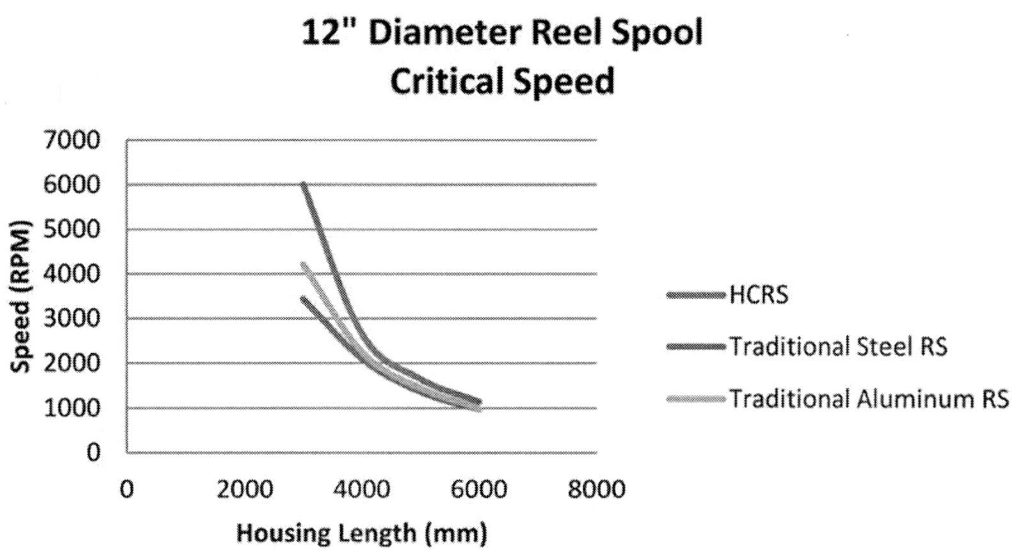

Reduced Shaft Weight – wind larger rolls while keeping hoisting requirements the same –

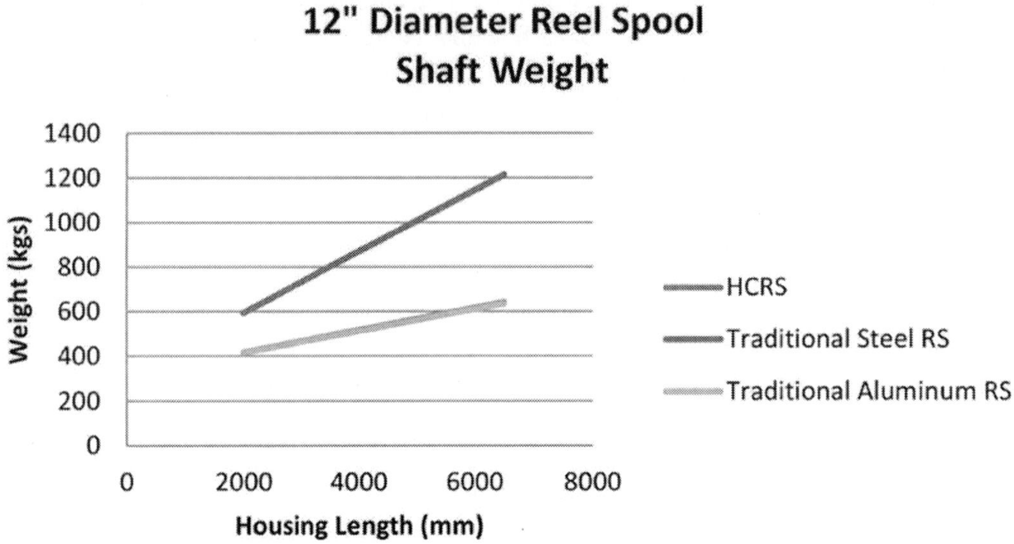

The most important innovation is the design and construction of the reel spools which allows us to produce them at a competitive cost to existing steel and aluminum shaft designs.

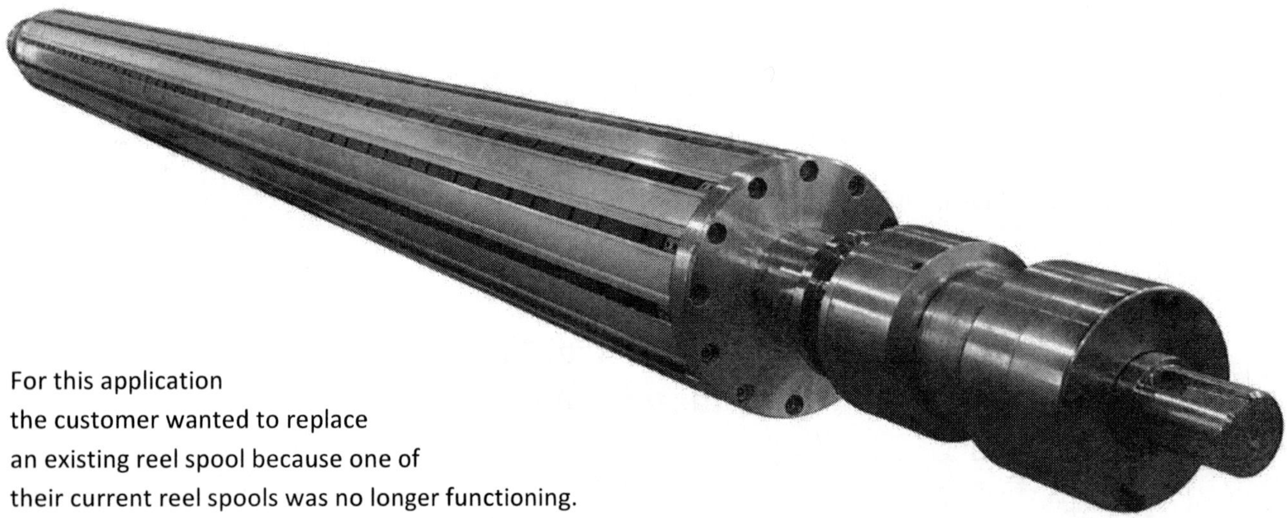

For this application the customer wanted to replace an existing reel spool because one of their current reel spools was no longer functioning.

They did not necessarily need the higher performance capabilities, but they would benefit from the higher critical speed in terms of less vibration and longer durability.

Using the HCRS-2000 technology, we were able to produce their shaft at a lower cost than traditional steel or aluminum reel spools as well as deliver sooner.

This is all possible due to the use of our proprietary hybrid carbon fiber/aluminum design and materials manufacturing technology.

CLOSING REMARKS

Using the unique properties of carbon fiber composite materials it is possible to solve critical tissue making problems that previously required much more invasive solutions. In most cases, existing equipment can be utilized or upgraded to improve productivity and reduce waste.

The advantages of this unique engineering material (high stiffness to density ratio) allow engineers to design shafts that maintain stiffness while increasing critical speed. Rotating inertia is also reduced allowing existing machines to handle lighter web materials without breakage.

Formatting Issue - Best Available Paper/Graphic

RYECO

Improve Operator Safety and Reduce Downtime with Automatic Web Marking on a Tissue Line

By:

Timothy Rye

RYECO Inc.

RYECO.COM

RYECO

Fact#1

- A missed ***web break, ply break, fold-over, tail,*** or ***large hole*** in a jumbo of tissue can cause major problems to downstream converting operations!

RYECO.COM

Formatting Issue - Best Available Paper/Graphic

RYECO

Fact #2

- The hand marking of moving tissue rolls with a crayon or tag is very ***unsafe, unreliable*** and ***only a guess***.
- Safety gates and light curtain interlocks are also making the practice ***impossible!***

RYECO.COM

RYECO

The Solution

- An ***Automatic Web Marking System*** is the only solution to improved operator Safety and higher Converting Line efficiency!

RYECO.COM

How are Web Markers used Today?

- In the Paper Industry Marking Systems are mainly used as Defect Markers for a Web Inspection System.
- Code Markers make synchronization marks for automatic unwind control.

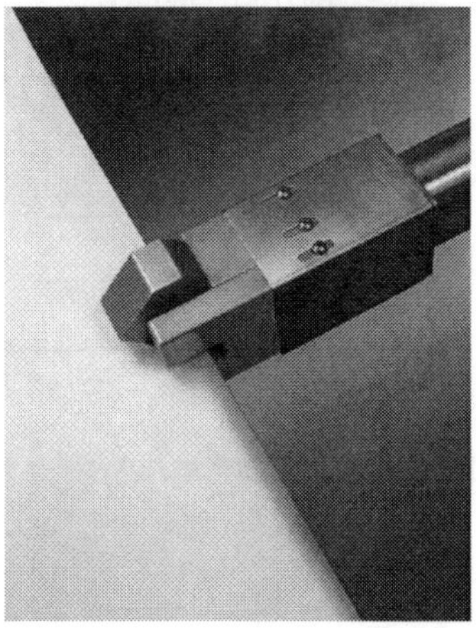

RYECO.COM

Tight Spaces of a Tissue Machine require flexible configurations

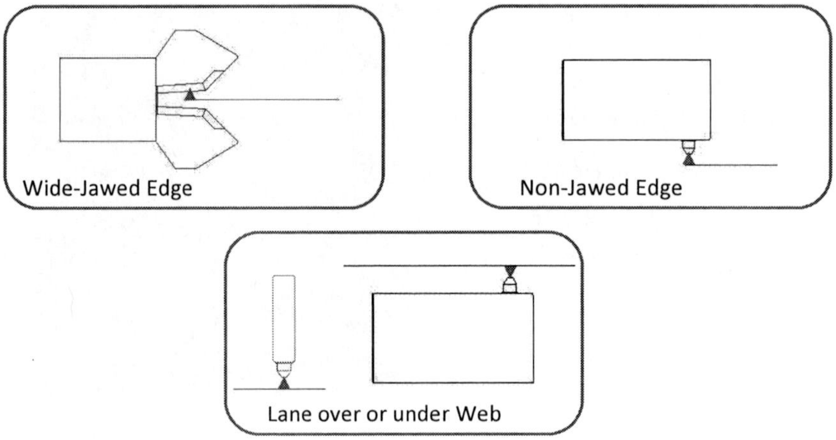

RYECO.COM

Formatting Issue - Best Available Paper/Graphic

From Machine to Winder or from Winder to Converting

- Place marks so they are seen by the downstream operator.

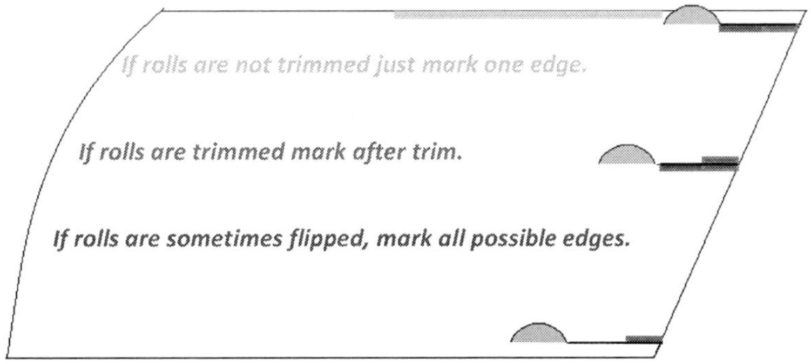

RYECO.COM

What Gets Marked?

Anything that needs to alert a downstream process

- On a Tissue Machine:
  - ***Defects*** seen by a Web Inspection System
  - Tissue Machine ***Breaks*** (Marked after sheet is restored)
  - ***Grade Changes*** or other ***off Quality***
- On a Winder or Rereeler
  - Winder ***Breaks***, ***fold-overs***, and ***tails***.
  - ***Ply Breaks*** or ***Web Splices***.
- Marking can also identify a unique Grade of product.

RYECO.COM

What Happens Next?

- When an operator in Converting sees a mark approaching he can take corrective action such as Slow, Crawl or Stop.
- Mark Detection sensors can be used to completely automate the stopping process.
- If using Code Marker, length is synchronized to the defect map of the Web Inspection System.

RYECO.COM

Conclusion

Safe, Happy Operators.

More and Better Product out-the-door.

RYECO.COM

Studies Reveal Costly Mill Leakage – The Unnecessary Loss of Fiber, Energy and Water

Rotary Drum Microfilter
Technology Applications in Tissue

Ross Elliot
Tecumseth Filtration Inc.

Presentation Outline

- Mill Leakage Issues
- Making Changes – What It Takes
- Mill Studies and Credibility – Rotary Drum Microfilters
- Business Drivers for Action – Risk and Uncertainty
- Anatomy of a Study – In-mill Evaluations
- Applications and Results
- Conclusions

Mill Leakage™

- Pulp & Paper mills must evolve in every part of the mill and organization if they are to be sustainable
- Society demands profitability and environmental sensitivity
- As a result, mills must continually adapt and improve
- Most operations pay little attention to leakage that drains profit from their mills every minute, every day
- Fiber, energy and water is needlessly being wasted as it exits the operation and the mill as effluent
- This unnecessary loss is Mill Leakage
- Rotary Drum Microfilter technology is a cost effective and efficient solution to reduce or eliminate leakage

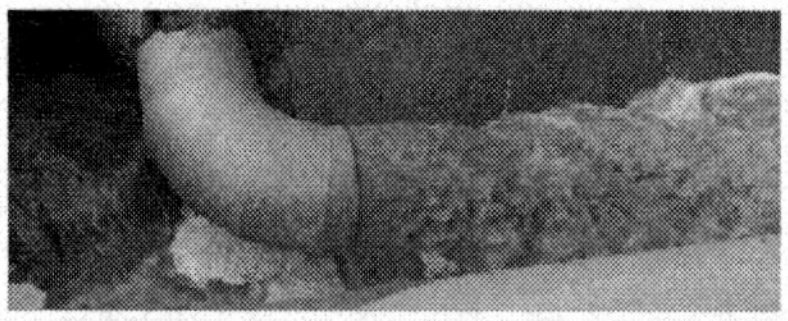

Example Problem and Solution

Problem

- Fresh water usage is high, it is expensive, and it must be heated to operating temperatures,
- Discharge volumes push clarifiers beyond their capability and mills face pressure from government on quality limits

Solution

- Use clarified process water on machine showers
  - Flood showers
  - Medium and high pressure cleaning showers
  - Felt cleaning showers
- Benefits: Reduced fresh water cost; no energy for heating; reduced flow to effluent and treatment
- Process temperature is higher – sheet dewaters faster in presses and dryers

Challenge Is To Get The Solution Completed

Bill Gates once stated

"I believe that if you show people the problems and you show them the solutions they will be moved to act."

- There are always lots of items on the 'wish list'
- There is usually no lack of ideas – good and bad
- The problem is to develop the appropriate action plan that will carry through to implementation
- There must be a champion who is willing to act and see it through

Why Most Ideas For Change Do Not Succeed

- Problem
- Idea
- Champion
- Benefits
- Plan

- Acceptance
- Funding
- Design
- Implement
- Audit

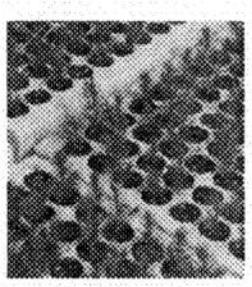

Risk

Reward

Uncertainty

Project Justification Requires Solid Results and Financials

Establish Project Proofs

- As Proofs, references are necessary but do not:
  - Demonstrate the technology or show the results
  - Provide details on the design requirements
  - Represent specific installation scenarios
- Effective studies provide information that is relevant to each application being considered:
  - Proof of concept and obtainable results
  - Realizable economic value
  - Operational issues and liabilities

Studies Can Provide Credibility

- Process issues identified:
  - Equipment sizing and install location
  - Support services required
  - Process design requirements or alterations
  - Application effectiveness and performance results
- Personal issues affected:
  - Career make-or-break decisions
  - Am I vulnerable? Can I get recognition?

Plausible Project Justification

Reasons For A Mill Leakage Study

Identify all the Options

- Test equipment over complete range of process conditions
- Exercise test equipment to limits of operating parameters
- Identify key issues that contribute to performance
- Collect sufficient data and observations – setup, tests
- Document cost and performance information

DATE	Filter Cloth	TIME	[illegible] GPM	[illegible] GPM	FEED	INLET mg/l	INLET % Consistency	REJECT GPM	REJECTS mg/l	REJECTS % Consistency	CLARIFIED GPM	CLARIFIED mg/l	Clarified % Consistency	Cloudy GPM	Cloudy mg/l	Cloudy % Consistency	RPM	Mg/l Reduced Inlet/Clarified	Comments	EFFICIENCY
Sep-15	30 Micron	9:05	30		CLOUDY	6080	0.6080		13340	1.3340		3520	0.3520				1.0	42.105263%		61.08%
		10:00	30		CLOUDY	6220	0.6220		11480	1.1480		3500	0.3500				1.0	43.729904%		62.95%
		11:35	30		CLOUDY	5960	0.5960		15020	1.5020		3140	0.3140				1.25	47.315436%	Sent sample to Irving Lab	59.82%
	120 Micron	1:00 PM	30		CLOUDY	7540	0.7540		16140	1.6140		3180	0.3180		2820	0.2820	1.25	57.824934%		72.01%
	120 Micron	1:45 PM	30		CLOUDY	9820	0.9820		6280	0.6280		3840	0.3840		2880	0.2880	1.25	60.896130%		156.73%
	120 Micron	3:00 PM	30		CLOUDY	7120	0.7120	7	13200	1.3200	20	3800	0.3800	2.7	2760	0.2760	1.25	46.629213%	Sent sample to Irving Lab	85.48%
	120 Micron	3:45 PM	30		CLOUDY	6160	0.6160		15180	1.5180		3400	0.3400		3080	0.3080	1.25	44.805195%		57.74%
Sep-16	120 Micron	8:30 AM	31.5		CLOUDY	6580	0.6580		16760	1.6760		4120	0.0412		2960	0.2960	1.25	37.386018%		96.10%
	120 Micron	3:55 PM	30	2.0	SWEETENER	8300	0.8300		27700	2.7700		1140	0.1140		1900	0.1900	1.25	86.265060%	CWW was 5680PPM	86.97%
Sep-17	120 Micron	11:40 AM	31.3	2.4	SWEETENER	8120	0.8120		25380	2.5380		1400	0.1400		1400	0.1400	0.50	82.758621%		87.71%
	120 Micron	12:50	32.9	2.1	SWEETENER	7940	0.7940		26480	2.6480		1140	0.1400		1400	0.1400	0.50	85.642317%		86.97%
	120 Micron	13:50 PM	32.3	2.4	SWEETENER	5020	0.5020		29620	2.9620		1100	0.1100		1640	0.1640	0.50	78.087649%	Lost Sweetener Feed	81.10%
	120 Micron	2:40 PM	32.7	2.6	SWEETENER	8940	0.8940		27700	2.7700		1400	0.1400		1620	0.1620	0.50	84.340045%	Sweetener Consistency 3.97%	88.83%
	120 Micron	3:15 PM	34.3	2.6	SWEETENER	6940	0.6940		23900	2.3900		1320	0.1320		1580	0.1580	0.50	80.979827%	Agat & Irving Samples	85.71%
	120 Micron	4:00 PM	34	2.7	SWEETENER	9020	0.9020		29120	2.9120		1400	0.1400		1340	0.1340	0.50	84.478936%		88.75%
Sept. 20	120 Micron	10:05	37.8	4.3	SWEETENER	11480	1.1480		28560	2.8560		920	0.0920		1100	0.1100	0.75	91.986063%	Increased amount of sweetener	95.09%
Sept-20	120 Micron	11:15	31.8	4.4	SWEETENER	10240	1.0240		24800	2.4800		1100	0.1100		1220	0.1220	0.75	89.257813%	Sweetener Consistency 2.74%	93.40%
	120 Micron	13:40	35.1	4.6	SWEETENER	9820	0.9820		28660	2.8660		1260	0.1260		1800	0.1800	1.25	87.169043%	CWW flow increased	91.18%
	120 Micron	14:15	35	5	SWEETENER	22840	2.2840	12	22840	2.2840	20	1020	0.1020	9	1800	0.1800	1.25	95.534151%		100.00%
	120 Micron	15:00 PM	35.2	6.3	SWEETENER	[illegible]	1.9600		29920	2.9920		980	0.0980		1400	0.1400	1.00	91.074681%	Increased Sweetener	96.27%
Sep-21	120 Micron	8:30 AM	49 ?		SWEETENER	11360	1.1360		29020	2.9020		1180	0.1180		7320	0.7320	1.80	89.792388%		93.80%
	120 Micron	10:00 AM	39.3	6.7	SWEETENER	12360	1.2360		25240	2.5240		1100	0.1100		7660	0.7660	1.20	91.100324%		95.25%
	120 Micron	10:30 AM	31.6	6.2	SWEETENER	13180	1.3180		24980	2.4980		1400	0.1400		7420	0.7420	0.98	89.377845%		94.88%

Design Of A Mill Leakage Study

Anatomy of a Study Report – The Table of Contents

- Executive Summary – quick read of key points
- Overview – study objectives and the setting
- Technical Discussion – process, testing procedures, equipment operation
- Summary of Findings – what happened, what it means
- Recommendations – action to follow
- Economics and Cost Analysis – savings, costs, finances
- Conclusions – key findings, recommendations, risk analysis
- Appendices: Support documentation and literature

Microfilter In-mill Study Purpose

Typical operational issues being addressed

- High fresh water usage
- Can use more polished recycled water
- Valuable solids and chemicals in effluent
- Clarifier is over-loaded
- Reduce discharge to municipality – volume, solids, BOD

What can be investigated

- Whitewater before or after Disc Saveall/DAF/other filter
- Effluent flow to clarifier
- Solids discharge flow out of primary treatment
- Screw press filtrate
- Vacuum pump sealing water

Rotary Drum Microfilter In-mill Study

Equipment set-up and operation

- Filter medium to be used – 120μ or 80μ or 30μ fabric
- Flow at inlet, outlets (flow balance)
- Solids measurements on flows (mass balance)
- Possible additions – sweetener stock, flocculation polymer
- Max/min throughput for various speeds of rotation
- Clear and Super-clear polished water fractions

Calculate The Savings

Savings Calculator: realistic projection of $2.23 million/yr

Includes:

- Fiber recovery
- Energy saved
- Fresh water reduced
- Production value
- Chemical reduction
- Discharge cost reduction
- Disposal cost factor

Filter savings calculation sheet

Fill in the boxes. The output is displayed in the red boxes.

Flow (l/min):	7700	Raw material costs ($/tonne):	200
Days of full production (days/year):	350	Power costs ($/kWh):	0.05
		Price of Steam ($/GJ):	2.00
TSS inlet:	400		
TSS outlet from Algas Microfilter:	100	Cost/tax on raw water ($/l):	0.0002
		Cost/tax on disposed water ($/l):	0.0001
Temperature water in production (oC):	55		
Temperature raw water (oC):	30		
Ash content in recovered material (%):	0%	Profit on paper for sale ($/tonne):	0
Polymer (mg/l):	0	Polymer costs ($/kg):	0
Bleaching (mg/l):	0	Bleach costs ($/kg):	0
Polymer for waste management (mg/l):	0	Polymer for waste management ($/kg):	0
Water content in waste for disposal (%	50%	Freight costs ($/tonne):	10
		Disposal cost/tax ($/tonne):	0
		Currency converter ($/x):	1
Total savings (US$/day):		**6389**	
Total savings (US$/year):		**2231363**	

START / Fibre / Energy / Water / Production / Chemicals / Disposal

Economics and Cost Justification

- Project cost includes equipment, engineering, construction
- Total project investment estimated to be $980,000
- A Return On Investment Summary indicates:

FINANCIAL ANALYSIS (in $1,000)

Category		Year 0	Year 1	Year 2	Year 3	Total
Savings Realized ($)		-	1,144	1,144	1,144	**3,432.2**
Investment	Expense ($)	(500.0)	-	-	-	**(500.0)**
	Capital ($)	(750.0)	-	-	-	**(750.0)**
	Total ($)	(1,250.0)	-	-	-	**(1,250.0)**
Depreciation ($)		-	(250.0)	(250.0)	(250.0)	**(750.0)**
Net Equipment Value ($)		750.0	500.0	250.0	-	**1,500.0**
Net Profit Before Taxes ($)		(500)	894	894	894	**2,182.2**
Taxes ($)		250	(447)	(447)	(447)	**(1,091.1)**
Net Profit After Taxes ($)		(250)	447	447	447	**1,091.1**
Discounted Cash Flow ($)		(250)	406	369	336	**861.7**
Economic Value Added (EVA)		(400)	406	369	336	**711.7**

Net Present Value (NPV)	1,112	**Operating & Cost Data**	
Internal Rate of Return (%-IRR)	170	Energy Saving ($/Day)	3285
Return on Investment (%-ROI)	69	Operating Period (Days/Yr.)	340
		Raw Material Cost ($/T)	200
Discount Rate (%)	10	Manufacturing Cost ($/T)	200
Tax Rate (%)	50	Sales Price ($/T)	550
Cost of Capital (%)	12	Fibre Saved (T/Day)	49.4

NPV	**1,112**
IRR %	**170**
ROI %	**69**

Example Study Applications

Tissue Mill – Consumer & Industrial Products

Issue: overflow of whitewater directly to effluent
- loss of fiber – mill uses 100% virgin fiber
- cost to discharge to municipality – volume and solids

Objective: reduce effluent flow and show filtrate suitable for showers

Study: machine whitewater was fed to microfilter
- 30 micron filter medium yielded best results
- Efficiencies in excess of 95% fiber removal could be sustained
- no solids in filtrate capable of plugging shower nozzles
  - 3rd party analysis of solids size
- Estimated yearly savings >$450,000

Example Study Applications

Tissue Machine – Industrial Products

Issue: DAF does poor job of clarifying discharge to municipality

Objective: remove solids (virgin fiber furnish) from effluent and retain filtrate for reuse

Study: excess process whitewater fed to microfilter
- Inlet solids feed to DAF and Microfilter ranged from 60 to >1000 ppm (grade/color change)
- DAF clarified accepts ranged from 20 ppm to >600 ppm during study
- Microfilter filtrate solids ranged from 1 ppm to 7 ppm
- Average solids removal was over 99%
- Filtrate acceptable for process reuse

Conclusions – Mill Leakage Studies

- Use of Rotary Drum Microfilter equipment provides critical insight into the advancement of process design
- In-mill studies overcome uncertainty and risk in making equipment and design decisions
- Studies provide an understanding of the technology and verifiable improvement opportunity:
  - Management, operations, maintenance, engineering examination of equipment and its operation
  - Results show production cost reduction, raw materials savings, environmental impact
  - Financial analysis provides the economic sense for proceeding
- Studies build the bridge over the uncertainty gap
- Mill Leakage is a rewarding target for cost reduction

Tecumseth Filtration

Thank you!

Ross Elliot
Tecumseth Filtration Inc.
www.tecumsethfiltration.com

GUARDING OPERATIONS

SEACON Engineering Associates, Inc.

Discussion by
David Kirchharr
Bill Lauber

*Background

World wide economic uncertainty forced many tissue manufacturers to postpone replacing existing process lines electing to maintain current operations.

Others are recommissioning idle lines to meet current demands for economy products.

Safe operations remains a key concern for manufacturers.

OSHA regulations, insurance pressures and desire to protect both people and equipment assets have resulted in increased emphasis to improve safety.

Improperly engineered guarding applications have resulted in conflicts with operations bypassing of guarding attempts creating ineffective safety and wasted expenditures

*Safety Studies

- These findings are based on Paper Mill studies published by the American Society of Safety Engineers in their Fall 2007 paper of *The Journal of SH&E Research*
- Studies show Safety training can lower the incident of injuries.
- The extent of Safety improvements from training personnel widely vary.
- Results in improving Safety are largely based on management, supervision and operator attitude, Knowledge, and Behavior.
  - Resources spent on safety training
  - Incentives for safe & penalties for unsafe practices
  - On the job operator attentiveness

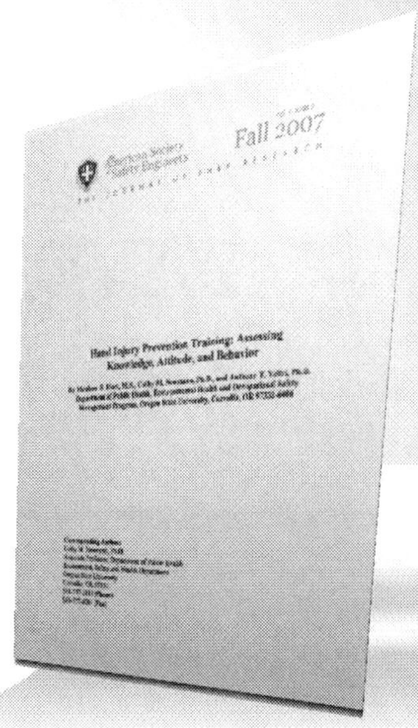

- Guarding systems do not suffer from fatigue
- Guarding systems do not experience moments of inattention
- Guarding systems perform the same way every time 24 hrs./day 365 days/year
- It's not the 99 things an individual does right that causes harm ... it is the 1 thing overlooked

*Benefits of Safety Guarding

*Project Approach

- The following slides outline step by step how successful Safety Guarding projects are engineered from start to finish
- Conduct on going risk assessment
- Determine access points
- Establish sectional areas
- Evaluate current process automation

*Interview & Observe Operators

- Part of on-going risk assessment
- Determine accessibility Requirements for start up & normal operation

*Design Guarding Access

- Determine fence location
- Establish gate locations
- Based on operator accessibility needs

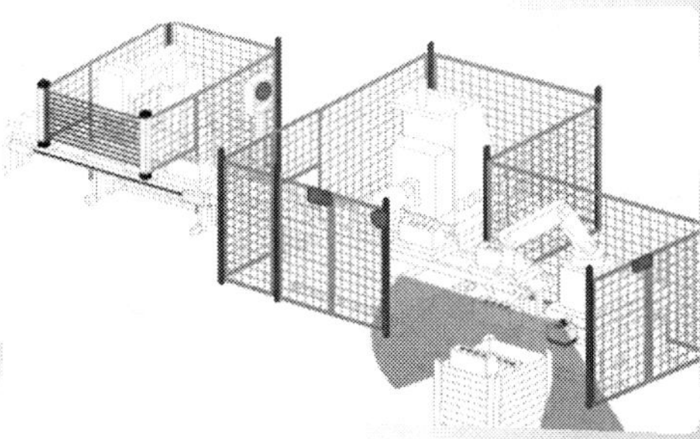

*Develop Safety Logic

Establish gate to process operation interlocks

Determine process areas to halt when interlock tripped

Develop logic sequence overrides for process startup

Design integrated safety and operational controls

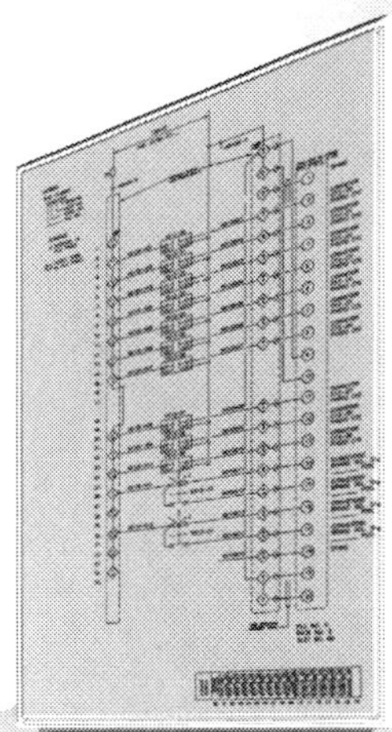

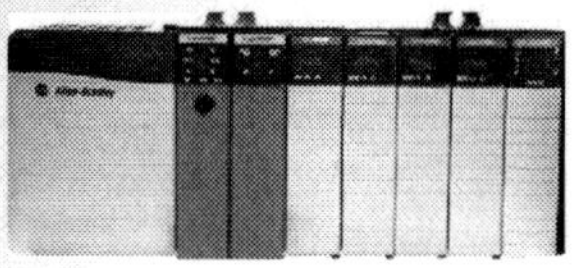

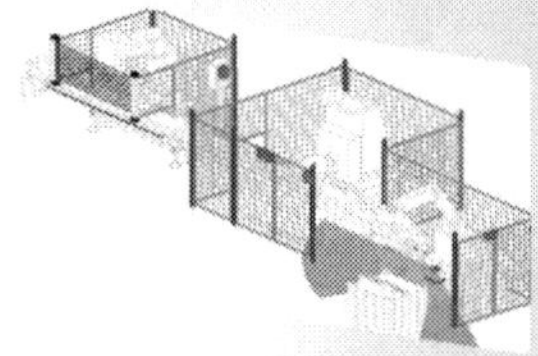

*Select equipment

- Safety controller
- Fence supplier
- Gate locks
- Safety sensors

*Implement solution

- Purchase equipment
- Stage safety system
- Install guard fence
- Install locks, sensors & controller
- Interface safety system with process controllers
- Start up, check out & train operations, maintenance & engineering

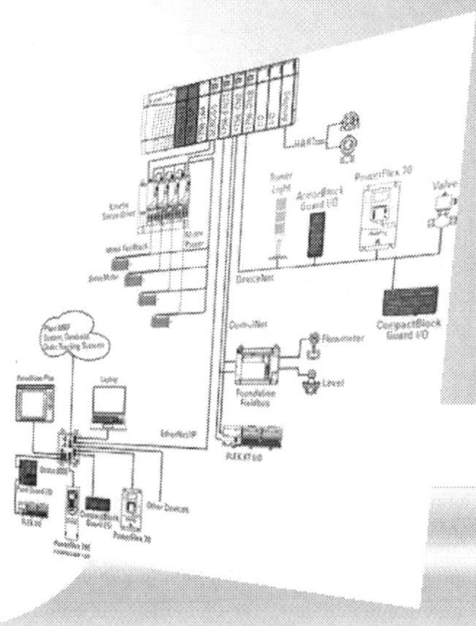

*Rest easy ...

Your process and personnel are safer with a properly automated guarding solution protecting them

*Guarding Operations

GUARDING OPERATIONS

SEACON Engineering Associates, Inc.

Abstract

Recognized Standard Locations for Airborne Dust Concentration Testing

While there are industry standard testing methods for concentration levels of fugitive and respirable dust, there is no standard for location of sampling filters. Recently some manufacturers are requiring acceptable levels within the machinery. Dust control systems have three main objectives;

- prevent respirable dust levels from exceeding permissible exposure limit (PEL)
- prevent fugitive dust levels migrating away from the source and accumulating to levels that become fire and explosion hazards
- improve final product quality.

Areas within the machine frame boundaries such as slitter banks, perf heads, doctor blades, and ply-bonding generate high amounts of dust. Regardless of the dust control system effectiveness, concentrations in these areas will most likely exceed any levels established by manufacturers. While numbers exceeding requirements may infer ineffective dust control, they may in fact indicate a higher level of dust control. Requirements below PEL levels should be reserved for areas outside the machine where operators and other personnel are exposed to respirable dust and where fugitive dust can accumulate. Standard test locations should be established to provide meaningful data.

Historical Perspective of Testing and Regulation

Analysis and sampling of airborne dust began in the 1920's to address the high rate of lung disease among mine workers. Coal Workers Pneumoconiosis (CWP)[(1)] was discovered in the early 19th century by doctors treating mine workers for what is known as black lung disease, miner's asthma, or CWP. Studies performed in the United States and Great Britain, discovered that different dusts had different levels of risk. In 1939 a report by the United States Public Health Services came up with a method of determining a worker's exposure to dust which related their time exposure at different dust levels and is known as a Time Weighted Average.[(2)] TWA is used today by OHSA and NIOSH to set personal exposure limits PEL for a variety of chemicals, fumes, dusts.

Silicon dioxide (quartz or silica) and asbestos are the most serious substances that cause lung pneumoconiosis. Silica has the power to kill white blood cells and causes the formation of fibrosis which prevents the removal of the particles from the lungs.[(3)] Asbestos cannot be removed by white blood cells and causes fibrosis to develop.[4] It is not yet understood why coal dust, without the presence of silica, also develops into severe fibrosis in a small number of cases.[(5)]

That was the birth of airborne dust testing. During that time many methods of testing were developed, but most of them had a specific goal of determining how dust caused lung disease. What they discovered was that dust particles in the size range of 0.5 to 10μm could reach deep in the lung into the alveoli. Alveoli are the final branches in the lungs where the gas exchange takes place. Dust particles larger than this are trapped by the nasal passages and bronchial tubes that expectorate the particles.[(6)] Smaller particles are either exhaled or are not in sufficient concentrations to be considered harmful.[(7)]

Fortunately, paper dust is not a severe health risk on the order of coal dust, silica or asbestos, but high concentrations of any dust should be avoided. Low risk organic dusts can cause lung problems in high concentrations. Other chronic lung diseases are Byssinosis caused by cotton or flax dust and Farmer's lung caused by moldy agricultural products.[(5)] This is why there is a category of dusts called Particulates Not Otherwise Regulated (PNOR) by OSHA and Particulates Not Otherwise Specified (PNOS) by the American Conference of Governmental Industrial Hygienists (ACGIH.)[(8)] Cellulose dust is in the PNOR category and has Personal Exposure limits of 15mg/m^3 total dust and 5mg/m^3 respirable dust.[9] ACGIH is more stringent and recommends levels of 10mg/m^3 and 3mg/m^3 respectively.[(10)] The difference between total dust and respirable dust is particle size. Total dust measures the total dust concentration in the air regardless of particle size. Respirable dust separates out particles larger than 10µm and the collection media only captures particles below this size.

OSHA has standards called Action Levels for certain substances that are regulated for worker exposure limits. The action level is 50% of the personal exposure limit (PEL) for dusts that are designated as having a ceiling limit.[(8)] The idea of the action level is to ensure that a worker has a very low exposure to concentrations above the PEL. That is to say that testing methods only give a snapshot of the average concentration level during a test interval and that there are likely peak concentrations above the PEL even during the tested period. The substances that have action levels have ceiling limits where exposures above the PEL should be avoided. Cellulose dust does not have a ceiling limit and only requires that workers average exposure for an 8-hour TWA should not exceed the PEL. Applying action levels for cellulose dust is not necessary, but measures should be taken to lower the concentration level as low as possible.

Paper Industry Guidelines

The target should be to have the total dust concentration level at or below the more stringent ACGIH respirable level of 3mg/m^3. Respirable dust concentrations will most always be less than the total dust since a cyclone is used to separate and reject airborne particles larger than 10µm.

Lung disease is not the sole reason to perform ambient dust concentration sampling surveys. One reason is to determine the effectiveness of dust control systems. Dust control systems are a must and are required for dusty processes in order to be compliant with NFPA standards and OSHA's housekeeping regulations. A dust explosion hazard survey is another reason to use sampling surveys. It can identify areas of the facility that are at risk of dust explosion and help determine the level of mitigation controls that need to be employed.

A dust system has three major functions: prevent dust concentrations from reaching worker exposure limits, minimize dust explosion hazards, and to maintain the quality level of the final product. A high amount of dust in the finished product can be transported throughout the mill via conveyors or forklifts that move the product from the machine to other processes or storage areas.

One reason to reduce the concentration levels far below the requirements for worker exposure is to minimize the risk of creating a dust explosion hazard due to accumulated dust. Particles below 50μm have a very low terminal velocity at 0.592 feet/min in still air.(6) Natural air currents from rotating equipment, ventilation inlets or outlets, humans walking, forklift traffic, etc. create air velocities higher than the terminal velocity of very small particles. For this reason dust can be carried great distances away from the producing source and settle in areas where air currents are relatively stagnant. Overhead areas are such places, since there is less activity to create strong air currents. Dust that escapes dust system control zones are considered fugitive dust. According to Drinker & Hatch spherical shaped particle sizes from 50 to 100μm will settle with a terminal velocity of 14.8 feet/min and particles greater than 100μm fall at a rate of 59.2 feet/min in still air.(6) This dust can travel quite a distance from the source point in the presence of strong air currents. A light breeze of air that can just be felt on the skin is 30 to 60 feet/min. So any light breeze that can be felt that is in the upward direction can lift even the larger particles of dust to higher elevations.

NFPA has standards regarding dust accumulation levels that would qualify an area as a dust explosion hazard.(11) In these areas special requirement are called for regarding electrical equipment, housekeeping plans, mitigation of ignition sources and fire, and explosion suppression devices. Testing of the accumulated dust should be done at every facility with either the "worst" dust tested or to have dust tested from different process areas in order to assess the actual dust accumulation thickness before housekeeping needs to be performed. The standard has a calculation that determines the dust layer accumulation thickness that would constitute a hazard. The formula uses 75 lbs/ft^3 as the basis for the bulk density of the dust but the equation allows adjustment for lighter dusts.

Today there is a large amount of literature that can be found on the topic of dust sampling. Much of the information explains what test methods and sampling equipment are available, but very little is written about where sampling equipment should be placed. The reason is that most of the literature encompasses the whole realm of airborne contaminants that includes gas, vapors, fumes, smoke and dust. For the dry end of paper machines and converting equipment, cellulose dust is the primary concern. Every dust has different characteristics and health risks. There are many different occupational processes that create different concentration levels of airborne dust. Because of the differences there is no one size fits all strategy for airborne dust concentration sampling across all industries.

NIOSH has developed test methods 0500 and 0600 which are the most common ways to test for total and respirable dust.(10) There are several sources for testing equipment that can either be purchased or rented. There are also several independent laboratories that can supply the filter cartridges and analyze the tested samples. These are methods accepted by OSHA to test for worker exposure to nuisance dusts. A pre-weighed 37mm diameter 5μm pore size PVC filter is assembled into a cartridge that has an inlet and outlet. A calibrated sampling pump is attached with tubing to the outlet of the filter. The pump then draws air at a rate of between 1 l/m and 2 l/m through the filter cartridge, where ambient air enters the inlet and any dust in the air stream will be deposited on the filter media. Concentration is then calculated by dividing the mass of the collected dust by the volume of air drawn through the filter. These calculations are performed by an AIHA accredited laboratory.

One test that is not discussed is the determination of the sedimentation rate. It is near impossible to derive an equation that could relate dust concentrations and sedimentation rates due to the many variables encountered. Both theoretical and experimental data depend on still air or very long settlement chambers with low horizontal velocities and spherical shapes. Paper dust is not spherical, being fibrous in nature, which impacts the terminal velocity. To further complicate the relationship, it would also be necessary to know the particle size distribution. In Table 1, it shows approximate size distribution from some collected samples. Samples D and E came from the same mill but are from different process areas. The same raw material producing different particle size distributions would also have different sedimentation rates. Terminal settling velocities that have been determined are for spherical dusts but other shaped particles have slower terminal velocities.[5] Illustration 1 shows a picture of some tissue paper dust next to a straight pin. It can be seen that the dust particles are random in shape and size. The air in an open room will have air currents from many different sources which also slows the sedimentation rate.

Particle Size microns	Sample A	Sample B	Sample C	Sample D	Sample E
<250	90	97	49	25	83
<125	79	80	33	14	62
<75	61	60	14	3	28

Table 1. *Percentage by weight particle distribution of five paper dust samples.*

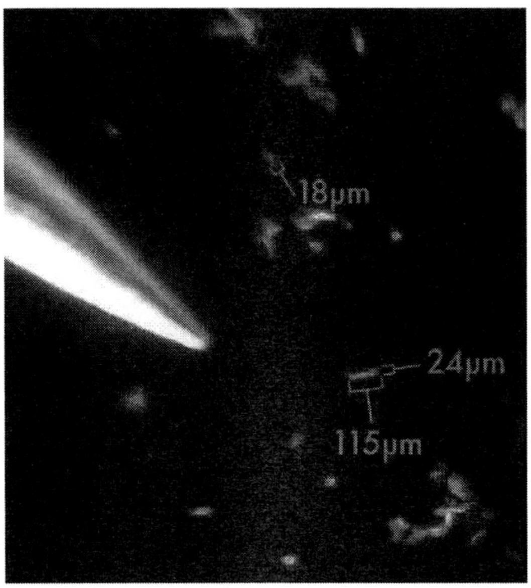

Illustration 1. *Tissue dust particle size comparison.*

microns	inches
1	0.00003937
10	0.00039370
18	0.00070866
24	0.00094488
50	0.00196850
100	0.00393701
115	0.00452756
250	0.00984252
420	0.01653543

Table 2. *Microns to inches conversion*

The value of knowing the accumulation rate is so that housekeeping procedures and schedules can be set. A sedimentation plate can be made of a 5 inch (127mm) square plate with sides high enough not to let airflows keep dust from settling at the edges, but not so high as to make it difficult to evaluate or remove dust. Determination of the sedimentation rate can be calculated by multiplying the plate area by the depth of the dust for a given time. This can also give you the bulk density of the accumulated dust that can be used to calculate the allowable dust layer thickness before an explosion hazard exists.

Simple instruments can be used to determine the bulk density and sedimentation rate. A scale that can measure to the tenth of an ounce or to a gram is required along with a sedimentation plate or any other type of container of a known volume. It is best to use the settled dust from the sedimentation rate test to find the bulk density. Paper dust when disturbed will agglomerate and become compacted when disturbed. If a sample of dust is collected by other means it will likely have a higher bulk density. Illustration 2 shows a simple method in determining the bulk mass and settling rate.

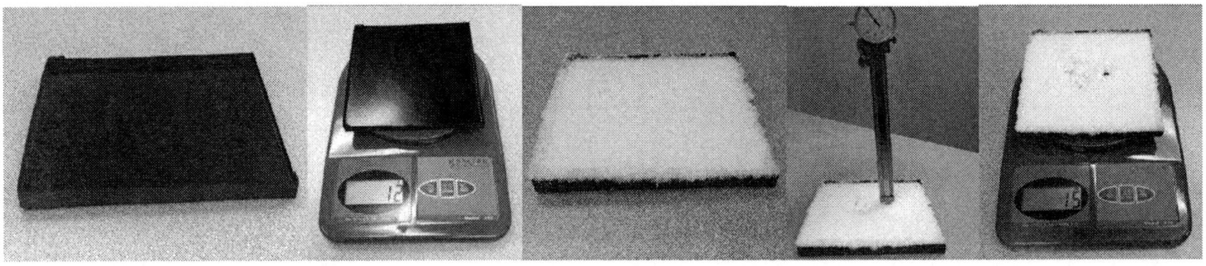

Illustration 2. *Sedimentation tray, scale, collected sample and vernier caliper.*

Most of the literature for dust sampling is concerned with occupational disease due to contaminant exposure. NIOSH has produced publication number 77-173 titled Occupational Exposure Sampling Guide. It is an excellent resource to use for development of a standard sampling strategy for tissue production operations.[8] It is 150 pages long and encompasses chemicals, fumes, and dusts. Only a portion of the guide is applicable to the testing of nuisance dusts. It also does not give specific directions on where place the sampler. It is very important to place the sampler near areas where a person is likely to be exposed to dust. It does however state that samples should be taken in the breathing zone. The normal configuration that is used is to place the cartridge filter with the inlet pointing downwards at a height of 5 feet. This would place the sampler nearly at the same conditions of human breathing zone.

The manual suggest some sampling strategies regarding the time periods and number of samples. Some of the suggestions for the ideal sampling using the methods 0500 and 0600 in regards to taking a full 8-hour period single sample are confusing. The methods give minimum and maximum flow rates and volumes through the filter cartridge that will not allow a test to be longer than 2.2 hours for total dust and 2.67 hours for respirable dust. The maximum volumetric flow through the filter is based on not loading the filter to its capacity of 2mg of dust during a test period at a concentration of 15mg/m^3 for total and 5mg/m^3 for respirable. The full 8-hour single sample period is the second best ideal method with a full 8-hour consecutive sampling being regarded as the best. This requires a large number of samples to be analyzed which has negligible benefit over the use of a single sample for a full period. However, if the dust concentration is above 4mg/m^3, the filter cartridge will become loaded with the limit of 2 mg before the test is completed.

Longer test periods can be selected and are allowed by the instructions, but this requires pre-testing to estimate the concentration of dust in the air in order to calculate how long the test can be conducted without collecting more than 2mg of dust. Portable real-time particulate concentration monitor can be used to pre-test but most only have a range of 0 to 50µm. Those capable of larger particle sizes cost over $20,000. Table 1 shows a majority of the particles are larger than 75µm; therefore a portable monitor is not an economically feasible pre-test. If the process area has never been tested previously, or has had a process change, use of a longer test is not recommended. Subsequent tests can use the previous test data to calculate the flow rate and period time so the filter loading will not exceed 2mg. See Table 3 for dust concentration levels.

Machine	TM "A"		TM "B"	
Web **inches (mm)**	130" (3,300)		130" (3,300)	
Speed **fpm (mpm)**	3,400 (1,036)		3,900 (1,189)	
	Before	***After***	***Before***	***After***
Overall Average **(mg/m^3)**	10.5	0.68	11.7	0.65
4 locations above web **(mg/m^3)**	18.5	0.92	18.8	0.68
4 locations outboard of machine **(mg/m^3)**	2.5	0.44	1.98	0.70

Table 3. *Before and after dust control comparison on two different tissue machines.*

There is another drawback to using a full period consecutive sample strategy. During the 8-hour test period a sample would include a shift clean-up where a large amount of dust is suspended by the use of compressed air to sweep the accumulated dust. The sample during this period will have a much higher concentration. Since one of the goals is to evaluate the dust control system, this higher average will reflect less effective dust control during this period.

Chapter 4 of the NIOSH manual gives the procedures to calculate the statistical upper and lower confidence levels. A single sample for the upset condition will only give the exposure for that particular day but on other days higher concentrations may occur. Some of the variables that could affect this are the number of hours between clean up periods and the tissue grades run on the machine. The upset condition sample is only necessary for evaluating the employee exposure and should be excluded when analyzing the effectiveness of the dust control system.

The NIOSH manual also allows for partial period sampling but advises that the confidence level is lower since the average is only valid for the period tested. There is latitude, using professional judgment to make inferences for the untested time period. Since most machines run at steady state with short intervals on converting equipment, for roll changes and sheet breaks, and on tissue machines, for parent roll changes and sheet breaks, it would be valid to estimate the untimed period as similar to the tested time period. Observations should be documented during the full 8 hours for roll changes, sheet breaks, grade changes, machine speed changes, and any other abnormal running condition. This will help in assessing whether it is a valid assumption that the untested time period is similar to the tested time period.

Locating Dust Testing Points on Paper Machines and Converting Equipment

There are three major goals for the dust testing:

- ensure that the worker exposure is below the recognized industry standards
- determine the effectiveness of the dust control system
- determine an accumulation rate that can be used to develop a housecleaning schedule to prevent a dust explosion hazard.

Location of the sampling media needs to first be selected. They should be within 3 feet of the outside of the machine frame and set at 5 feet high. Samples should be taken at the drive and tending sides at every process change that generates a significant amount of dust; typically the unwind stands, sheet combining, calendering, ply bonding, slitting, perfing, and re-winding with the number of locations being determined by which of these processes are used on the machine. A tissue machine's dust producing processes are the Yankee doctor, stabilizer foils, slitters, reel section, and the parent roll. Illustrations 3 & 4 show suggested locations for sampling filters.

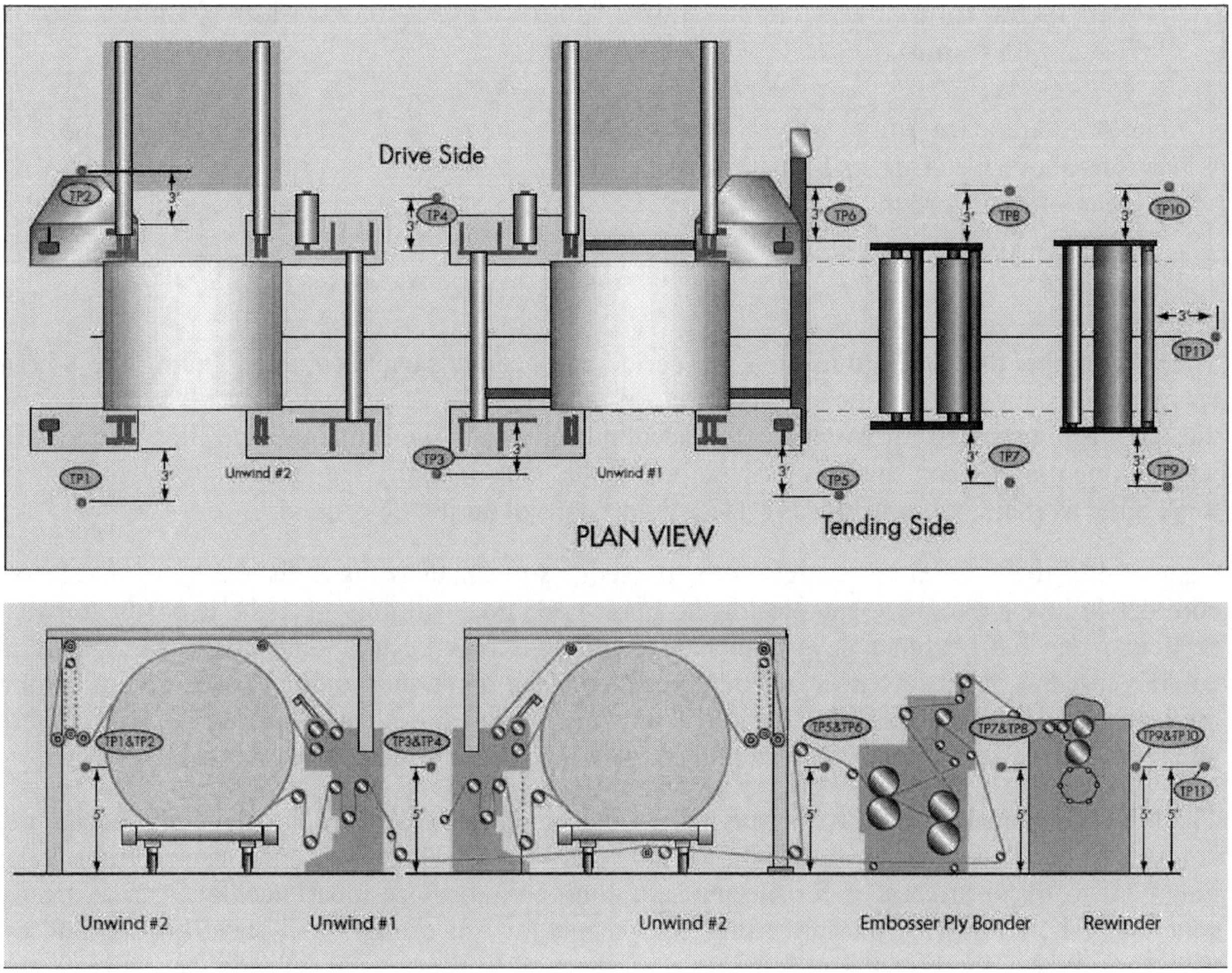

Illustration 3. *Suggested sample locations converting equipment plan and elevation views.*

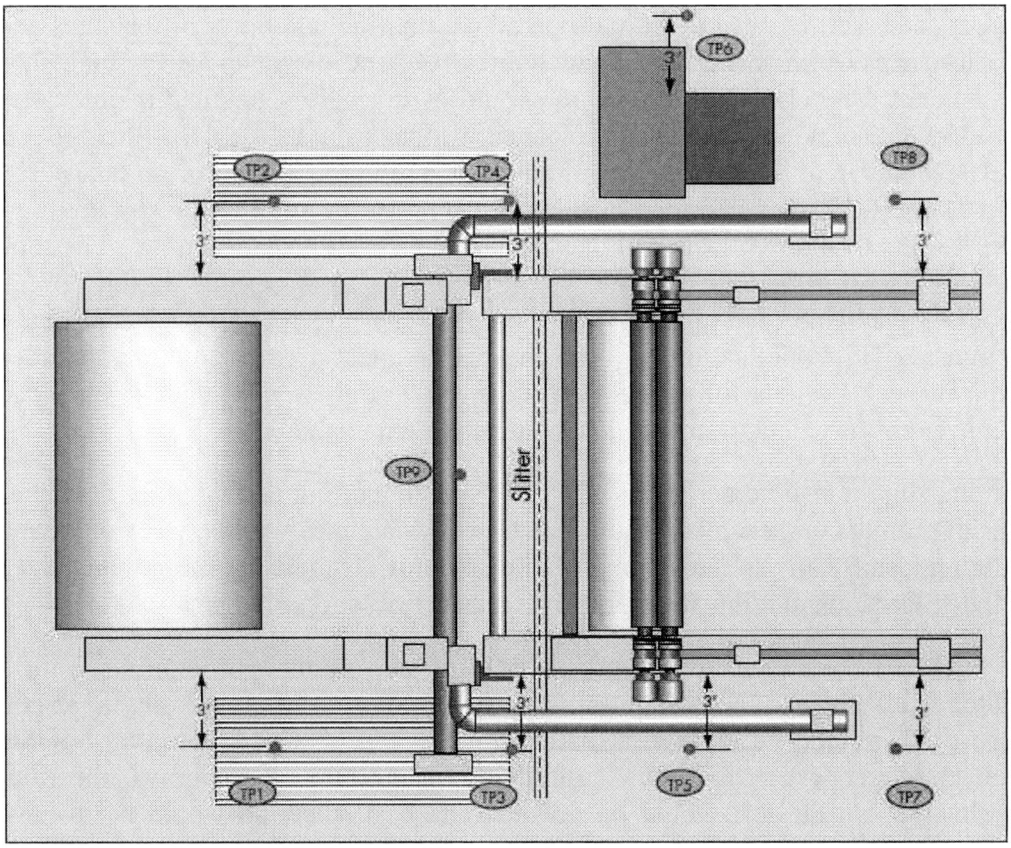

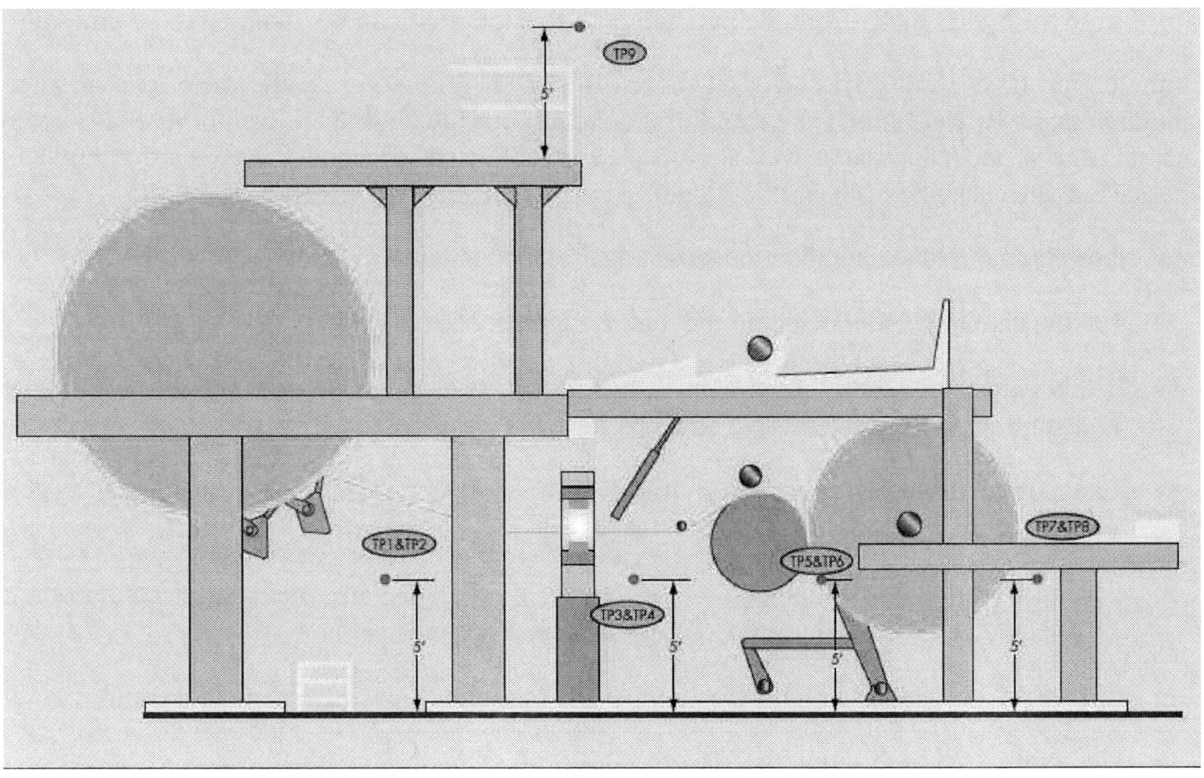

Illustration 4. *Suggested sample locations for tissue machines plan and elevation views.*

At least two test periods should be set for normal operations that include only roll changes and sheet breaks. Two test periods should be set to include a machine clean up where high dust concentration levels occur. One clean up test period should be during a scheduled maintenance down where the entire machine is thoroughly cleaned which would likely be the highest concentration levels encountered. The test period should last as long as the clean up takes. The second clean up test period can be performed at the same time as the total and respirable dust testing takes place. Results of the two tests can then be analyzed for exposure limits. The total tested time should be at least 70% to80% of an 8 hour shift. The test periods during normal operation should be for the same interval of time to simplify the upper and lower confidence level calculations in the NIOSH publication 71-173. The clean-up test period should be included in the average exposure calculations for worker exposure limits only and should be excluded from the average concentration calculations for dust control system evaluation.

Respirable dust testing should also be performed. Most of the converting and winding machines have no more than seven operators; each one should be tested. Machines with between eight and eleven operators during a shift can exclude testing of one operator. If there are more than eleven operators refer to the NIOSH manual for the appropriate number of workers to be tested.

According to the NIOSH test method 0600 the flow rate for an aluminum cyclone used for a 37mm cartridge filter is 2.5 l/m and the maximum volume for concentrations of $5mg/m^3$ is 400 liters. The cartridge filters should not exceed a loading of 2mg. As stated before, the NIOSH instructions allows for longer test periods but if the dust concentration is unknown, the filter could become overloaded and the test would be invalid. The best selection would be to have three samples that would cover a period of 7.5 hours and assuming that there will be at least 1 hour of time for breaks away from the machine; the break period can be assumed to be $0mg/m^3$.

Sedimentation trays should be placed at three different heights, one on the floor, one at 5 feet high, and one at 10 feet high. They should be done at both the tending side and drive side of the machine. Areas under the paper web should also be collected, especially in areas that build up dust. Several areas can be tested without adding to the cost since the trays are reusable.

Some other parameter that should be observed during the testing are:

- Testing should be conducted on the dustiest grades of paper;
- If other dust producing equipment is in the area background levels should be obtained and subtracted from the sample analysis for evaluating dust control system effectiveness;
- Test period during machine or room clean-up not to be used for dust control system effectiveness evaluation;
- Careful observation of machine down times, roll changes, and other upset conditions to be documented for the duration of the 8 hour test period;
- Use an AIHA accredited laboratory to perform gravimetric analysis on the samples;
- The total tested time should be 70% to 80% of the 8 hour work shift.

Table 4 shows a matrix of test types and suggested time periods.

Time	Total Dust Tests	Time	Respirable Dust Tests	Time	Sedimentation Rate
7:45	15 min. Total Dust Test w/o Dust Control	7:45	15 min. Total Dust Test w/o Dust Control (optional)	7:45	
8:00 10:10	130 min. Total Dust Test	8:00 10:35	155 min. Respirable Dust Test	8:00 3:00	7 hour Sedimentation Test
10:10 12:20	130 min. Total Dust Test	10:30 1:05	155 min. Respirable Dust Test		
12:20 2:30	130 min. Total Dust Test	2:30			
3:00		3:00			
4:30 5:00	Test Period for Length of Shift Clean-Up	4:30 5:00	Test Period for Length of Shift Clean-Up minimum 30 min.	4:30 5:00	

Table 4. *Test type and time period matrix.*

Some mills are assigning guaranteed concentration limits within the machine frames. This provides no useful data with regards of the three main goals for dust sampling. Effective dust mitigation requires control zones within the machine frames. Control zones help contain the fugitive dust in areas where the airflow will carry it into exhaust hoods. Sampling in these areas would have a velocity difference between the ambient air and the inlet air to the filter which would impact the collection efficiency of the filter; true results would not be obtained. Another obstacle with this strategy is that very high concentrations would be encountered due to the convergence of the air containing dust to a narrow slot at the collection hood.

Dust is generated in tissue processes where the sheet is disturbed by turning rolls, ply separation, calendering, doctoring, slitting, etc. When the dust is dislodged from the sheet it prefers to stay in the boundary layer of air traveling with the sheet. Centrifugal forces around turning rolls, or wherever there is a change in the direction of the web, cause the dust particles to exit the boundary layer. Some areas inside the machine frame are natural to have high concentrations of dust. Consideration must be given to the design and placement of hoods to avoid creating sheet effects. Maintenance accessibility also may prevent the mounting of hoods in the ideal location. For these reasons, high concentrations of dust within the machine frame cannot be totally avoided. Control zones defined by baffling will prevent dust escaping outside of the machine frame where worker exposure is a concern. Because the dust is confined to the control zone, concentrations will be higher than outside the machine. Any dust that escapes from within the machine frame will mix with the room air and the concentration will be diluted. Sedimentation rate tests can be conducted in areas within the machine frame where hood mounting is difficult in order to determine the required frequency for machine clean-ups. With the involvement of mill operators, mill hygienists, safety compliance officers, dust sampling consultants, and dust control equipment consultants a consensus standard could be issued that could be used for paper machines, tissue machine, and converting equipment in the paper industry.

Works Cited

1. Derickson, Alan. *Black Lung: Anatomy of a Public Health Disaster.* Ithaca: Cornell UP, 1998.

2. United States. National Institute of Health. United States Public Health Service Division. *The Significance of Dust Counts.* By J. M. Della Valle. Center for Disease Control and Prevention, June 1939.

3. United States. Department of Health and Human Services. National Institute for Occupational Safety and Health. *Occupational Respiratory Diseases: Section II Pneumoconiosis Silicosis.* By John M. Peters, Center for Disease Control and Prevention, Sept. 1986.

4. United States. Department of Health and Human Services. National Institute for Occupational Safety and Health. *Occupational Respiratory Diseases: Section II Pneumoconiosis Asbestosis.* By John M. Dement, James A. Merchant & Francis H. Y. Green. Center for Disease Control and Prevention, Sept. 1986

5. Dorman, Richard George. *Dust Control and Air Cleaning,* Oxford: Pergamon, 1974.

6. Drinker, Philip, and Theodore Hatch. *Industrial Dust, Hygienic Significance, Measurement, and Control.* 2nd ed. New York: McGraw-Hill, 1954.

7. Roach, S. A. *The Industrial Environment, its Evaluation and Control: Sampling Air For Particulates.* U.S. Department of Health, Education, and Welfare, NIOSH 1973

8. United States. U.S. Department of Health and Human Services. National Institute for Occupational Safety and Health. *Occupational Exposure Sampling Strategy Manual.* By Nelson A. Leidel, Kenneth A. Busch, & Jeremiah R. Lynch. Center for Disease Control and Prevention, Jan 1977.

9. United States. United States Department of Labor. Occupational Health & Safety Administration. *1910.1000 Table Z-1.*

10. United States. U.S. Department of Health and Human Services. National Institute for Occupational Safety and Health. *Test Methods: Particulates Not Otherwise Regulated 0500 Total/ 0600 Respirable.* Center for Disease Control and Prevention

11. "NFPA 654: Standard for the Prevention of Fire and Dust Explosions for the Manufacturing, Processing, and Handling of Combustible Particulate Solids." *National Fire Protection Association* (2006).

Maintaining Optimum Performance of Tissue Machine Clothing

Tissue World 2012

John Schwamberger
Paper Chemicals Product Line Manager
DuBois Chemicals

john.schwamberger@duboischemicals.com

DuBois

Challenges in Maintaining Performance of Tissue Fabrics

- High speed
- Low basis weight
- Recycle fiber contaminants and supply
- High number of daily fabric nip cycles
- Functional chemistry challenges
  - Strength additives, Yankee chemistry, softeners
- Water system closure
- Energy cost increases
- Environmental/regulatory restrictions on chemistry
  - VOC limits, FDA restrictions, etc
- Increasing product quality demands

Benefits that can be gained by Improved Fabric Performance

- Improved runnability, production rate
- More stable operation, fewer web breaks
- Improved product quality / uniformity
- Better efficiency, particularly at the end of a press fabrics life
- Ability to use different or more economic wet/dry strength agents, without filling press fabric void volume
- Longer fabric operating life, resulting in lower costs
- Less downtime production losses for batch washing
- Eco-efficiency
  - Water and energy savings
  - Reduced effluent impact

Why Target Fabric Contaminant Reductions
Potential Savings Opportunities

Suppose:

- 220 ton/day hypothetical tissue/towel grade with profit of $300/Ton*
- Press fabric life
  - 30 days, $25,000/felt*
  - 12 outages, scheduled to felt life, 10 hrs each
- Four downtime batch washes monthly
- Sheet breaks
  - 45/month, 15 min/break
  - Many breaks occurring towards end of felt life or when a batch wash is needed
- Machine slows down 200 fpm (61 meters/min) at last 3 days of felt life

* Tissue profit and fabric pricing is hypothetical to illustrate potential benefits

Why Target Fabric Contaminant Reductions
Resulting Benefits

Consider if:

- Felt life was extended 5 days and outage schedule changed to correspond to new felt life
- No slow-down at end of life
- Batch wash reduced to 1 month on average
- Breaks reduced 15% by better fabric performance throughout its life

	Before	After	Production Savings	Financial Savings
Reduced Breaks	900	810	1620.0	$ 74,250
	Lost Time (min./month)		Time Saved (min/yr.)	
Reduced Batch Washing	240	60	2160.0	$ 99,000
	Lost Time (min./month)		Time Saved (min/yr.)	
Increased Machine Speed	6000	6020	256.7	$ 77,000
	Average Speed (ft./min.)		Increased Production (tons)	
Increased Felt Life	30	35	1.7 Felts Saved/Year; 1020 (min/yr)	$ 87,500
	Average Life (days)			
			Total Process Savings	$ 337,750

Wire and Fabric Cleanliness

Fabric contaminant control is accomplished by a variety of mechanical devices, fabric design and chemical cleaning methods

- Mechanical Tools
  - Showers
  - Uhle & suction boxes
  - Vacuum systems
  - Doctors
- Fabric Designs
  - Materials, treatments
- Chemical Applications
  - Preventative – Surface treatment chemistries
  - Cleaning techniques

Showering Considerations
Most Important Showering Aspects for Fabric Cleanliness

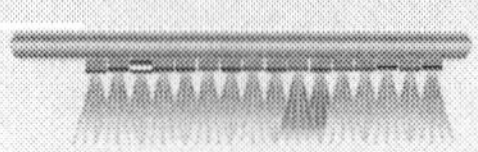

- Oscillating HP Needle Showers
  - Typically used on both forming & press fabrics
  - Provides effective mechanical cleaning proportional to applied kinetic energy - Set-up and maintenance are critical
  - Damaging to fabric if pressure is too high
- Flooded Nip Showers
  - Set-up and flow rate important
  - Water quality & temperature are important factors
- Traversing Shower
  - Spot cleaning, 1-4 nozzles common, low frequency coverage
- Chemical fan cleaning showers
  - Traditionally dedicated full width chemical fan showers

Chemical Contaminant Control
Fabric Surface Treatment Technologies

- Cationic passivator or barrier treatment
  - Traditionally most common
  - Based on charge control
- Non-ionic wire passivation
  - Newer technology, works by making the fabric surfaces less attractive to bonding by "stickie materials"

Chemical Contaminant Control
Cationic Barrier Treatment Technologies

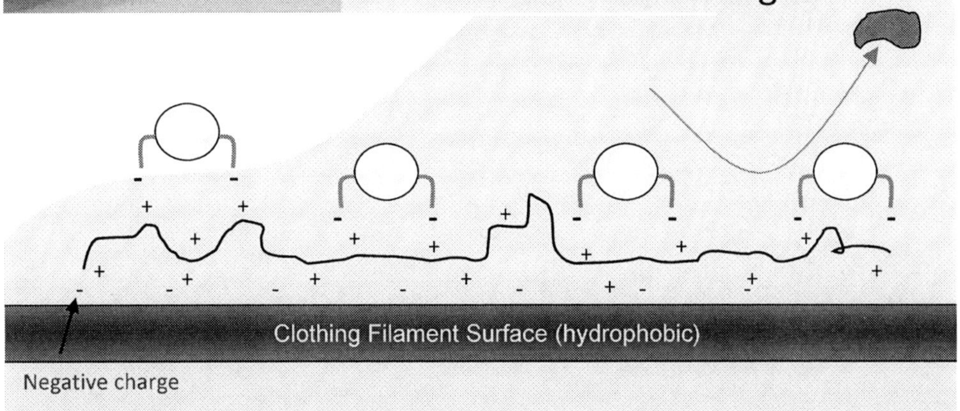

1. Cationic polymer attaches itself to negative charges on fabric surface
2. Anionic/Hydrophilic trash attaches itself to cationic charges of polymer
3. Hydrophilic trash repels stickies

Chemical Contaminant Control
Why Cationic Treatments Build-up on Fabrics

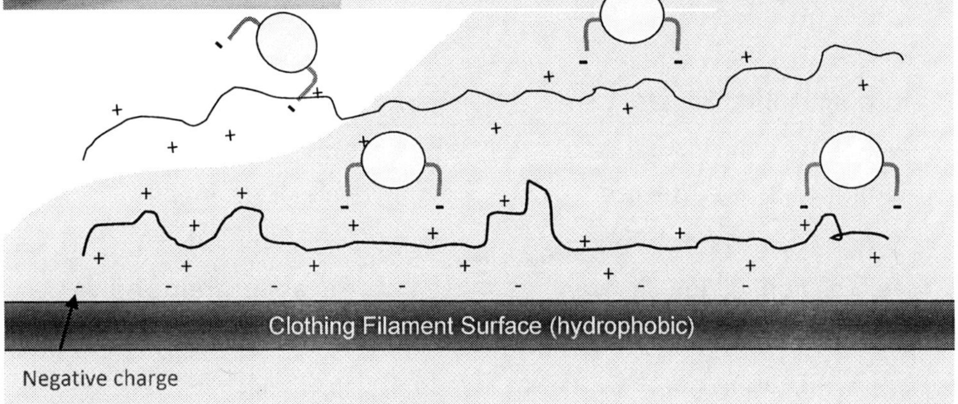

1. Anionic trash leaves
2. New cationic polymer attaches itself
3. More anionic trash attaches itself

Chemical Contaminant Control
Non-ionic Surface Treatment Technologies

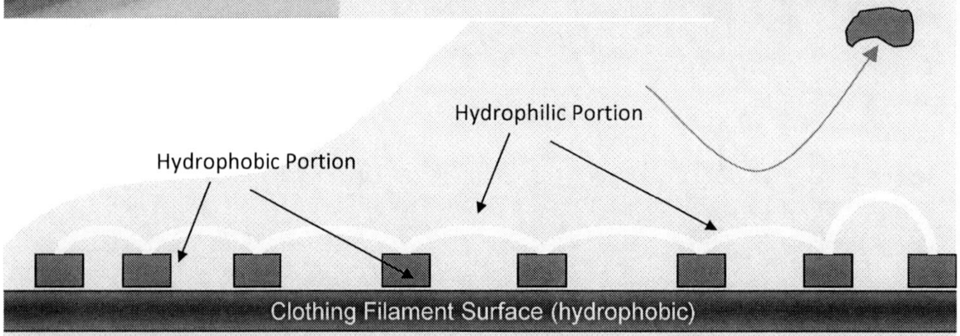

1. Non-ionic technology has hydrophobic portion and hydrophilic portion
2. Hydrophobic portion of polymer adheres to hydrophobic portion of clothing
3. Hydrophilic portion of polymer repels stickies

Chemical Contaminant Control
Fabric Surface Technologies Comparison

Cationic Chemistries	Non-ionic Chemistries
When wire performance or drainage is not effected	When cationic affect performance or drainage
When fresh water is available for the application shower	When filtered whitewater is used for the application shower
When downtime for stripping barrier coating is not an issue	When no downtime wire stripping is advantageous
When cationic effectively prevent stickies	When cationic are ineffective

Traditional Chemical Contaminant Removal
Cleaning Technology for Tissue Fabrics

- In many cases, fairly basic strategies
- Often viewed as a remedial procedure used after runnability has deteriorated – often with substantial hidden costs
- Batch Cleaning
  - Requires down-time
  - Often use chemistries with undesired side-effects
    - Sodium Hypochlorite (potential press felt damage)
    - Hydrocarbon solvents (marginal results, VOC)
- Spot Cleaning
  - Traveling Nozzle or manual applications

Traditional Press Fabric Cleaning Methods

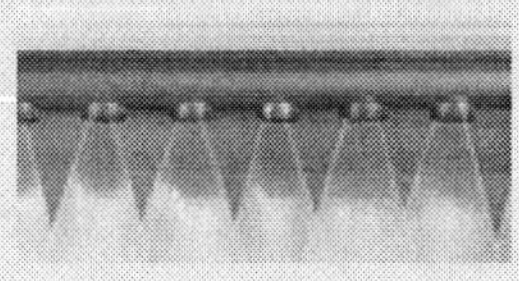

Continuous Conditioning / Cleaning

- Low concentration application continually
- Limitations:
  - High chemical cost for marginal results.
  - May fail to reach threshold product concentration for effective cleaning
  - Often utilizes dedicated showers

Batch on the Run

- Intermittent cleaning, higher concentrations than continuous cleaning
- Limitations:
  - Can upset machine chemistry or inhibit felt water transport during cleaning
  - Reduced overall press stability during cleaning
  - May impact Yankee coating

Keys to Effective Chemical Cleaning of Press Fabrics

- Understanding Contaminants
  - Used fabric analysis
- Understanding machine conditions & process additives
- Designing a method of cleaning to compliment mechanical cleaning equipment and fabric designs
- Effective Chemistries are only one piece of the puzzle
  - Dwell time, temperature, chemical concentration, rinsing and mechanical action all critical parameters
  - Application safety profile very important consideration

Understanding Source of Contaminants

- Many potential contaminant soils impact fabric performance
- Most common problematic soils from analyzed residual press fabrics
  - Calcium Carbonate
  - "Stickies" in recycle grades. (Dominant classes are PVA, Styrene Acrylic, SBR Latex)
  - Wet and Dry Strength Aids
  - Yankee Chemistries
  - Defoamers – Esters and Oil Based
  - Fines

Pulsed Method of Fabric Cleaning

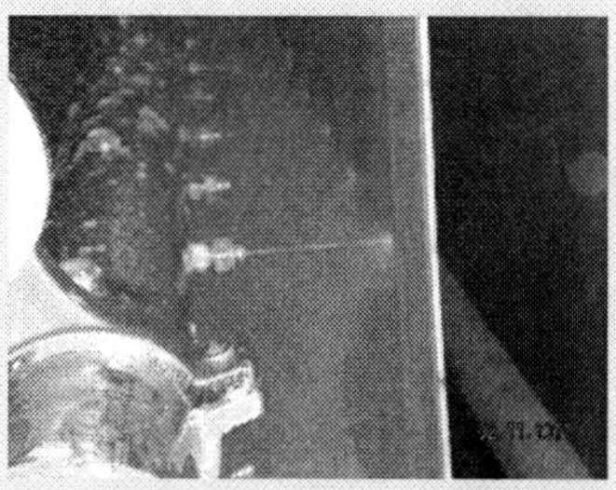

- Patented unique method of felt cleaning
- Utilizes H.P. needle showers to deliver cleaning chemistries during normal paper production
- Achieves effective concentrations only in small sections of the felt
  - Press section stability
  - No negative effects of gross chemical addition to entire fabric width
- Concentration only a fraction of a percent
- Minimized risk of impact Yankee chemistry
- Mechanical & chemical cleaning synergy

U.S. Patent s # 7,597,782; 7,850,824; 7,918,968

Cleaning Chemistry – Specific Problems and Solutions Wet Strength Resin

Problem – Wet Strength Resin Deposition

- Can form gels that swell in press fabric and fill void volume
- May cause need to take downtime and batch wash with sodium hypochlorite – significant risk of fabric damage

Solution

- Stabilized peroxygen detergent used in combination with an alkaline cleaner
- Allows for intermittent cleaning during production

Results Demonstrated

- Extended fabric life
- Ability to provide substantial saving by using CMC to supplement more expensive proprietary wet strength resins
- Eliminate downtime batch washing with sodium hypochlorite

Cleaning Chemistry – Specific Problems and Solutions Stickies Deposition on Wires and Fabrics

Problem

- Recycle fiber stickies are main source of difficulty in wires and felt
- Traditional solvents
- VOC
- Limited Effectiveness

Solution – Engineered Solvent

- Targeted to Stickies
- Aqueous, dilutable
- Reduced solvent use

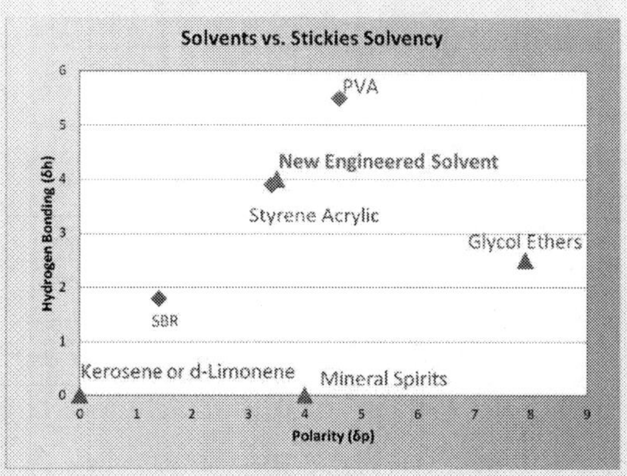

Conclusions

- Effective chemical cleaning and contaminant soil prevention should be looked at as part of the complete solution to maximize the performance of tissue machine clothing
- All the mechanical and chemical tools available should be used in conjunction with the best fabric design to provide for the safest, most effective contaminant control program.
- Substantial financial return can be gained when all available tools are used in together in most effective manner

Yankee Surface Chatter –Mechanics, Monitoring and Methods to Minimize Impact

By Sam Archer, Gary Furman, William Von Drasek, Nalco Company, Naperville, IL

ABSTRACT

Chatter is one of the most insidious issues that can develop on the Yankee dryer. It is a term that is often used to describe observations made about sheet, coating and Yankee surface defects. Its presence can lead to waste, lost time and result in increased maintenance and capital expenditures. Conditions that lead to the appearance of chatter are often traced to doctor tip vibrations and are frequently related to mechanical issues or process operating conditions. Over the last decade the appearance of chatter on Yankee dryers has increased dramatically and most of the issues seem to be related to the management of the creping process.

Once conditions exist that can cause the tip of the doctor blade to vibrate, the amplitude can be intensified by resonance responses with other machine components. This review focuses on process driven chatter and discusses three areas surrounding this phenomena:

1) Mechanics (how it forms)

2) Monitoring (how to develop early warning)

3) Methods (practical ways to eliminate and/or avoid the appearance of chatter).

INTRODUCTION

In the day to day operation of the tissue machine operators are continually challenged to achieve higher levels of productivity and quality. The quest for these improvements often results in a process that strays toward the outer edges of the operating window. Excursions beyond the limits of the process can result in creping conditions that lead to onset of chatter. By understanding the mechanics of the phenomena and the current best practices for monitoring, a set of protocols can be developed that lead to avoidance of chatter in the coating and on the Yankee surface.

During the last two decades there has been a marked improvement in the ability to monitor the mechanical integrity of the tissue machine. It is a rare occurrence that the tissue machine mechanically fails. What has been observed with increasing frequency is a functional failure around the creping process. The creping process is a complex transformation involving mechanical, chemical and operational elements. When these elements are fully functioning they define the operating window of the process. The best way to avoid chatter is to understand the different elements and their operational limits. A functional failure is an excursion beyond the capability of the process and does not imply that an element is broken. Often processes in a functional failed state can lead to a mechanical failure, a loss of process continuity and increased waste. This condition is referred to as a "process driven" failed state.

MECHANICS OF PROCESS DRIVEN CHATTER

A good way to eliminate chatter is to first start by understanding it. The formation of chatter on the Yankee surface is a process anomaly. A number of reviews have been presented that help the tissue maker understand how chatter occurs (Corboy 2003, Alessandrini 2003). The following discussion builds on this previous work. It focuses on the tip of the creping doctor blade and the influence of process conditions that can lead to onset of chatter.

Since chatter manifests at the tip of the doctor blade it is important to understand the interacting forces at this location. A simplistic, but understandable model has been previously presented (Corboy 2003, Apple 2007). This model attempts to describe forced vibration of the total mass between the strong back loading cylinders and the creping doctor blade. A modified recreation of this model is shown in Figure 1. If consideration is given to the relationship of higher frequency vibration and the occurrences of chatter, a special focus should be given to understanding motion of the doctor blade tip (See Figure 2). Equation 1 (Corboy 2003) provides a general mathematical description for motion within these systems but does not address what initiates the movement.

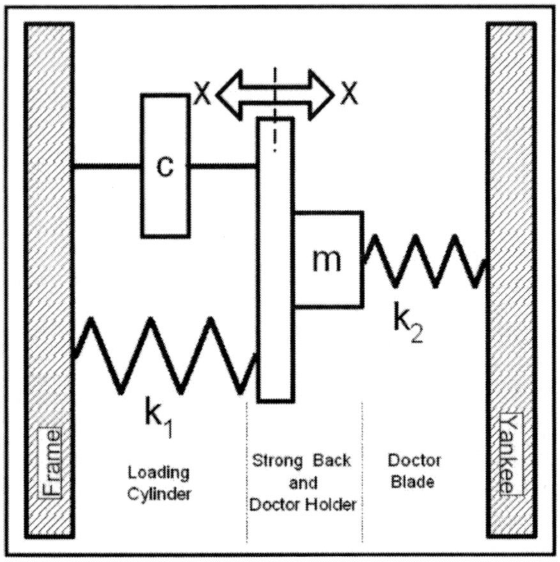

Figure 1 – *Dampened spring system used as basic model for describing forced doctor blade vibration. (Adapted from Corboy 2003).*

$$Fsin(\omega t) = m(dx^2/dt^2) + c\ dx/dt + kx \qquad (Eq.\ 1)$$

Fsin(ωt) = *Motion of a the mass within the system*
m = *Mass of the system*
c = *Dampening Constant*
k = *System stiffness factor* (k_1 *and* k_2)
x = *Displacement from the origin*
t = *Time*

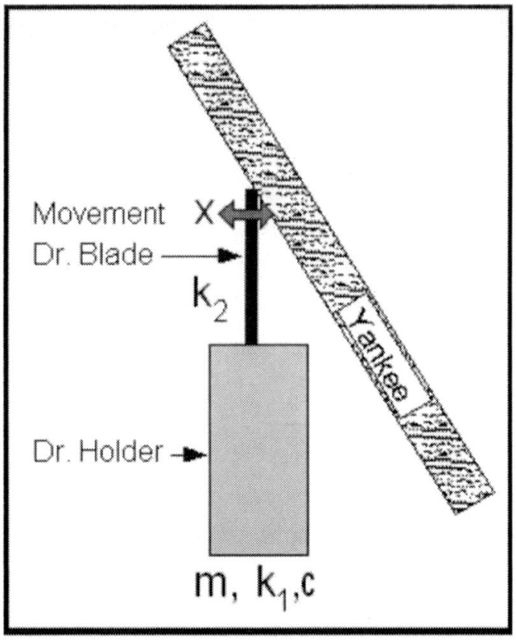

Figure 2 – *Mechanical factors that affect the movement of the doctor blade tip.*

Previous work (Archer 2008) further described process driven chatter as sheet, coating and Yankee surface defects related to a phenomena known as stick-slip (See Figure 3). The phenomenon causes excessive vibration at the tip of the doctor blade and increased out of plane movement. Excessive out of plane vibration can lead to chatter (See Figure 3). In this work a number of forces and creping elements are defined (See Figure 4). All of these elements are part of a system that defines the creping operating window.

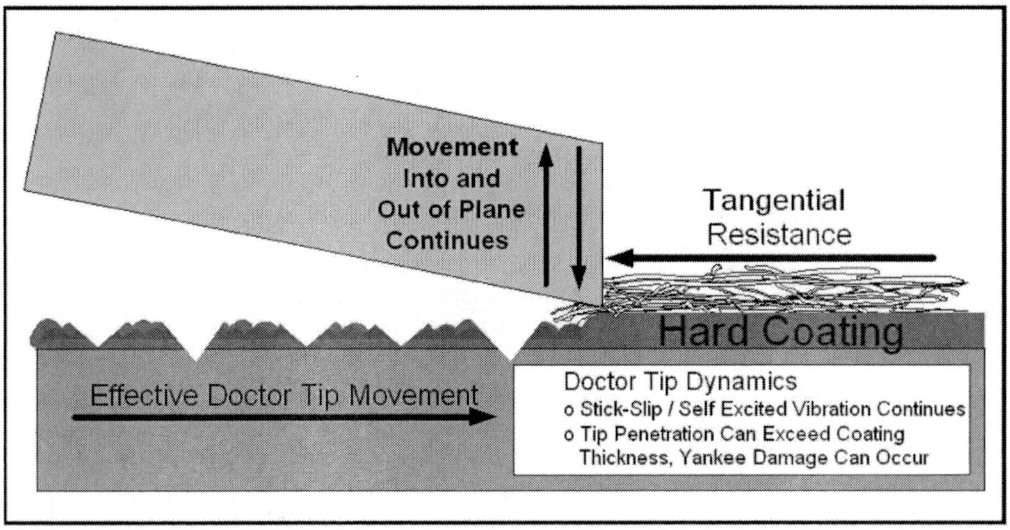

Figure 3 – *A demonstration of out of plane movement. and chatter that can be formed.*

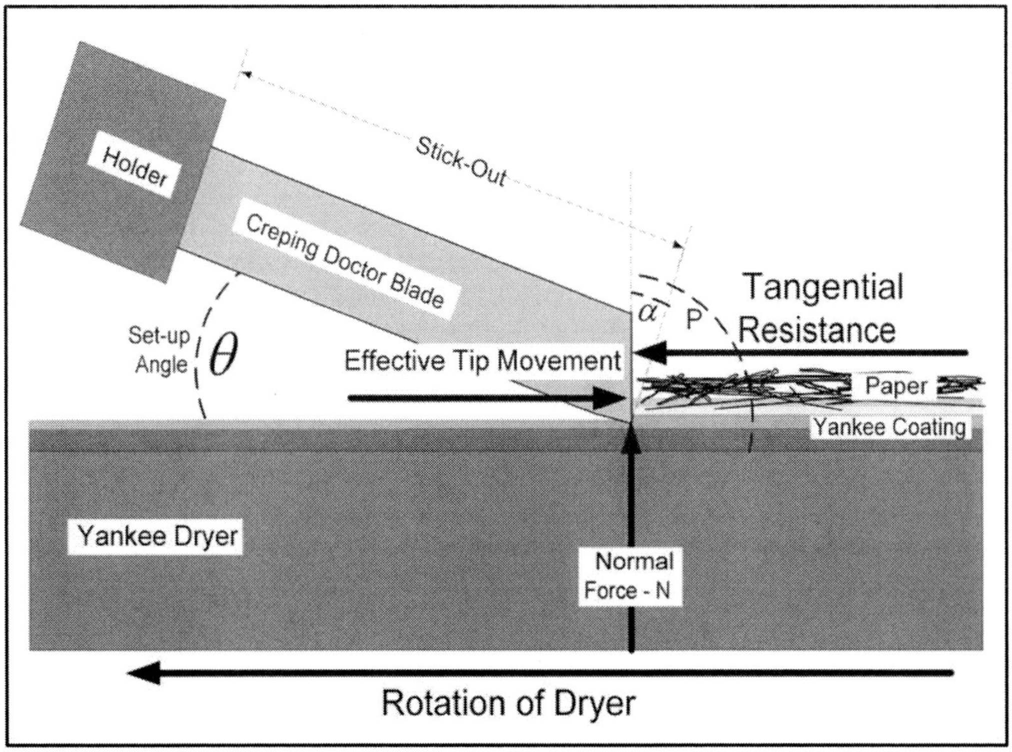

Figure 4 – *The Creping System- Mechanics, Chemistry, Operations.*

TANGENTIAL RESISTANCE

A critical force within the system, as defined in Figure 4, is tangential resistance. A review of numerous chatter events would indicate that this is the main factor that has increased over the last two decades. In an effort to improve productivity or quality, mill operations have driven for more and more adhesion of the sheet to dryer. This desire for more adhesion has resulted in thicker, harder coatings that cause greater tangential resistance at the doctor blade tip. Often so much resistance develops at the tip of the doctor blade that it can no longer stay within the boundaries of the coating and lifts up (out of plane). This is the beginning of the stick slip phenomena, which leads to increased doctor tip vibration and ultimately chatter.

The amount of tangential resistance is directly related to many factors. Figure 5 is a closer view of the area around the tip of a creping doctor blade. A functional expression for tangential resistance is shown in Equation 2.

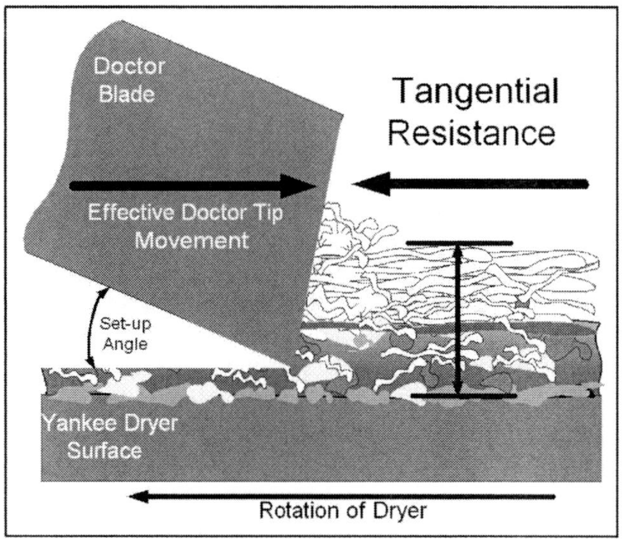

Figure 5 – *Forces and coating characteristics around the tip of the creping doctor blade. The coating/fiber mass (m) and shear modulus (G'), can significantly influence the total tangential resistance.*

Tangential
Resistance $\Rightarrow f[(\mu_k F_N), (m\Delta v), G', Cx, P]$ *(Eq.2)*

Where:

- $-F_N$ = *normal force of the doctor blade tip*
- $-\mu k$ = *coefficient of kinetic friction, dependent on surface characteristics.*
- $-m$ = *mass of coating material, mc, and sheet, ms, on the dryer. The mass associated with the sheet is a function of basis weight and a sheet adhesion factor, A, (0 to 1).*
  $m = (m_c + m_s A)$
- $-v$ = *surface velocity of Yankee moving the coating and sheet into contact with the tip of the creping doctor blade*
- $-G'$ = *elastic behavior of the coating developed on the Yankee at the current velocity.*
- $-C_x$ = *tendency of the coating adhesive to become hard or react with other chemistries in the process.*
- $-\theta$ = *Set up Angle*
- $-P$ = *pocket angle,* $P = (90 - \theta + \alpha)$.

One of the most important items leading to increased tangential resistance is the coating that develops on the Yankee surface. The amount of, m, and the physical characteristics, G', will dramatically influence the resistance that the doctor blade will encounter. It should be remembered that coatings on the Yankee surface are not pure substances, but composites made of many different materials (Sloan 2006, Furman 2007, Boudreau 2009). A conceptual drawing of a Yankee coating is shown in Figure 6.

As would be expected with a composite, the characteristics are usually significantly different than the neat binding material. In the case of the composite that develops on the Yankee surface the characteristics are defined by all of the included materials, the type of binding materials, any modifiers and the operating environment. Some soluble process chemistries react with the adhesive binder while others precipitate and form solid structures within the matrix. Non-soluble fibrous cellulosic materials and other inorganic solids tend to act as mechanical reinforcing structures significantly affecting the shear modulus (G') of the coating. All of these components act to make the coating significantly harder than the adhesive binder. Figure 7 illustrates the impact of different included materials on a commercial Yankee coating adhesive. The higher the shear storage modulus, G', the harder the coating will be.

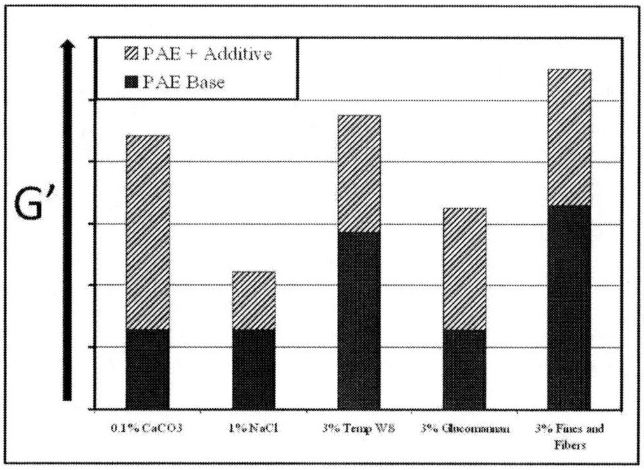

Figure 7 – *Impact of various components commonly found in Yankee coatings on shear storage modulus, G'.*

The environment around the creping doctor blade is very dynamic. There are a number of conditions that can develop that lead to hard, tangentially resistant Yankee coatings. Many of the interactions have been documented in the past (Archer 2008, Grigoriev, 2008) and are summarized in Table 1. As the items in the table demonstrate most tissue making process changes or chemistries result in a thicker, harder more tangentially resistant coating.

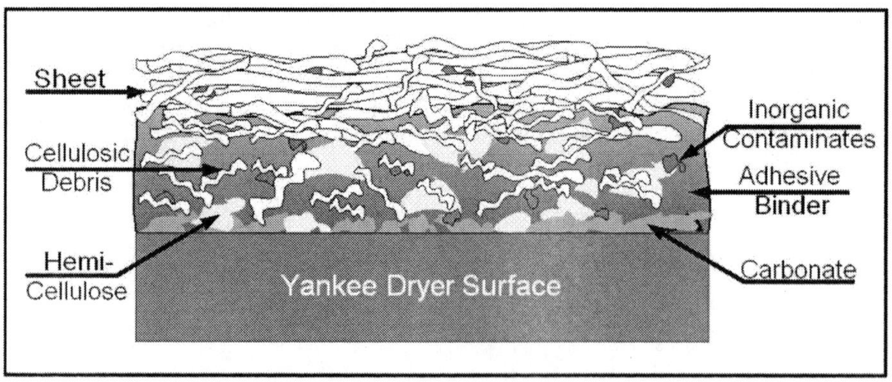

Figure 6 – *Conceptual drawing of a Yankee coating.*

Table 1 – *Typical impacts of process conditions and additive usage on tangential resistance. (Note: Based on qualitative and quantitative observations.)*

Process Condition or Additive	Impact on Composite Coating Shear Modulus – G'	Impact on Binding Adhesive	Coating Hardness	Tangential Resistance	Potential Issues
Lower Moisture Creping	-Increases	-Drier - Lower Adhesion -Stronger internal binding	-Harder	-Higher	-Picks and breaks -Short operational Dr. life -Chatter
Strength Additives	-Increases	-Potential higher MW -Some chemically crosslink with coating adhesive	-Harder	-Higher	-Picks and breaks -Short Dr. life -Chatter -Loss of sheet Control
Increased Yankee Speed	-Increases	-Less time for binder and composite to absorb energy of blade	-Harder	-Higher	-Loss of sheet control -Loss of effective Adhesion
Debonder	-Decreases	-Lower adhesion -Interferes with binder	-Softer	-Lower	-Loss of sheet control -Loss of adhesion
Hemi-cellulose	-Increases	-Lower adhesion -Interact with binder	-Harder	-Higher	-Loss of sheet control -Loss of adhesion
Fines	-Increases	-Lower adhesion -Interact with Binder	-Harder	-Higher	-Loss of sheet control -Loss of adhesion
Calcium Carbonate	-Increases	-Lower adhesion -Interact with binder	-Harder	-Higher	-Loss of sheet control -Loss of adhesion
Spray on Softener	-Decreases	-Lower adhesion -Interferes with binder	-Softer	-Lower	-Loss of sheet control -Streaky coating
Increase Total Add-on	-Increases	-Higher Adhesion	-Harder	-Higher	-Picks and breaks -Short Dr. Life -Chatter
Increased Adhesive to Release Ratio	-Increases	-Higher Adhesion	-Harder	-Higher	-Picks and breaks -Short Dr. Life -Chatter
MAP	-Increases	-Higher Adhesion -Lower G'	-Harder	-Higher	-Picks and breaks -Short operational Dr. life -Chatter

MONITORING THE PROCESS

Controlling the process to avoid chatter almost always involves frequent monitoring. Those that have been successful are involved in both qualitative and quantitative data acquisition and decision making. Qualitative information is usually obtained through time based inspections of the process. Quantitative condition based monitoring of the system yields real time information about changes that are occurring in the tissue making system. Neither of the monitoring technologies will guarantee the elimination of all damage or lost time but they will provide operators with useful information for productive problem solving and process interventions. Tables 2 and 3 below should help operators identify out of norm process excursions that may need to be addressed.

Table 2 – *Qualitative time based observations of Yankee surface.*

Bright Light	Strobe Light	Yankee Edges
Check for thick or thin coating.	Check for appearance of cracked glass → Hard	Check for appearance of edge deposit → Must work to remove. Leads to chatter.
Check for streaky coating	Check for appearance of corrugation → Chatter	Check for indication of wear on doctor blades on Yankee surface.
	Chatter.	

Table 3 – *Quantitative continuous monitoring of tissue making process.*

Moisture Profiles	Yankee Load	Vibration Monitoring
Monitor both CD and MD profiles.	Check for increased Yankee drive load that can indicate excessive tangential resistance.	Identify no chatter baseline vibration characteristics.
Variation can lead to uneven coating characteristics, higher tangential resistance and chatter.	Increased load may be due to either coating characteristics or doctor blade wear.	Check for résistance and excursions in both frequency and amplitude.
		Identify and eliminate source of vibration.

Vibration Monitoring – "Best Available Technology"

Continuous on-line vibration monitoring, to detect the onset of chatter, provides improved runnability and asset protection. Chatter that is developing in the coating is often corrected by adjusting chemical add-on rates and ratios or reconditioning the Yankee surface with a doctor blade change. In cases where coating build-up and chatter in the coating is severe, the dryer surface may require short-term maintenance that involves a shutdown and sanding of the dryer. The sanding removes residual coating and smoothes out the surface. Early detection of the onset of chatter and execution of these corrective steps reduces the risk of damaging the metal dryer surface with chatter marks. Once the metal dryer surface becomes damaged, it can only be repaired through the costly operation of regrinding the surface. Regrinding also accelerates asset depreciation due to a reduction in the dryer wall thickness that can negatively affect the vessel pressure rating.

Utilization of an on-line monitoring system to detect doctor blade vibration is possible using piezoelectric accelerometers strategically mounted near the creping doctor blade. These sensors have a small footprint and are hermetically sealed making them well suited for the dusty moist environment. Direct mounting of the accelerometers on the doctor blade is not practical because of the geometric constraints and limited service lifetime of the blade. Therefore, indirect monitoring is practiced by mounting the sensors on the doctor holder as shown in Figure 8. In this configuration, vibration originating at the doctor blade propagates through the doctor blade holder to the sensor. At least two sensors are recommended in order to observe side to side variation in the vibration levels that may be present across the Yankee width.

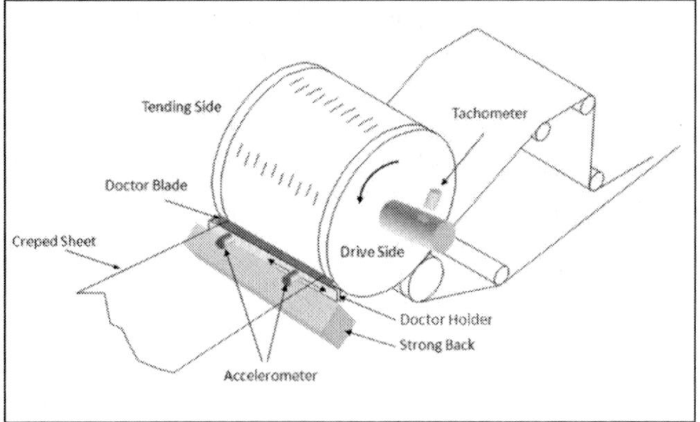

Figure 8 – *Accelerometer mounting option on the crepe blade holder for chatter monitoring.*

Accelerometers are point measurements that integrate vibration propagating through the process to the sensor. Discrimination of vibration frequencies originating from other sources such as fan pumps, screens, felts, bearings and rolls is made by monitoring higher frequency bands associated with doctor tip vibration and potential coating chatter. An example, illustrating frequency and amplitude changes that occur from chatter is shown in Figure 9. The images show a series of vibration frequency spectra collected over several days. The left hand image shows vibration frequency data collected under normal operating conditions when there was no evidence of coating chatter and no product quality issues. Data collected during this period represents a baseline condition. The characteristic features in this set of frequency spectral data show similar band structures and amplitudes. Conversely, the right hand image shows significant changes in amplitude of selected frequency bands as well as the appearance of new high amplitude bands. These new bands appeared when chatter was observed in the coating. In practice, alarm levels are set on selected frequency bands to alert operators when the amplitude exceeds a set level. When an alarm condition arises, the operator can follow a decision tree protocol to determine the appropriate course of corrective action.

Another example illustrating useful information obtained from on-line vibration monitoring is the variation in the doctor blade energy resulting from changes in tangential resistance. In this case, it is convenient to monitor the RMS (Root Mean Square) amplitude of the time waveform data collected from the accelerometer. The RMS reduces the complex time waveform data to a single value useful for trending overall doctor holder vibration energy. The trend plot presented in Figure 10 shows the dynamic behavior of the vibration energy for creping and cleaning doctor blade changes. When a doctor blade is changed the RMS value drops dramatically. This is due to significant removal of coating material from the Yankee surface with a new, sharp tipped, blade and resultant lower tangential resistance. As the blade ages and the coating begins to build-up, the tangential resistance increases causing the RMS to increase. The rate of change for the RMS increase can vary as well as the maximum value reached before a blade change. It has been observed that the rate of change and peak value can be dramatically affected by adjustments to the Yankee coating and the mechanical set-up and operation of the doctor blade and holder.

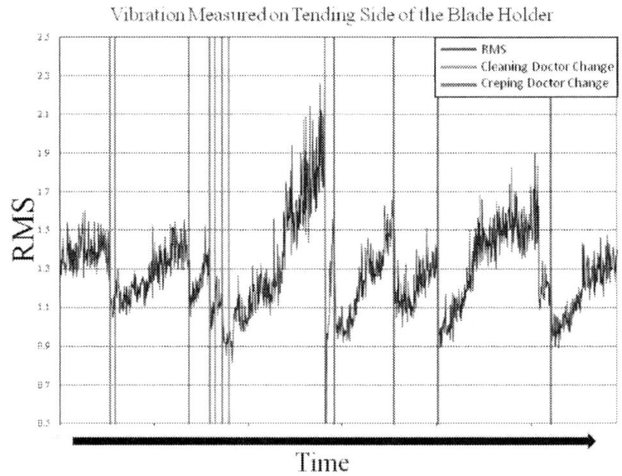

Figure 10 – *Real time RMS run chart showing increased amplitude over the life of a doctor blade. Note the drop in RMS when doctor blades are changed.*

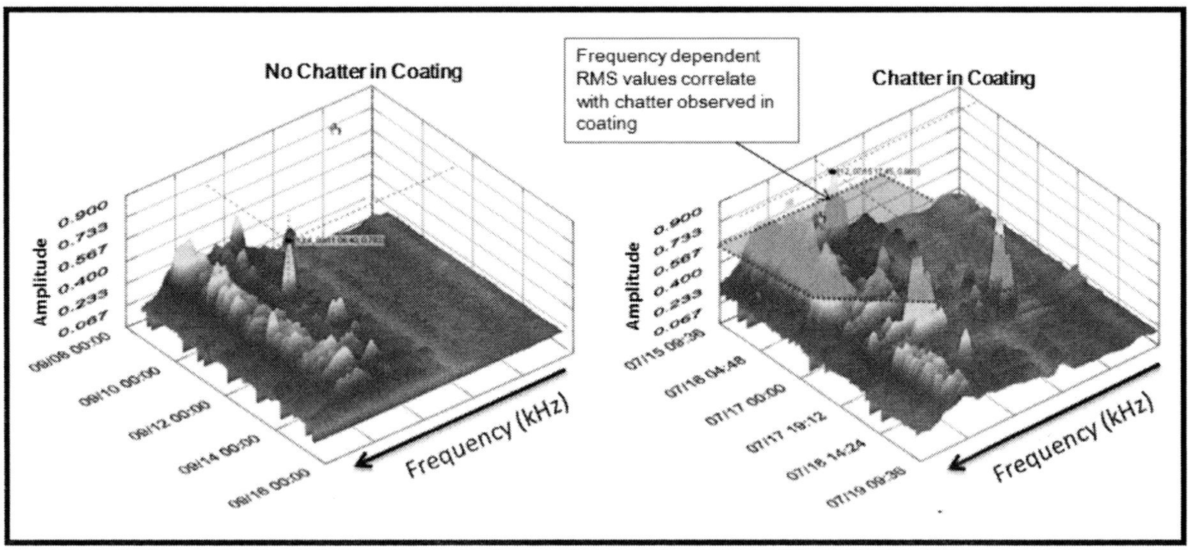

Figure 9 – *Vibration frequency spectra data collected over several days for coating with and without chatter based on visual inspection. When chatter is present the amplitude of the high frequency bands increases as well as the appearance of additional excited bands at other frequencies.*

METHODS TO MANAGE THE PROCESS

If active monitoring of the Yankee surface indicates chatter conditions are developing, process interventions can be made in a systematic, timely way to abate the condition. Since chatter is a system failure, all the critical system elements around the tip of the doctor blade need to be considered (See Figure 4). The most probable condition that will be encountered is a hardening of the coating resulting in an increase in tangential resistance. This resistance must be either mechanically addressed to maintain stable operations or the resistance due to the coating must be reduced.

Mechanical Intervention

Management of composite coatings that are harder and have greater tangential resistance requires a change in doctor blade management strategies. Typically this means the doctor blade tip loading, kg/cm, has to increase. Successfully increasing doctor blade tip kg/cm requires a basic understanding of blade loading mechanics. No matter what system design is utilized to provide force to the doctor blade the effective tip kg/cm is a function of the set-up angle (as shown in Equation 3 and Figure 11).

$$\text{Doctor Blade Tip } kg/cm = F_N \sin\theta \qquad (Eq.3)$$

F_N =*Normal force at tip of blade*

θ = *Set-up Angle*

With an understanding of the mechanics of tip loading it is possible to consider different doctor blade operational strategies. Typically, operators will try to control thicker, harder, more resistant coatings through three distinct approaches.

1. Increase the pressure to the strong back loading cylinders.
2. Decrease the stick-out (free-height) of the doctor.
3. Increasing the set-up angle through adjustment of the strong back/doctor blade holder when the machine is down.

Chemical Interventions

Modification of harder resistant coatings has been successful in the past. Utilization of surfactant type releases or modifiers have provided the tissue maker with a tool to help keep the coating soft and manageable (Furman 2004). Traditionally these materials help reduce coating thickness, hardness and resistance to movement of the doctor blade tip. At the same time the coating is being softened a loss of adhesion is often observed. The extent of the reduction is determined by type and amount of adhesive and modifier chemistries applied to the Yankee dryer.

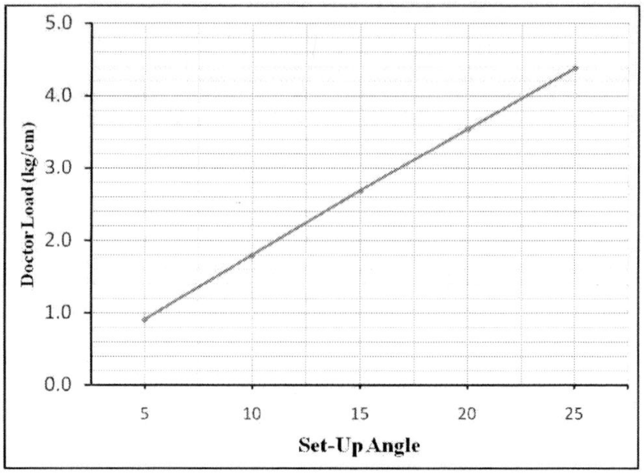

Figure 11 – *Demonstrates the change in effective pli as set-up angle changes. (Note: This Figure is based on a system with cylinder loading of 3 bar, 15 degree set-up angle that resulted in a 2.7 kg/cm tip loading.)*

Newer systems are currently being commercialized that characteristically deliver very high sheet to Yankee adhesion (Grigoriov 2009). With these systems it is possible to deliver the required adhesion at extremely low add-ons rates. Yankee protection and reduced blade wear is assured by the uniform coating composite that develops on the surface of the dryer. Through proper application of both adhesive and release it is possible to create a coating composite that is manageable with existing doctor blade loadings and geometries while still delivering improved productivity and quality.

Systematic Approach to Chatter Abatement

A systematic chatter abatement program utilizing many of the approaches defined above is presented in Figures 12 and 13. The process first requires a determination as to the origin of the chatter – mechanical driven? Or process driven? To make this determination the tissue maker may need to complete a number of mechanical and process audits or solicit the aid of technical experts from the industry. Once this determination is made work will be defined that leads to a satisfactory abatement of chatter on the tissue machine.

If the source of the chatter is deemed to be mechanical in origin, expedited planning should be completed to repair mechanical elements that have failed. Should the source of the chatter be determined to be process driven a systematic series of activities should be followed to eliminate it. Initially, optimization of the coating package should be completed. This will be fol-

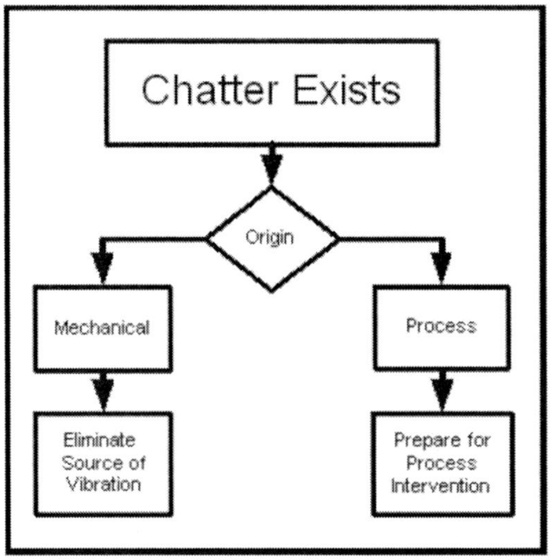

Figure 12 – *Basic definition of origin of tissue machine chatter.*

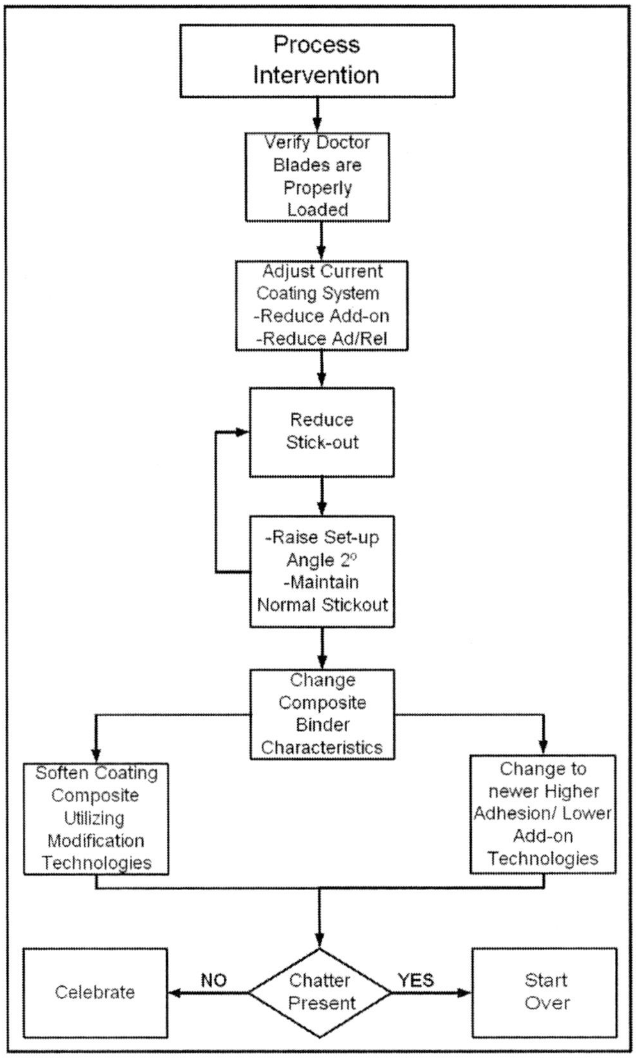

Figure 13 – *Systematic methodology to abate process driven chatter. (Note: If any of the interventions above result in elimination of chatter, further changes to the process should be suspended until a full evaluation can be completed.)*

lowed by focused attention on doctor blade mechanics. A number of loading, stick-out and set-up moves are recommended that will result in increased doctor tip penetration and ability to maintain effective movement through the coating. Chemical modifications will need to be made if mechanical doctor blade adjustments are unsuccessful. If none of the solutions outlined below eliminate the appearance of chatter on the Yankee dryer, operations will need to re-start from the beginning and test basic assumptions as to the origin of the chatter.

CONCLUSIONS

Chatter is a process anomaly that can negatively affect quality production and asset availability. The phenomenon has been tied to excessive vibration at the doctor blade tip. Many occurrences of chatter are directly related to process changes. In general the changes have resulted in composite coatings that are thicker, harder and more tangentially resistant. These more resistant coatings interfere with the ability of the doctor blade tip to move within acceptable boundaries of the Yankee coating. This often leads to out of plane movement, excessive vibration and the onset of chatter. Currently, there are operational and best available technologies that allow time based or condition based monitoring of the process. This information is critical to problem solving, decision making, and planning. Control and elimination of conditions that lead to excessive doctor blade vibration and surface chatter requires systematic process interventions. The approach integrates mechanical, chemical and operational changes to the system. With disciplined application of the suggested changes process driven chatter can usually be reduced or eliminated.

REFERENCES

Alessandrini, A., Pagani, P., "Chatter Marks: Origin, Evolution and Influence of the Creping Doctors", Tissue World, Nice, France, March 2003.

Apple, Chan Lok Shun, www.personal.cityu.edu.hk/~bsapplec,/forced.htm.

Archer, S., Grigoriev, V., Furman, G., Bonday, L., Su, W., Chatter and Soft Tissue Production: Process Driven Mechanisms", Tissue World Americas Conference, Miami, Florida, March 2008.

Boudreau, J., Hollmark, H., "Chemical and Morphological Analysis of Tissue Yankee Coating", Nordic Pulp and Paper Research J., Vol. 24, no. 1, 52 -59, 2009.

Corboy, W., "Vibration Induced Yankee Surface Wear – An Overview of Chatter," Tissue World Conference, Nice France, March 27, 2003.

Furman, G., "Effects of Modifying Agents on Adhesive Film Properties – Findings Toward Improved Yankee Coatings", Tissue World Americas Conference, Miami, Florida, September 2004.

Furman, G., "Yankee Coating Chemistry", Karlstad Tissue Conference, Karlstad Sweden, April 18, 2007.

Grigoriev, V., Furman, G., Frette, G., New High Adhesion Creping Technology for Improved Sheet Properties and Energy Savings", Tissue World, Nice, France, March 2009.

Grigoriev, V., Furman, G., Archer, S., Su, W., Kaley, C., "Wet-end Impacts on the Creping Transformation" PTS Tissue Symposium, Munich, Germany, March 2008.

Sloan, J., "Trouble Shooting Coatings", Tissue Runnability Course, Covington, KY, October 2006.

ACKNOWLEDGEMENTS

The authors would like to thank Winston Su (Nalco) for his contributions to this work.

TISSUE WORLD CONFERENCE
MIAMI, MARCH 20TH, 2012

Chatter Marks forecast and detection through the examination of the harmonic behaviour and evaluation of the friction coefficient

Piero Pagani, Oradoc S.r.l.

Chatter Marks are now one of the most important problems in the production of Tissue.

This type of damage has different features, depending on the origin and the working conditions.

For example, the following chatter marks are generated by the same phenomenon (vibration of the creping blade) in different conditions.

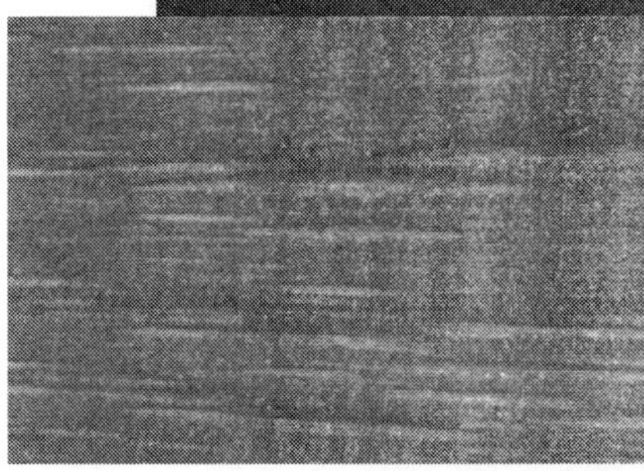

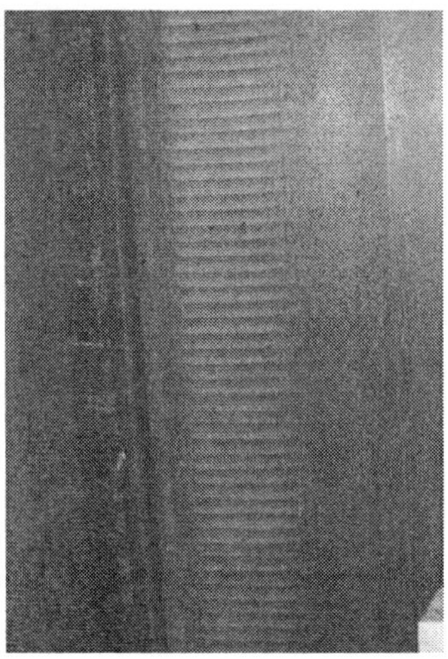

The chatter marks shown here are generated by:

- Fibers build-up at Yankee face sides
- Doctor back vibrations
- Damages coming from the impact between the cutoff blade and the Yankee
- Creping blade and its friction with the coating layer
- Low temperature of the Yankee cylinder
- Non uniform coverage of the spray boom
- Loss of distribution of the coating

Looking into the dynamic condition

The creping blade, if considered without load when it touches the Yankee cylinder, looks like this:

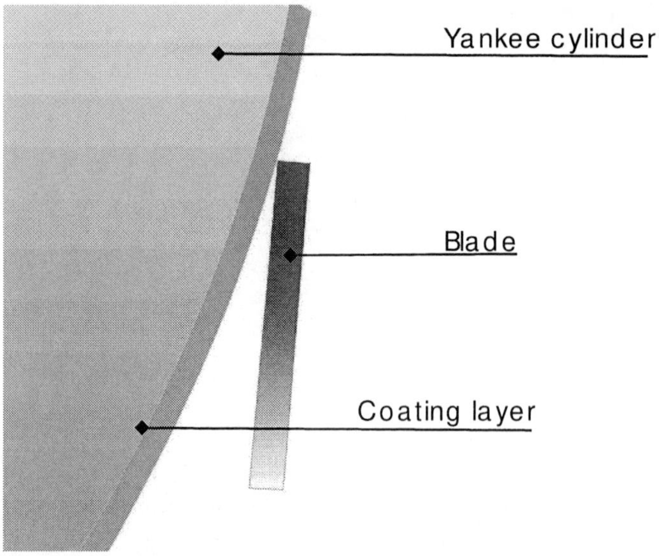

But when the blade is pressed against the Yankee cylinder with a force that is usually called blade load its shape changes.

This force bends the blade as shown here.

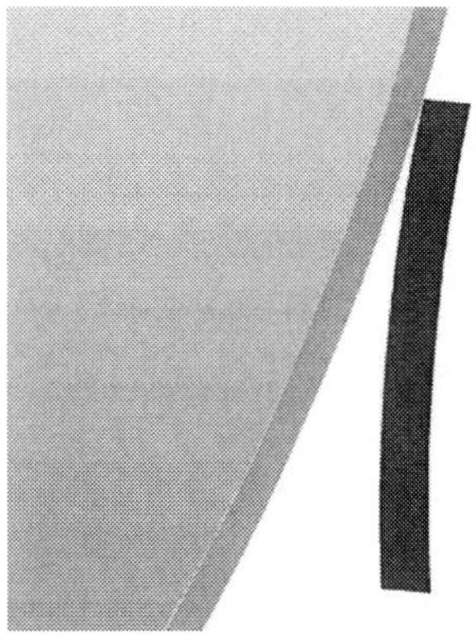

The shape of the blade is an arc of circle, and the geometrical parameters are given by the mechanical characteristics of the blade and its load, i.e.:

- The blade thickness
- The blade Young module
- The stick-out
- The blade load

This is the static condition, usually found during the doctor setup.

In working conditions the real shape of the blade tip is different, due to all the forces that are involved in the creping process.

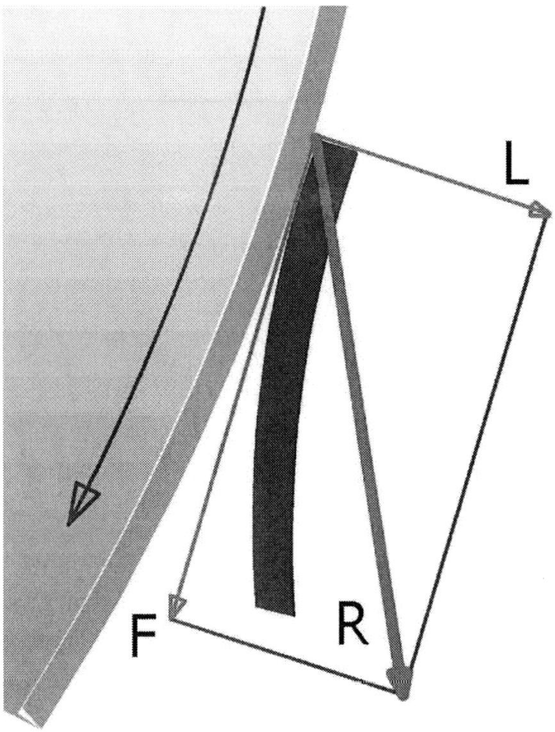

The vectors are:

L: Reaction of the blade load
F: Reaction of the friction
R: Resultant

The blade bent has a shape that differs from the previous one because it is close to the vertex of a parabola.

The vector L is constant, in modulus and direction, and it is given by the force that we apply to the doctor. i.e. the Blade Load.

The vector F has always the same direction (tangent to the Yankee surface) but its modulus varies with the friction coefficient.

This means that also the vector R has a modulus and direction that varies with the friction.

When the machine works, we have to consider that the surface of the coating that is in contact with the blade is normally softer than the blade itself.

This means that the blade interferes with the surface irregularly, and this fact gives a different reaction R during the work.

For instance, if the friction is low (F vector short), the reaction R will be shorter and farther from the Yankee surface, while with a high friction (F vector long) the reaction will be longer and closer to the Yankee surface.

It is clear that the harmonic behaviour of the whole doctor changes quickly during the operation, following the local conditions, i.e. the conditions of the friction coefficient of each point of the Yankee surface.

The figure on the right shows what happens with a soft coating formula.

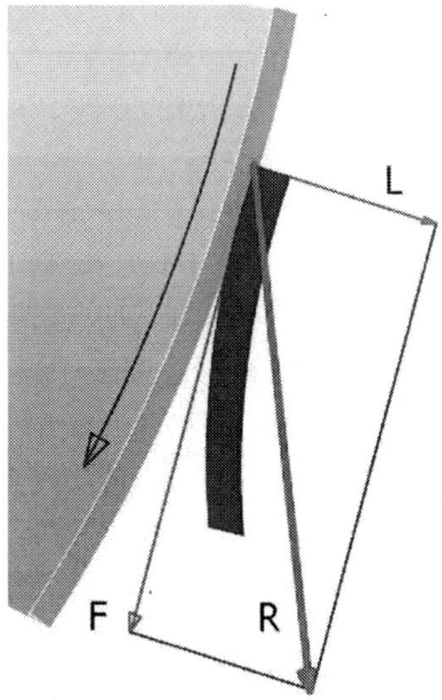

A longer F vector gives a longer R vector, and this shows that the direction of the reaction R is closer to the Yankee surface.

The same applies to a higher blade load, which results in a longer R vector, directed farther from the Yankee surface.

On the contrary, a lower blade load results in an R vector closer to the surface, with different harmonic working conditions.

This explains why, with a different blade load, the harmonic conditions (vibrations) change.

Influence of the friction coefficient on the chatter marks

We can divide the chatter marks into some types, each related to the origin of the vibration.

Machine and pure doctor back vibrations

The chatter marks generated in this case have a pitch of some centimetres, typically from 5 to 10 cm, depending on the machine speed.

The reason is the vibration of the doctor back or a vibration coming from the machine that the doctor back cannot absorb.

The position of this type of chatter marks is usually random.

It is possible to forecast the harmonic behaviour of the doctors if we know:
- The type of vibrations coming from the machine
- The transfer function of the doctor
- The machine working characteristics.

It is possible to have a forecast of the behaviour using the convolution theorem, which lets us give a prediction of the harmonic condition of the doctor back if we know the above elements.

The formula is simple:

$$x(t) = h(t) * f(t)$$

where x(t) is the harmonic response to the working condition h(t) of a system having a transfer function f(t).

Practically, this is what happens:

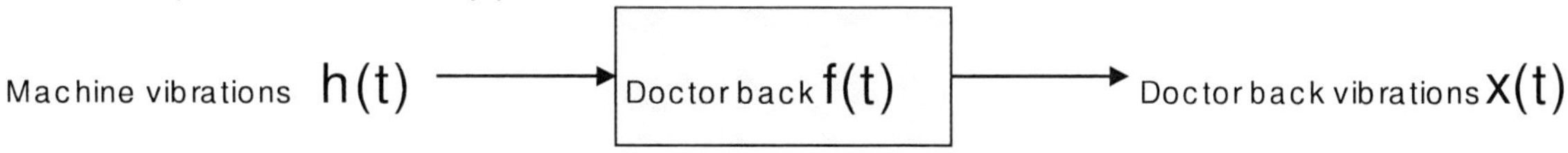

While it is possible to obtain the function h(t) (it is sufficient to record, process and analyze the harmonic behaviour of the machine), the transfer function is not easy to measure, nor calculate.

But when this is possible, it is also possible to have a clear idea of how the system works under the operating conditions of the machine.

On the contrary, it is possible to evaluate the transfer function f(t) if we know the operating condition h(t) and we can measure the harmonic response x(t).

In fact the critical and difficult operations are:

- to evaluate the transfer function f(t).
- to calculate f(t) knowing h(t) and x(t).

This is possible using the modern calculation systems, i.e. a computer equipped with a special dedicated software.

This type of behaviour can also be calculated for the sudden impact between the blade and the cylinder, like in the case of the cutoff doctor.
In this case as h(t) can be used the $\delta(t)$, where $\delta(t)$ is the Dirac impulse function.

This type of vibration has peculiar characteristics and it is not much influenced by the friction.

Vibrations coming from fibres build up and deposits.

These vibrations have a localized source, e. g. at the edges, mainly in Crescent Former machines.

The friction has a high influence on the behaviour, but the only way to reduce this problem is to avoid the fibres and deposits build up or to clean them locally.

Nevertheless, the vibrations coming from the edge can trigger off other vibrations in the remaining part of the doctor back, and this is another source of vibrations for 'tuning'.

Vibrations coming from friction along and around the Yankee surface.

The vibrations coming from this phenomenon have some peculiar characteristics:
- they start from a certain part of the surface.
- they have a very low pitch (some mm).
- they are sensitive to the coating and to the local surface conditions.
- They are highly sensitive to the friction.
- They are normally generated by the vibration of the blade.

Looking at many detailed analysis made over the past 8 years, we found that this type of chatter marks always have the same development:

1 - The friction between blade and coating increases.
2 - The blade starts vibrating, even at a low level.
3 - The coating surface starts to become rough.
4 - The blade vibration level increases and it affects the coating film more deeply.
5 - The damage of the coating increases until the blade 'goes through' the coating thickness and starts to interfere with the Yankee shell surface.

If we act no later than stage 4, we can avoid the irreversible condition of stage 5.

Use of a dedicated instrument

We have seen that chatter marks are always generated by a vibration or an impact. In certain cases the friction is the main responsible for this dangerous phenomenon.

If we are able to detect the vibration level, the type of frequencies related to the vibration and the friction coefficient, we are also able to forecast the risk of chatter marks.
Moreover, if we have enough knowledge of the specific doctor and we can isolate in the information the part of vibration connected to the vibrations of the doctor back (see above), we may increase the reliability of our prediction. This is possible mainly basing on a long experience in doctor backs design.

The operation described above is quite complicated, if made manually, but it is a simple work for a computerized dedicated instrument.

There is also the possibility to use a special sensor, called VFC Sensor (Variable impedance matching Friction Coefficient Sensor) that gives some additional information that allows to forecast the possibility, or the probability, of generating chatter marks.

As an example, here are two different graphs obtained from such an instrument:

A – Vibration diagram of a specific position and sensor

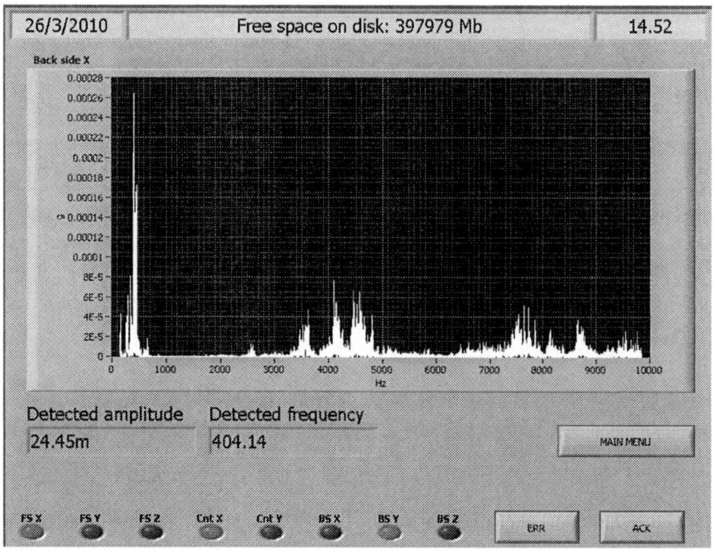

In the above diagram the instrument is able to detect at the drive side:

- The presence of chatter marks for fibres build up.
- The presence of a 'wet spot' midway with the centre (this comes from a comparison with the same type of sensor installed at the centre).
- The presence of chatter marks, still in the coating layer, due to high friction.

The red spots in the bottom part show that there is a critical condition that must be considered.

B – Friction coefficient evaluation

Following the considerations made during the above session, it is possible to have an evaluation of the friction coefficient in the area where we supposed to have a 'wet spot'.

We moved the VFC Sensor in the position where we supposed that the wet spot was present , and this was the diagram of the signal coming from the Yankee cylinder:

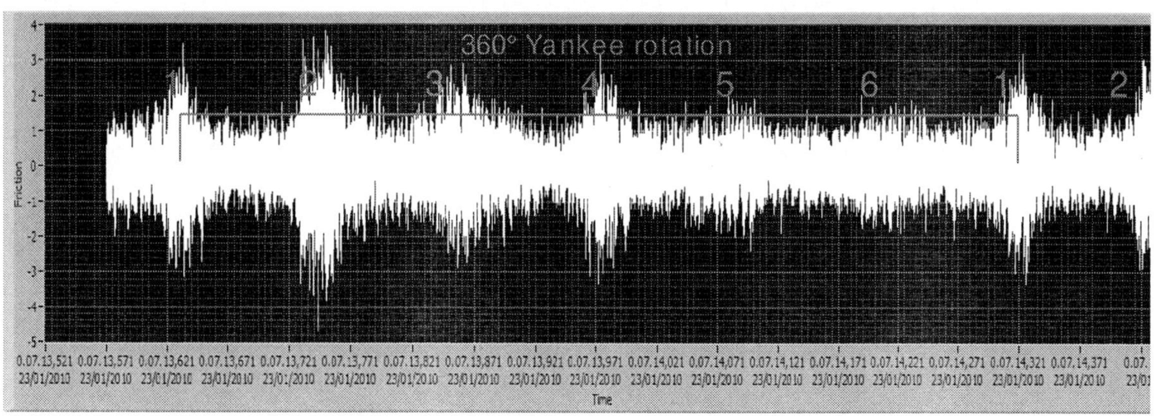

It is possible to see the rotation of the Yankee cylinder just looking at the waveform, on which we have marked the numbers of the straw pipes conventionally. We see that the area corresponding to the straw pipes marked 1 and 2 has a higher friction than, for example, the area of straw pipes 5 and 6.

This means that probably the conditions of the coating (in this case due to a difference of temperature) varies dramatically.

In order to appreciate the difference in the diagram when the working conditions are correct, we can show the same type of plotting recorded on a machine that has a good working condition.

The comparison with the signal of the VFC Sensor in a normal condition is quite clear here below. In this example the amplitude of the friction is good (about 1/100 of the above value) and the difference between the different areas is not significant.

In the following image we have left the Y-scale with its autoscale feature, to be able to appreciate the difference in amplitude between the different points.

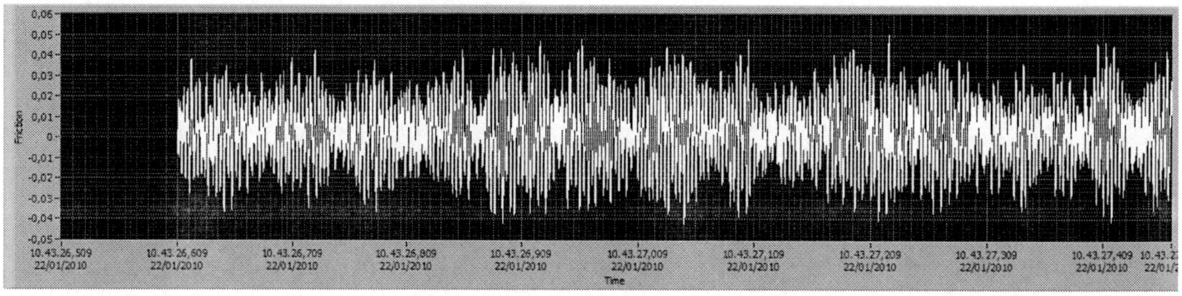

To appreciate the different conditions better, we have plotted the same waveform using the same Y-scale as the main example, and we can see that the tracing looks like a line.

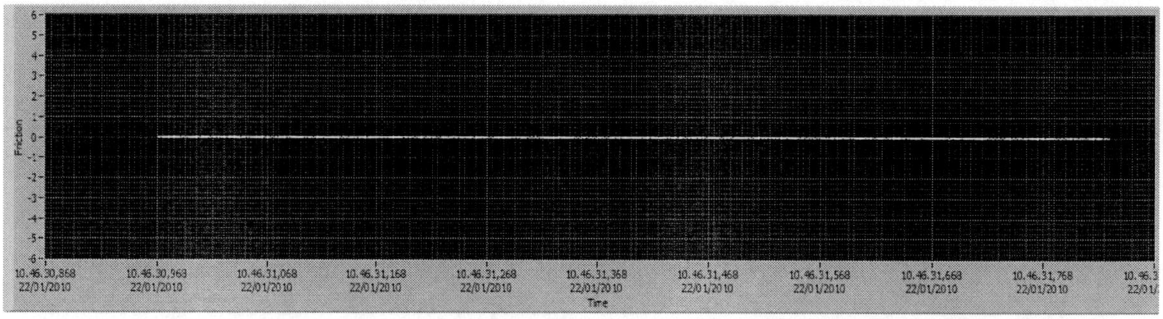

A flat line ensures that the friction is good and also that its value is constant.

We define the value of the friction vector (it is the vector F of the sketch at pages 3 and 4) studying the working conditions of the machine for a sufficient period (some days) and considering the comments of the Papermaker.

When the working conditions are defined as good by the Papermaker, we define the Friction Unit (it is called a Versor F in mathematical field) and from that moment all values of friction will be referred to this unit.

Remote assistance

If the instrument is connected to Internet it is possible to assist the Papermaker to analyze the information coming from the instrument correctly.

This assistance comprises:

- The possibility to read the same diagrams that the Papermaker can see on his screen
- The possibility to download the waveforms of the previous days, up to 1 month
- The analysis of the past events, with the aim to find when a certain phenomenon started
- A deep investigation of the situation, mainly looking at the working conditions, if the readout of the instrument remains not clear

Moreover it is possible to store the interesting waveforms coming from the sensors and to compare them with the same waveforms obtained in different conditions.

Practical use of the instrument

The instrument is used mainly to prevent, or counteract, chatter marks.
But it can also be used to compare the performances of different coating formulas, or to check the working conditions of the spray boom.

Another use is the continuous monitoring of the results of the steam/condensate system, as the working parameters can be evaluated looking at the friction between the blade and the Yankee surface.

The typical use of the instrument is to detect the dangerous conditions in the coating layer.

In the following page we have two examples, with the images coming from the instrument, in two different cases.
It is shown an example of a machine working correctly and another one of a machine having great troubles with coating generating chatter marks.

It is clear that, setting the instrument up in the right way, the risk of chatter marks results in a different visualization, that gives the possibility to act immediately.

MACHINE WORKING CORRECTLY

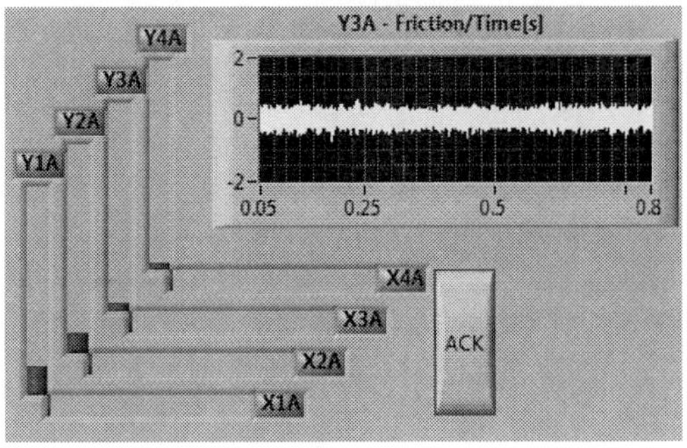

Harmonic behaviour: friction regular, no dangerous vibrations.

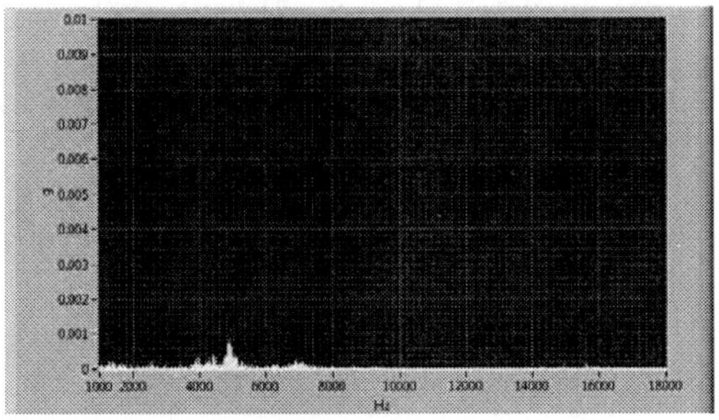

FFT diagram of the doctor: no important vibrations.

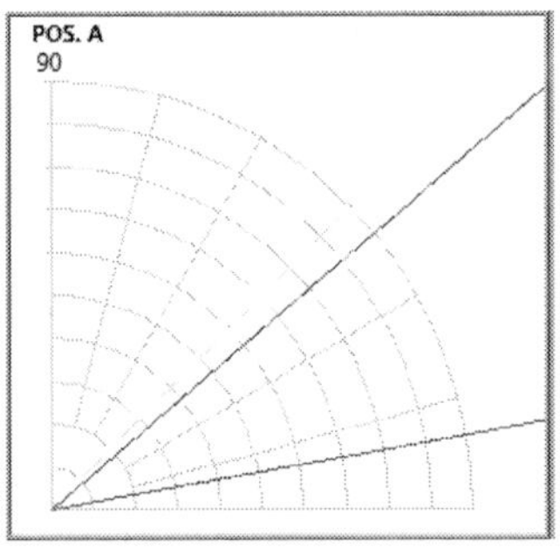

Friction polar diagram: the resultant R is short and directed along the tangent.

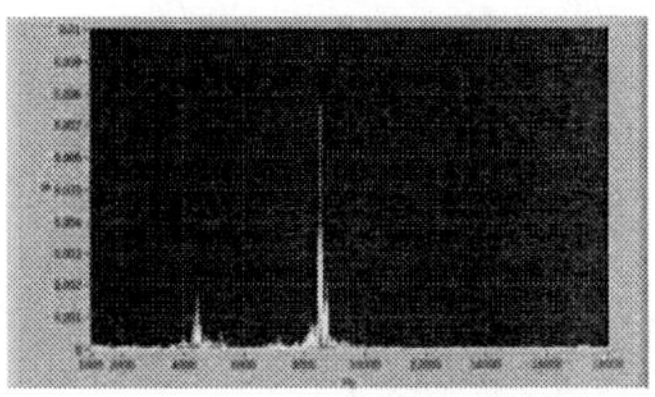

Blade worn: the second harmonic appears.

The harmonic energy is increased. The blade is facing a higher friction due to the worn bevel.

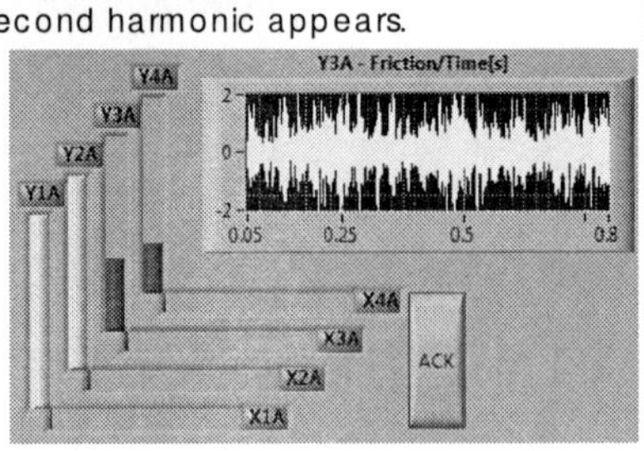

MACHINE WITH CHATTER MARKS

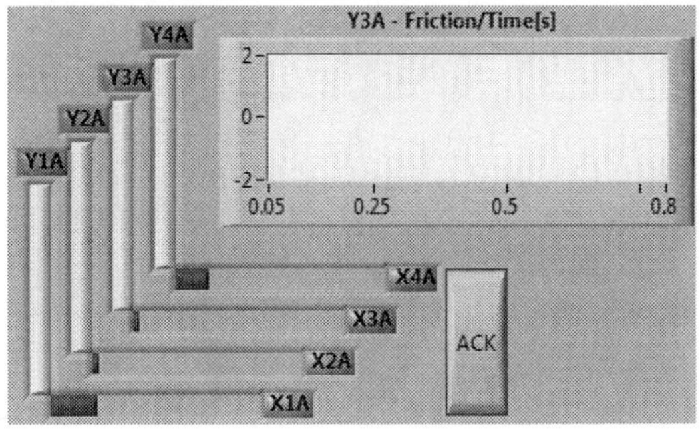

Harmonic behaviour: friction very high, alarms on all positions. This is a clear dangerous condition.

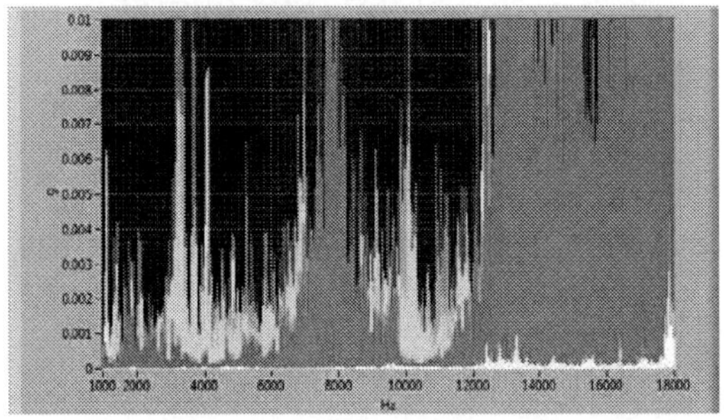

FFT diagram in the same scale as the example at left. The vibrations are extremely high.

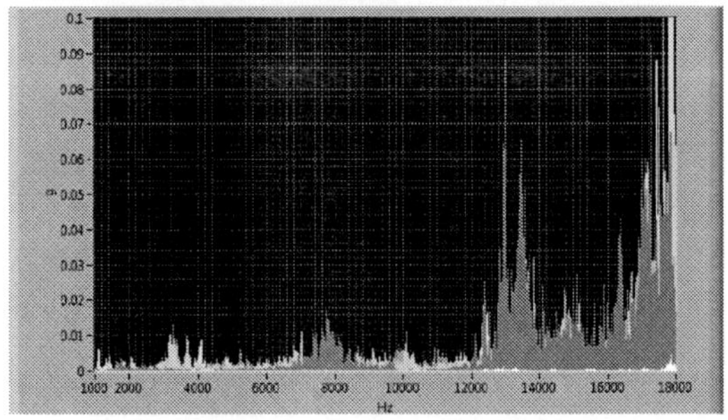

To read the real values we have to increase the scale of ten times (0.1g instead of 0.01g).
All vibrations modes are present. It is a clear demonstration of the Chladny theory.

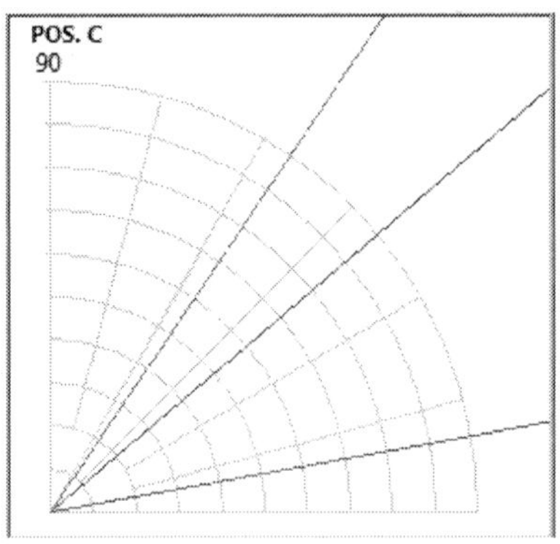

The output of the VFC sensor shows that both R vectors are very long and are located outside the maximum resistance section.

Where to install the instrument

This type of instrument gives its best results if the sensors are installed close to the gravity centre of the doctor back.

If this is not possible, a different location of the sensors can be chosen, provided that the output is corrected by considering the distance of their position from the gravity centre.

The main unit, as well as the output devices (screen, keyboard) are installed in the control room.

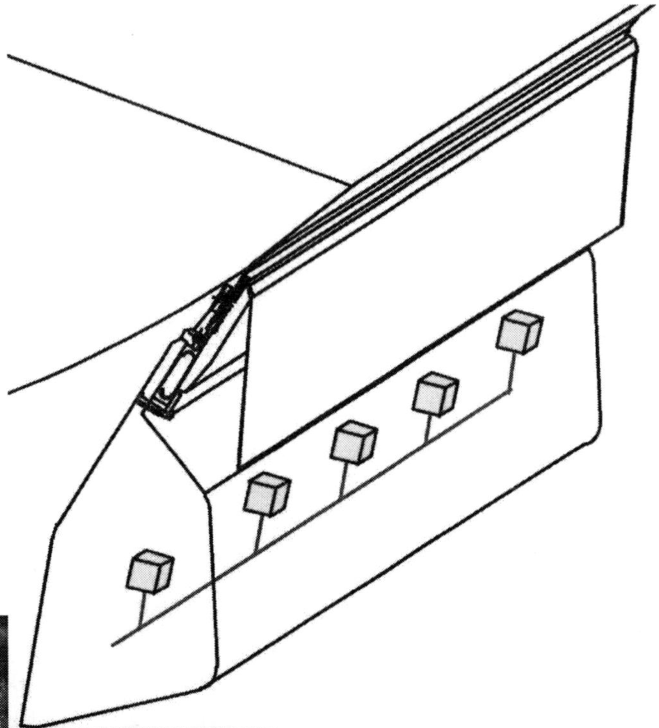

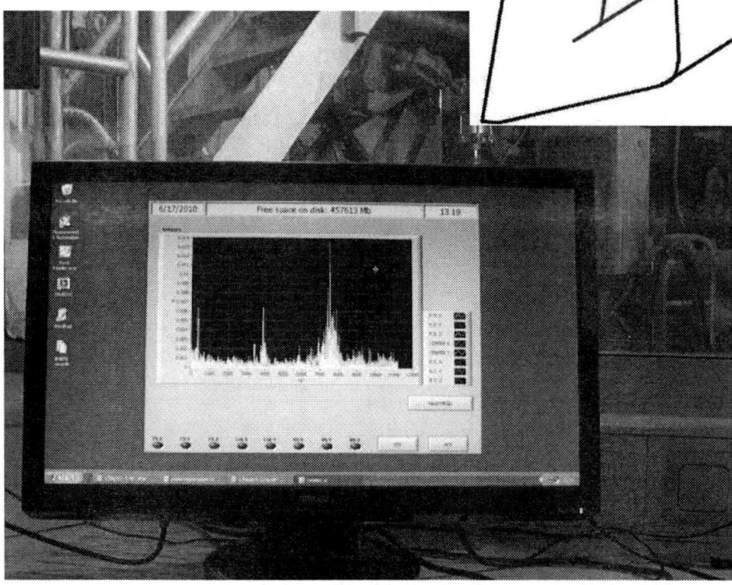

In case of need,
a normal PC can be used as an emergency backup unit.

BTG
RAISING YOUR PRODUCTIVITY

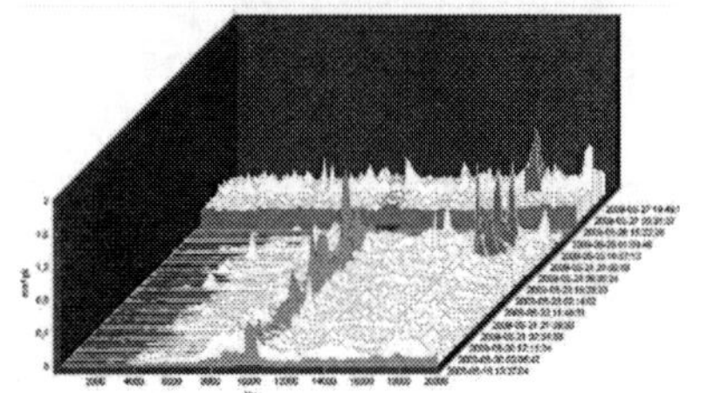

Practical Methodology for Handling Chatter Issues

Tissue World Miami ´12

Florent Bougerolle
Product Manager Creping – DUROBLADE

20th of March 2012

Operating at the edge…

BTG
RAISING YOUR PRODUCTIVITY

- High speed, high-performance machines
- Drive to increase productivity
- Drive to reduce costs

➡ **Machine instability exacerbated**
Critical operation situation

- But what if we get it wrong?
- Cost of chatter
  - Up to 400,000€ / year / machine

2

...with limited understanding

BTG RAISING YOUR PRODUCTIVITY

- Chatter is a complex phenomenon
- Interaction of many factors
- Analysis mostly based on engineered empirical approach

- However, we see better insight from other workers:

- Archer
- Furman
- Pagani
- Lennman

 3

Fundamental principles

BTG RAISING YOUR PRODUCTIVITY

- **Energy is transferred first from the Yankee to the blade (excitation)**
    - Involved: the coating and the paper

- Blade slipping resistance accumulates stress to be released

- This can manifest as **chatter**

- Danger is amplitude and energy

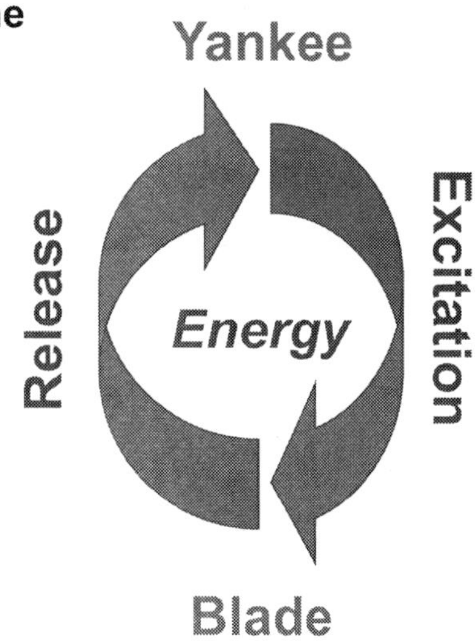

 4

Visible parts of the chatter iceberg

RAISING YOUR PRODUCTIVITY

- **The symptoms we see include**
  - CD marks on coating or Yankee surface
  - Broad frequency range of 0-25kHz
  - Unusual sound around Yankee
  - Chatter visible with strobe light
  - Vibration sensors

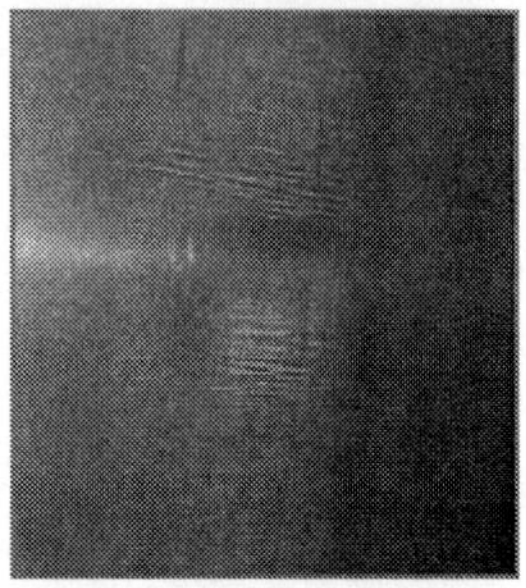

- We can focus on three main sources
  - **Mechanical blade environment**
  - **Edge contamination**
  - **Blade/coating interaction**

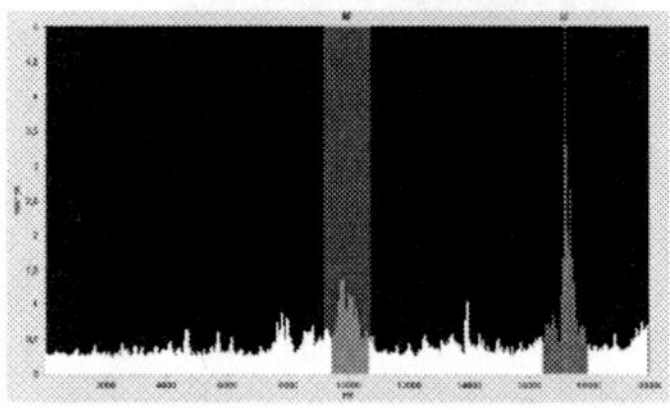

5

Track down any mechanical problems

RAISING YOUR PRODUCTIVITY

- Beam + BH features: settings & conditions usually critical

- **Uneven load all over Yankee surface**
  - Uneven specific pressure
  - Quality of blade load profile system
  - Unloaded edges
  - Cracks, tilt fingers, twist BH

- Weakness which allows freedom of movement

6

Blade holder focus

BTG RAISING YOUR PRODUCTIVITY

KEY THOUGHTS

- **Frequent BH cleaning & maintenance**
- **Blade holder angle around 18°**
- **Best practices regarding set-up and profiling**

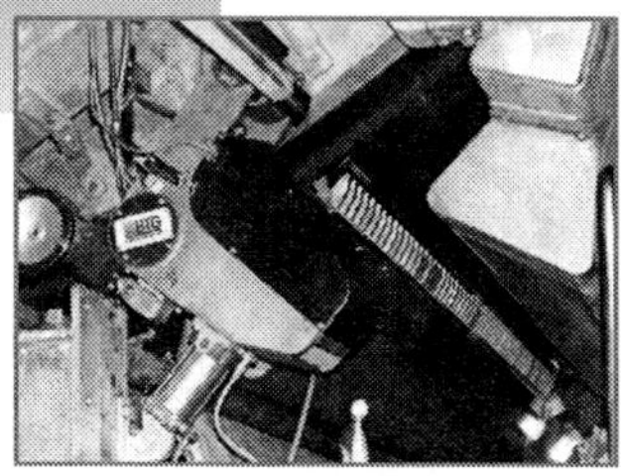

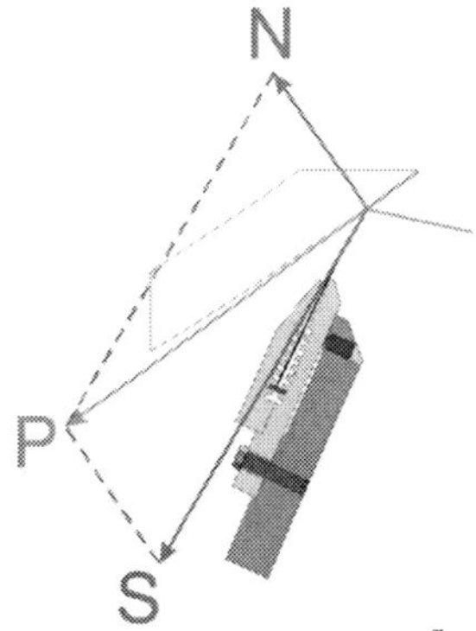

Edge contamination issues

BTG RAISING YOUR PRODUCTIVITY

- Fines, free fibers outside paper trim will deposit on over-heated coating
- Contamination on Yankee surface & between blade and Yankee
- **Problem**
  - Abrasive blade-life killer: decreases local specific pressure
  - Pushes blade away from Yankee surface: decreases local specific pressure
  - Induces stress on blade, upsets slips, stick and friction mechanics

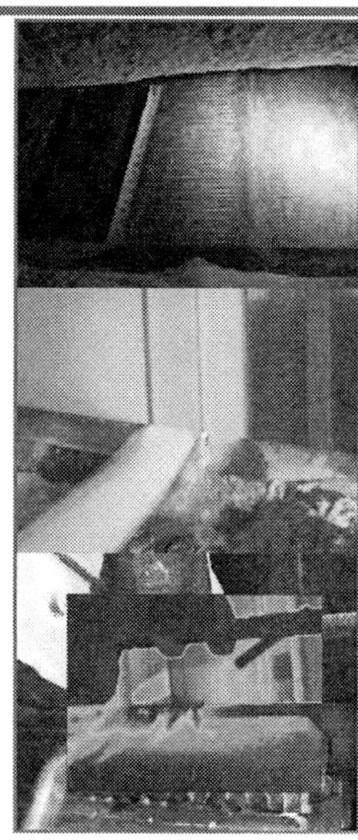

Maintain Yankee clean edges

BTG
RAISING YOUR PRODUCTIVITY

KEY THOUGHTS

- **Treat contamination from Wet End & Process**
- **Fine-tune spray boom configuration and critical settings**
- **Separated edge spray system**

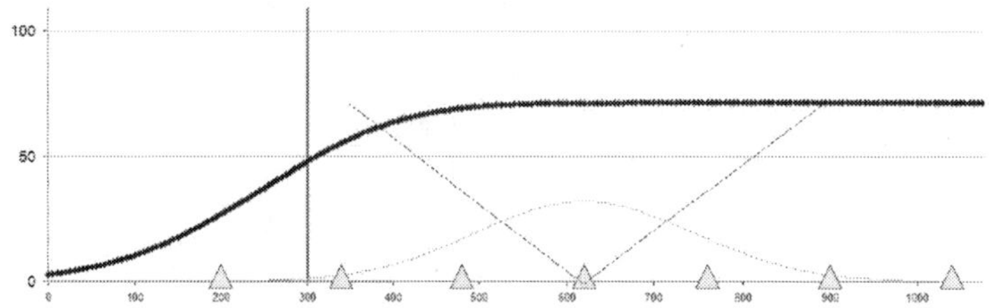

Penetration of blade in coating

BTG
RAISING YOUR PRODUCTIVITY

- **Vibration generated by brake effect which excites blade**
  - Process & paper parameters
  - Coating + blade interaction:

➡ **Properties of coating on Yankee**
Creping doctor configuration and settings

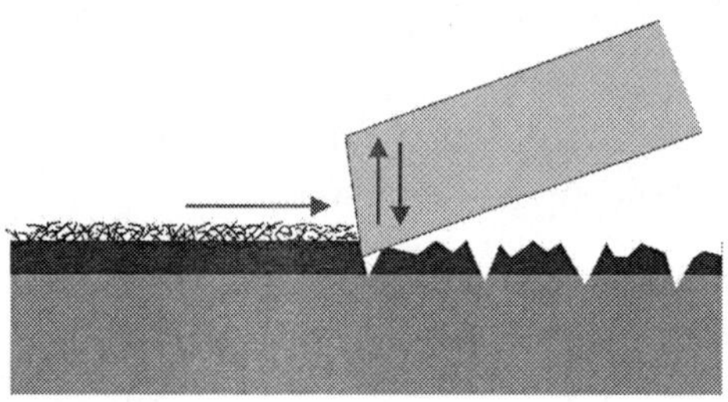

Keep coating away from disturbances

BTG RAISING YOUR PRODUCTIVITY

- **The coating is a critical item for tissue quality and process performance**
- **Change in behavior is bad for the operation**
  - Sensitiveness of coating on moisture
  - Release ratio
  - Add-on rate
  - Coating pollution (fines, etc…)
  - Grade change - WSA

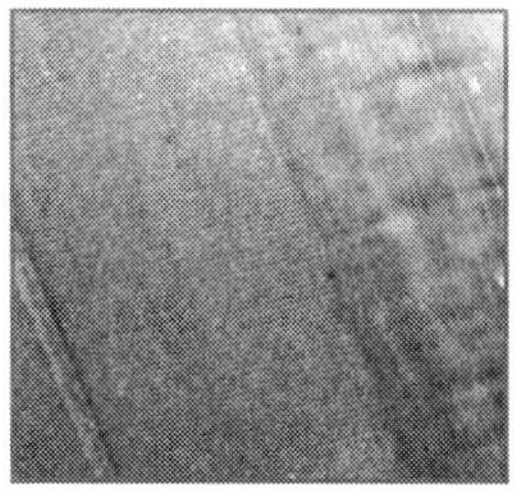

- **Local disturbance**
  - Low spot, moisture streak, higher add-on, etc..
  - Process- or equipment-related

11

Coating management

BTG RAISING YOUR PRODUCTIVITY

KEY THOUGHTS

- **Define the best coating strategy for quality and production performance**
- **Work on process to maintain stable behavior independent of changes**

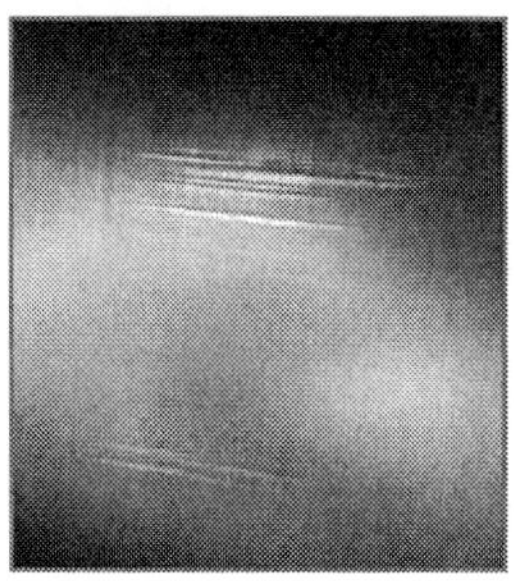

12

Engineered high-performance blades

BTG RAISING YOUR PRODUCTIVITY

- **An efficient engineered blade tip will:**
  - Reduce coating shavings and loss of lubrication
  - Refresh coating to avoid building up a too thick, too hard coating layer
  - Maintain reduced level of coating contamination

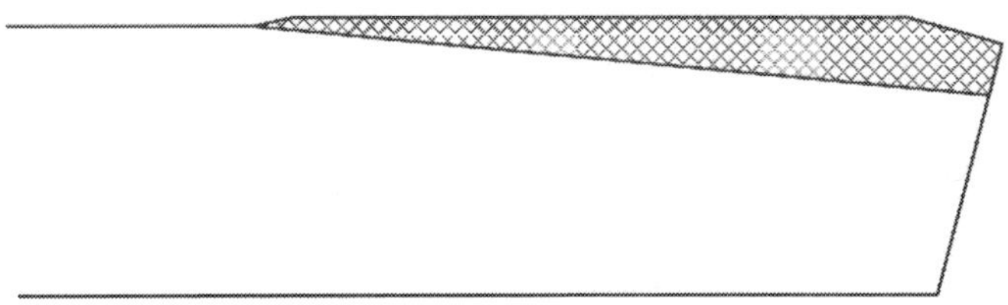

13

Slower wear, slower vibration development

BTG RAISING YOUR PRODUCTIVITY

- **Wear development increases vibration**
  - Diminishes coating control
  - Sliding wear progressively decreases the specific pressure
  - Impact wear closes pocket and increases the force pushing the blade away from the yankee

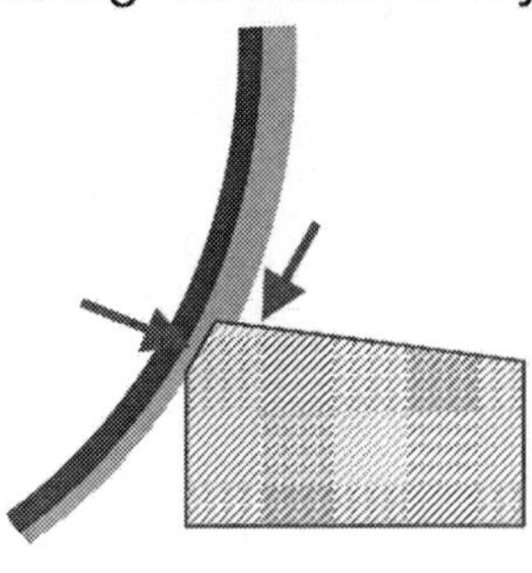

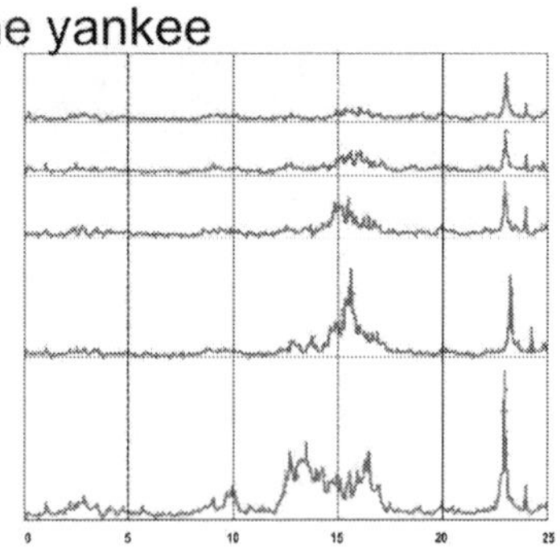

14

High-performance blade

KEY THOUGHTS

- Work with your blade supplier to enhance blade tip design
- Define with him the best material to reduce Yankee friction and maintain the blade tip efficiency

15

Additional considerations

KEY THOUGHTS

- Upgrade Yankee surface protection
- Inspect Yankee surface frequently
- As soon as you see deep chatter in coating, or light in Yankee surface, polish the surface
- Not to forget !
  - Cut-off blade
  - Cleaning blade

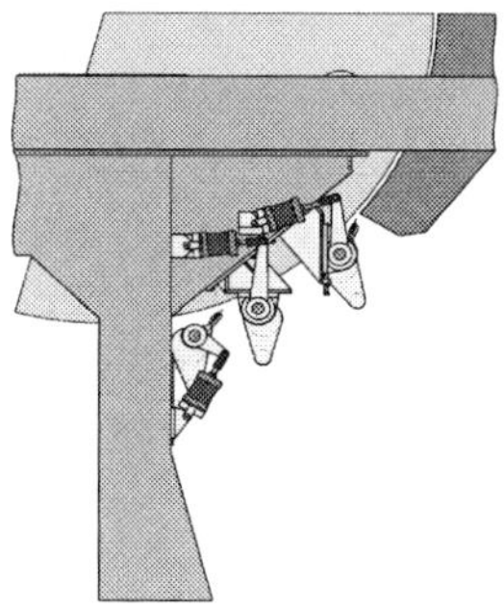

16

Chatter solution does not come from <u>one</u> factor

BTG
RAISING YOUR PRODUCTIVITY

Unaligned process

- CL/CR blade
- Blade material
- Blade geometry
- Blade holder profiling
- Blade holder alignment
- Yankee edge control
- Yankee coating
- Yankee drying control
- SPR: Yankee crown fit
- Felt dewatering
- Formation
- Wet fines management
- Wet end charge

Unstable process

Poor performance

High chatter risks

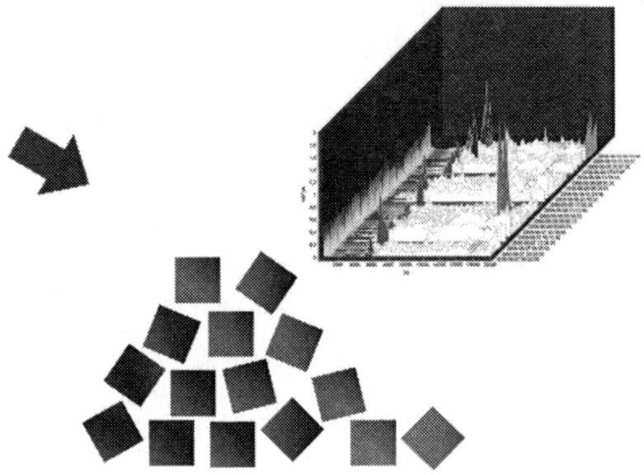

17

One Chatter Solution: Solves <u>all</u> factors

BTG
RAISING YOUR PRODUCTIVITY

Aligned process

- CL/CR blade
- Blade material
- Blade geometry
- Blade holder profiling
- Blade holder alignment
- Yankee edge control
- Yankee coating
- Yankee drying control
- SPR: Yankee crown fit
- Felt dewatering
- Formation
- Wet fines management
- Wet end charge

Process stability

High performance

Low chatter risks

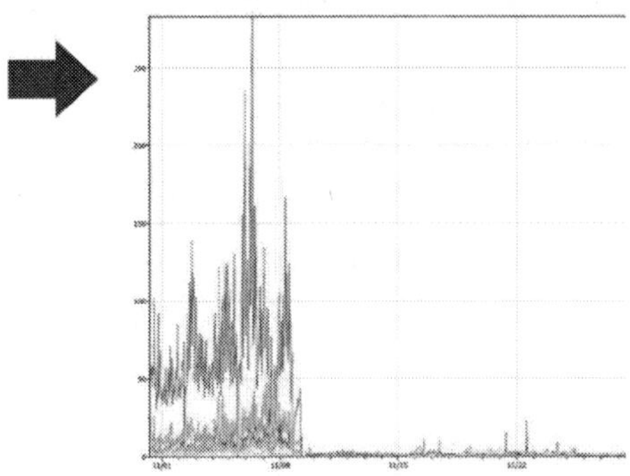

18

BTG
RAISING YOUR PRODUCTIVITY

For defining & implementing your customized chatter solutions, please contact me at:

florent.bougerolle@btg.com

Thank you for your attention…

Improving Yankee Reliability With High Speed Precision Thermal Imaging

PROdry Technology, Inc.

Productivity · Reliability · Optimization

PROdry Technology, Inc.
Vancouver, Washington USA

"Setting the standard in papermaking diagnostics"

1

PROdry

Presented at

TissueWorld 2012
MIAMI, FLORIDA

March 20, 2012
by

John A. Tyburski
President, PROdry Technology

2

PROdry

Outline:

- The Tool

- Blistering
- Chatter
- Buckling

- **Summary**

3

PROdry℠

HIGH SPEED PRECISION THERMAL IMAGING

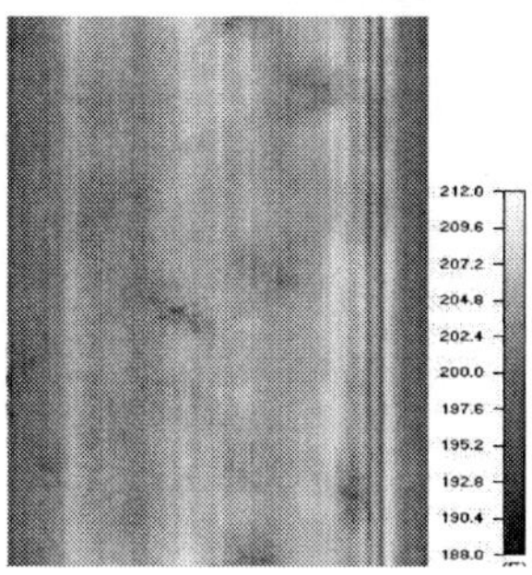

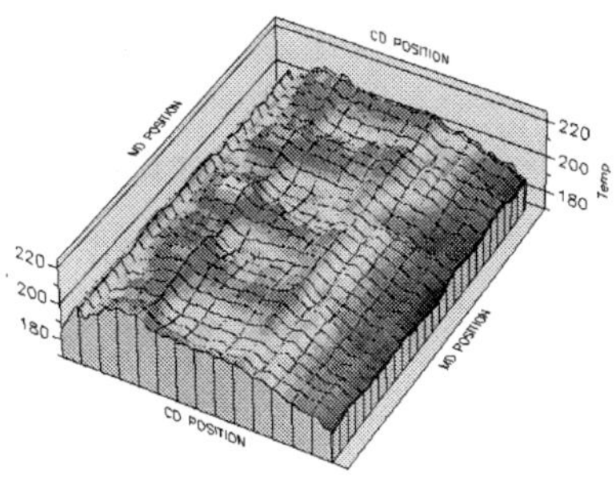

4

PROdry℠

High Speed Thermal Imaging Features

- BOTH FULL-FRAME AND LINE-SCAN IMAGING CAPABILITIES
- HIGH-SPEED CAPTURE OF FIELD OF SURFACE TEMPERATURES (Approx. 2600 HZ)
- ABLE TO "VIEW" ENTIRE DRYER THROUGH A NARROW SLOT VIA LINE SCAN
- ABLE TO LOCATE PROBLEM AREAS PRECISELY ON DRYERS

5

PROdry℠

Thermal Imaging Objectives

- DETERMINE IF INTERNAL PROBLEMS EXIST THAT EFFECT PAPERMAKING
- PINPOINT LOCATION OF PROBLEMS TO BE ADDRESSED DURING OUTAGE
- EVALUATE DRYING PROTOCOL
- IDENTIFY OTHER COMPONENTS IN DRYING SYSTEM CAUSING PROBLEMS
- CREATE ARCHIVAL RECORD TO FACILITATE LATER TROUBLESHOOTING EFFORTS ON THE MACHINE

6

PROdry℠

Example: **Comparison of Full-Frame vs. Line Scan IR**

Full-Frame infrared image presents a video image of the object at about 30 Hz.

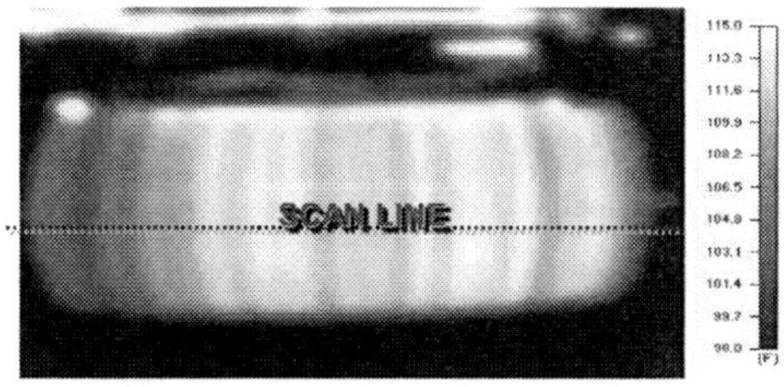

Line-Scan infrared records lines at over 2500hz, "extruding" an image of the rotating object for a pre-determined period of time, revealing hidden details.

7

PROdry℠

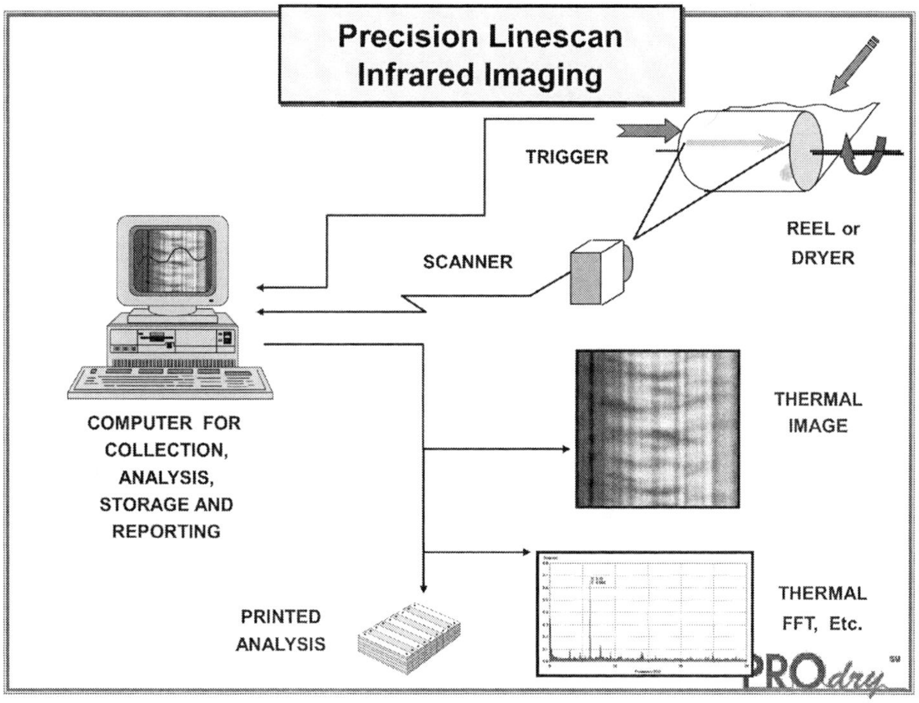

Formatting Issue - Best Available Paper/Graphic

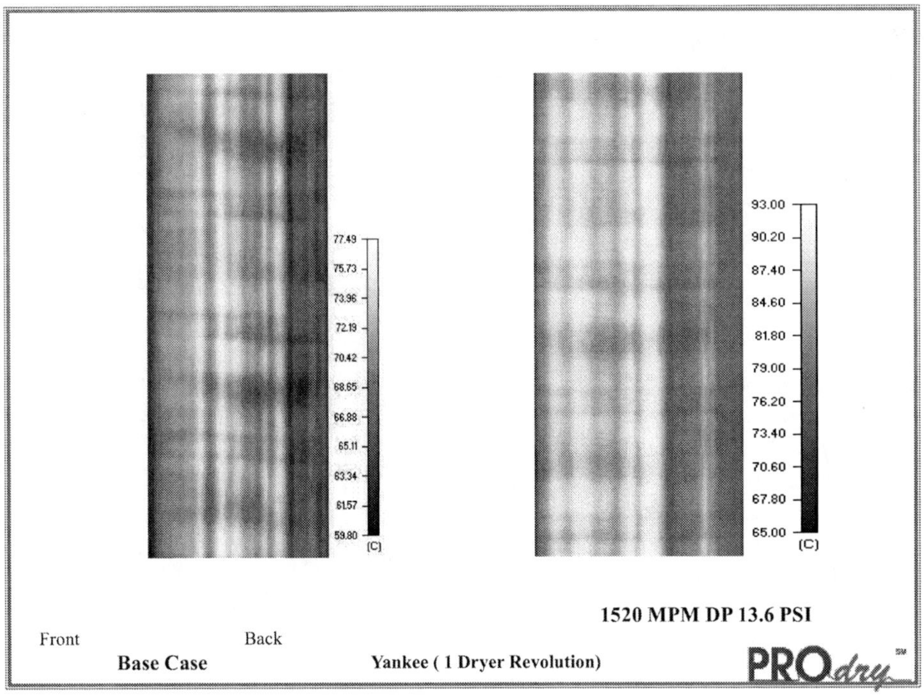

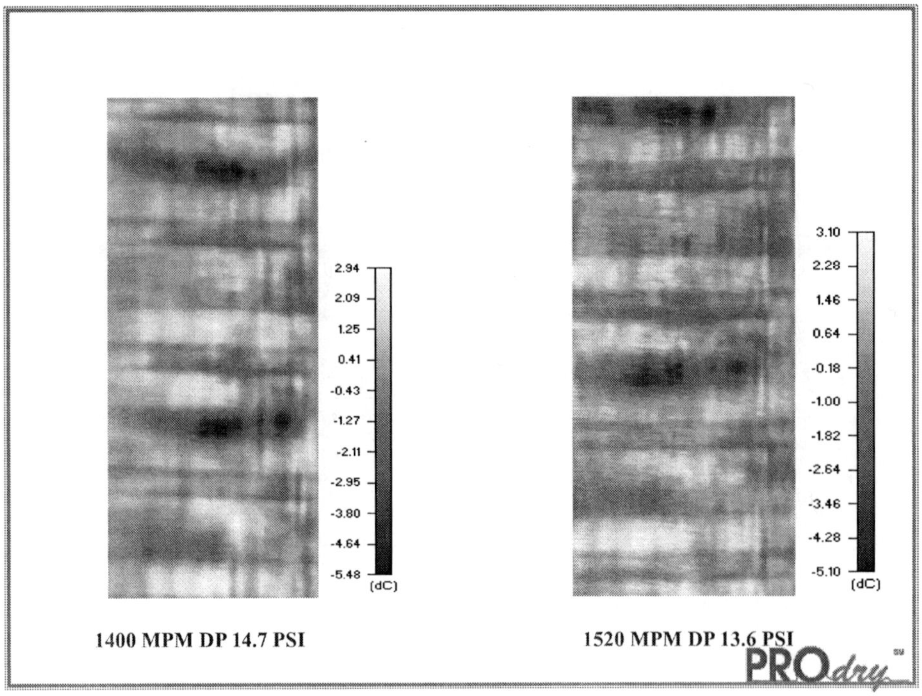

What is "Sheet Blistering"

- "Blistering" is a *common problem* on Yankee dryers
- *Called* "flashing", "stealing", "blistering", etc.
- *Progresses* from minor "holes" in the sheet to catastrophic "blow outs", interrupting operation
- More common on *sheet edges*, but also the interior
- Influenced by *many* factors encompassing both equipment and operational issues
- Often a *transient, local* condition on the Yankee dryer
- Usually *difficult* to predict and control

PRO*dry*℠

A Simple Formula for Sheet Blistering

"Wet" Incoming Sheet

\+ "Hot" or variably "hot" Yankee Dryer

\+ "Closed-Up" Sheet

= Sheet Blistering

PRO*dry*℠

Formatting Issue - Best Available Paper/Graphic

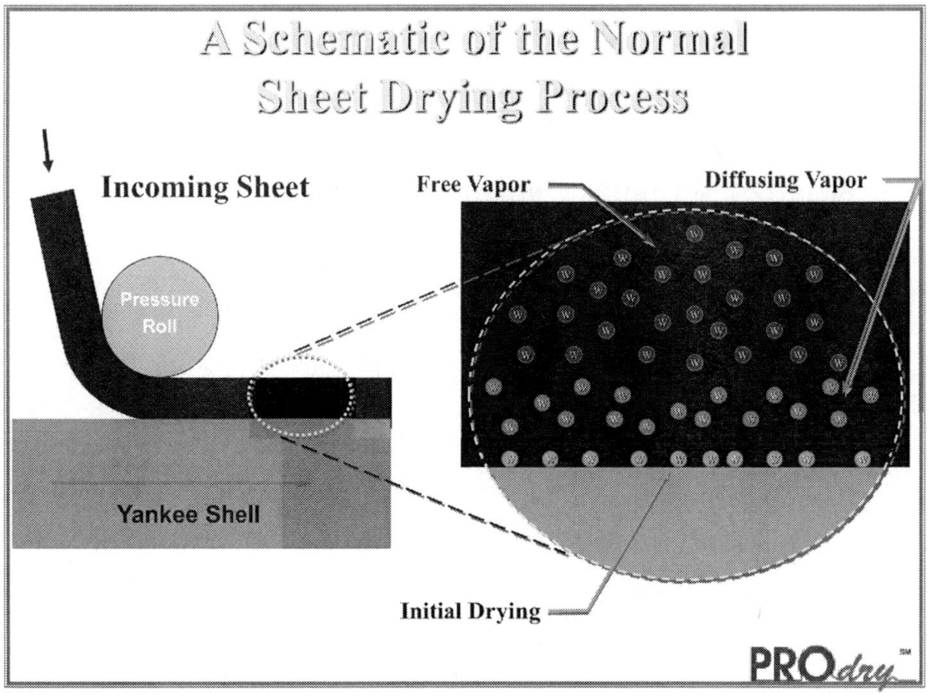

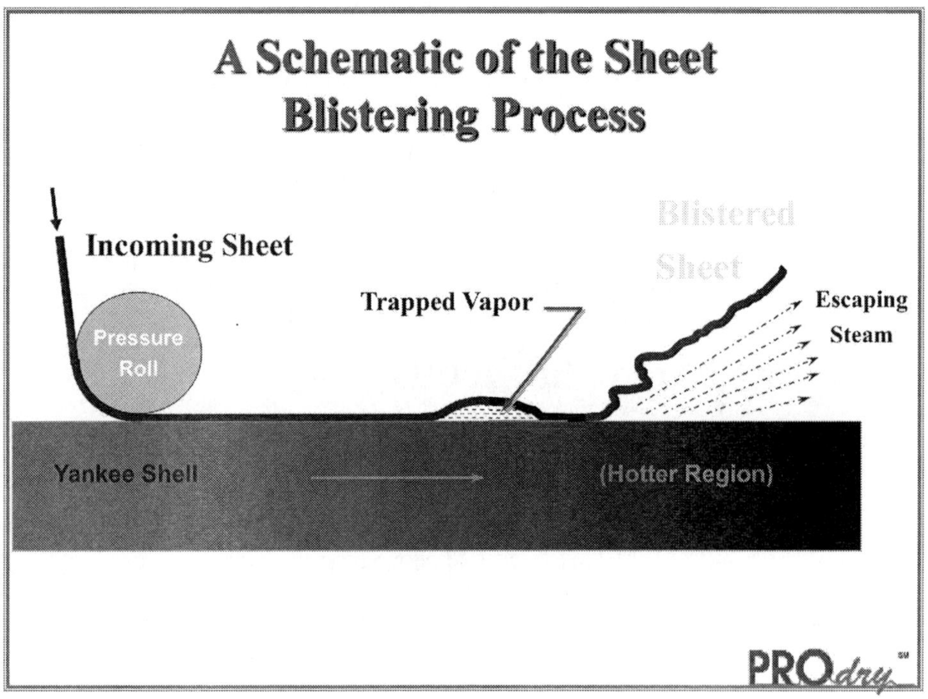

Some Other Observations About Sheet Blistering

- ✓ The disruption of sheet integrity on the Yankee dryer is *caused by an excessive rate of evaporation* and a wet, non-porous sheet that traps the vapor.
- ✓ Blistering is *usually a "point" phenomenon*, and can occur where a hot spot on the Yankee dryer meets a wet area of the sheet.
- ✓ Blistering *may not cause a sheet break* immediately, especially if the initial blister is in the trim or away from the sheet edge.

Some Other Observations About Sheet Blistering

- ✓ Eventually a spot on the Yankee dryer remains hot enough that subsequent sheet applications cannot cool it, and it becomes *progressively hotter.*
- ✓ Blisters then begin to *grow in size* and sheet breaks are inevitable.
- ✓ Productivity and efficiency *always suffer* when sheet blistering occurs on the Yankee dryer.
- ✓ *CAUTION*: Some efforts to control blistering can be *detrimental to Yankee dryer safety.*

Formatting Issue - Best Available Paper/Graphic

Development of a Sheet Blister at the Trim Cut on a Yankee Dryer

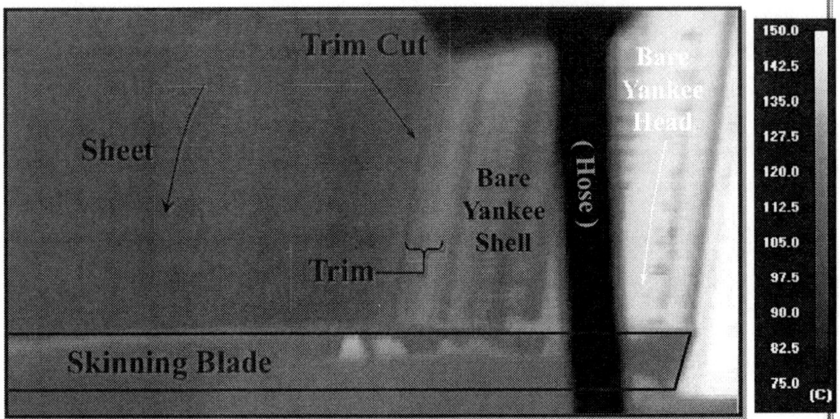

Machine Speed ~ 1440 meters/min.
Basis Weight ~ 20 grams/meter^2

Sheet Blister Development: Standard IR Images

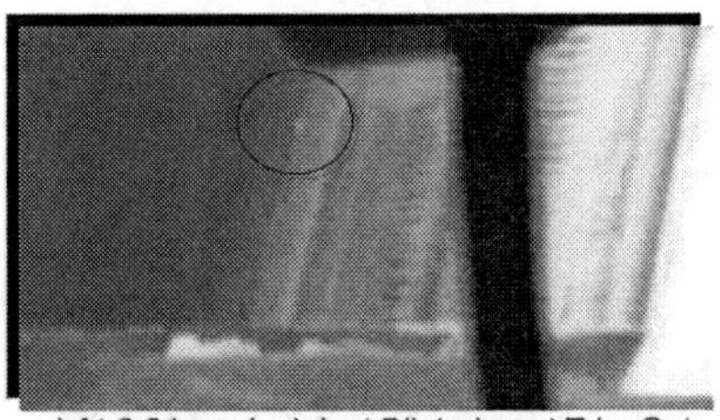

a) At 2:31pm, Incipient Blistering at Trim Cut

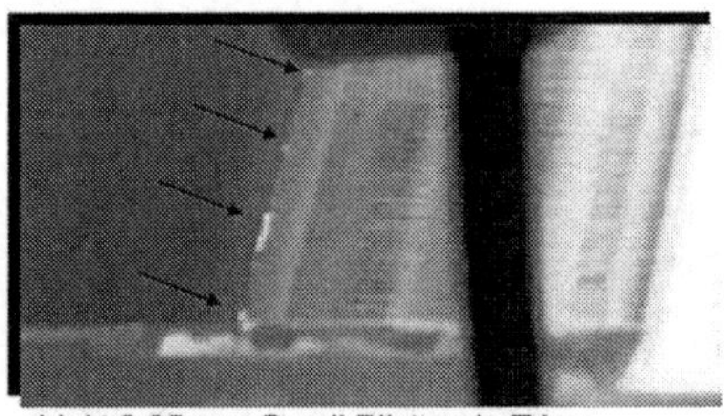

b) At 3:05 pm, Small Blisters in Trim

c) At 3:05 pm, Developing Blister in Sheet

d) At 3:10 pm, Fully Developed Blister in Sheet

Formatting Issue - Best Available Paper/Graphic

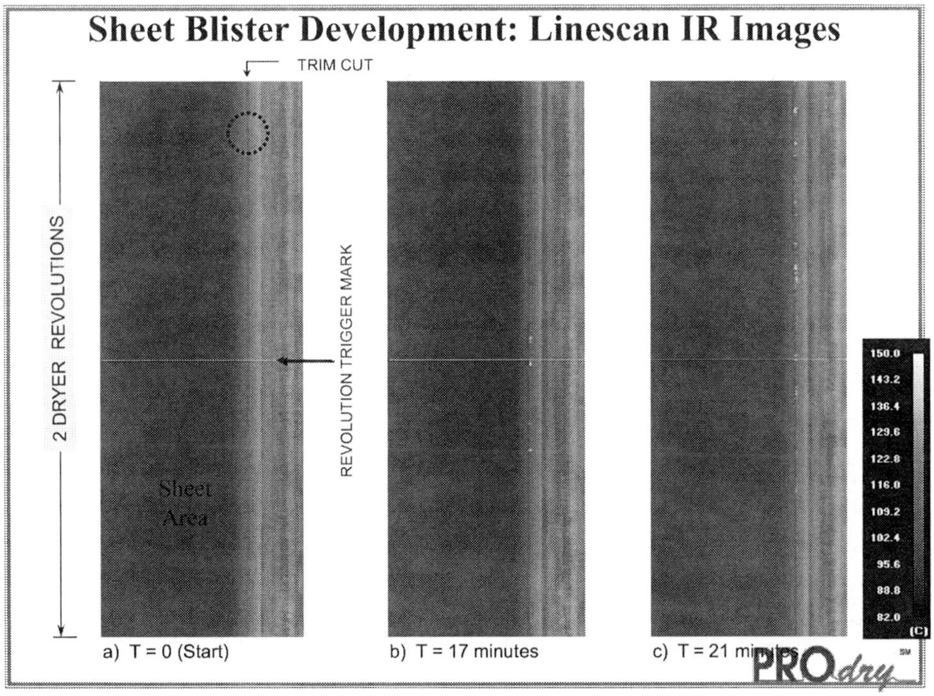

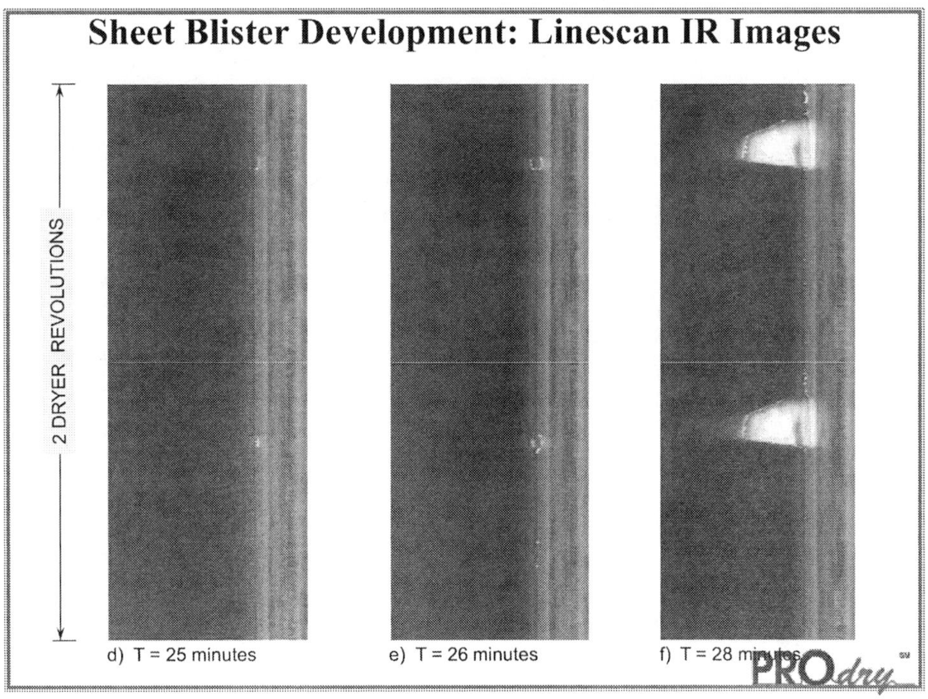

Formatting Issue - Best Available Paper/Graphic

Sheet Blistering: 3-D Infrared Images

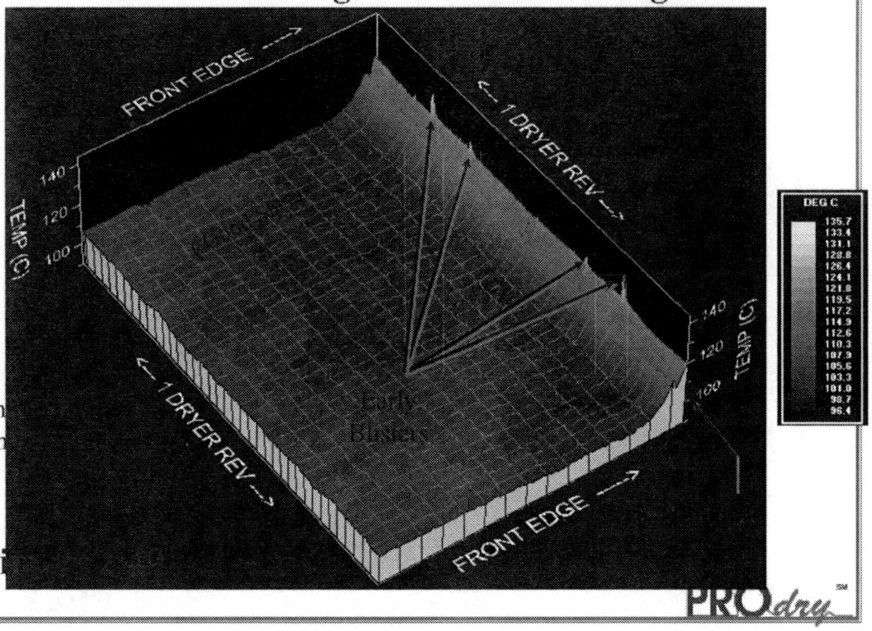

Sheet Blistering: 3-D Infrared Images

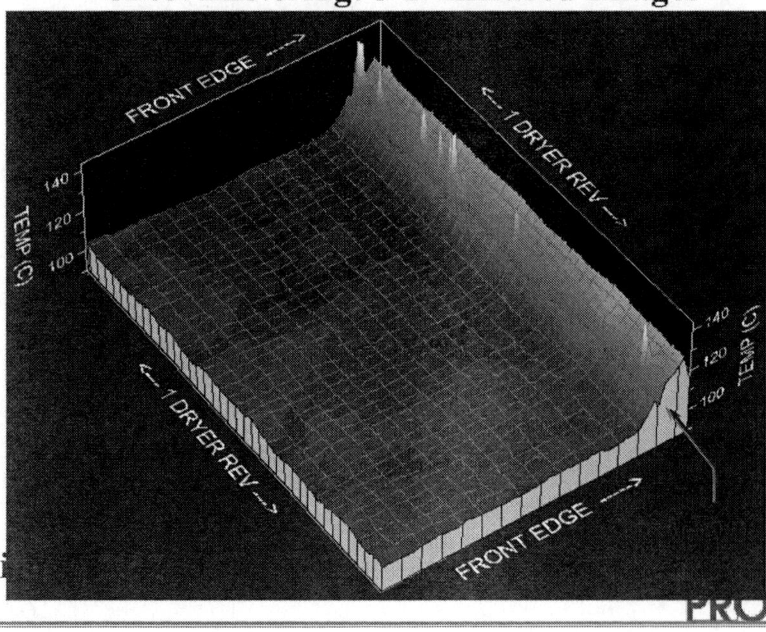

Formatting Issue - Best Available Paper/Graphic

Sheet Blistering: 3-D Infrared Images

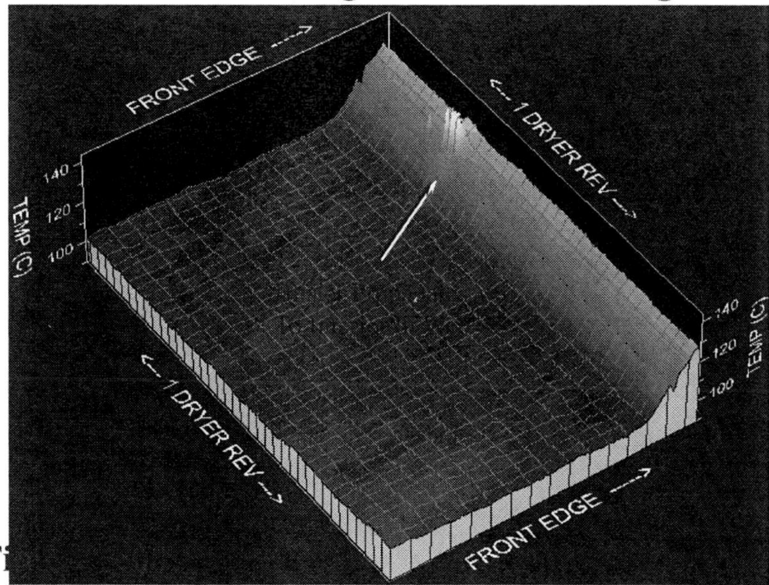

aks
me

Ti

PRO*dry*

Sheet Blistering: 3-D Infrared Images

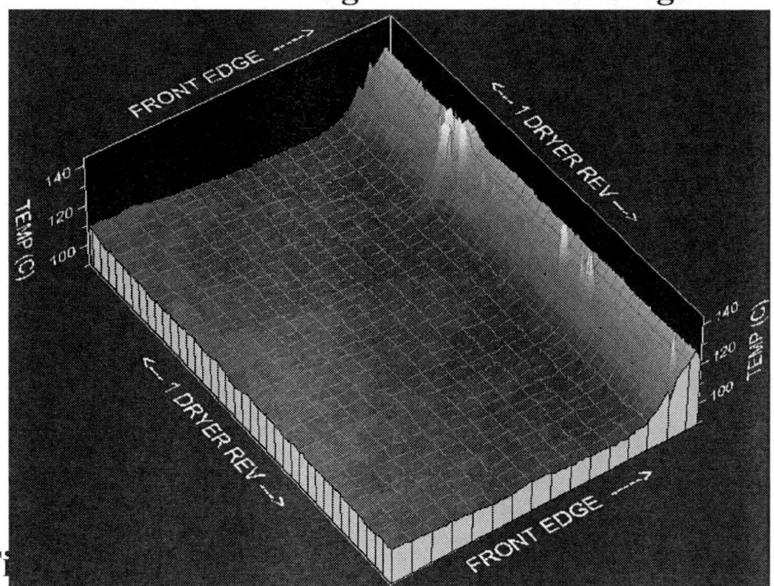

Ti

PRO*dry*

Sheet Blistering: 3-D Infrared Images

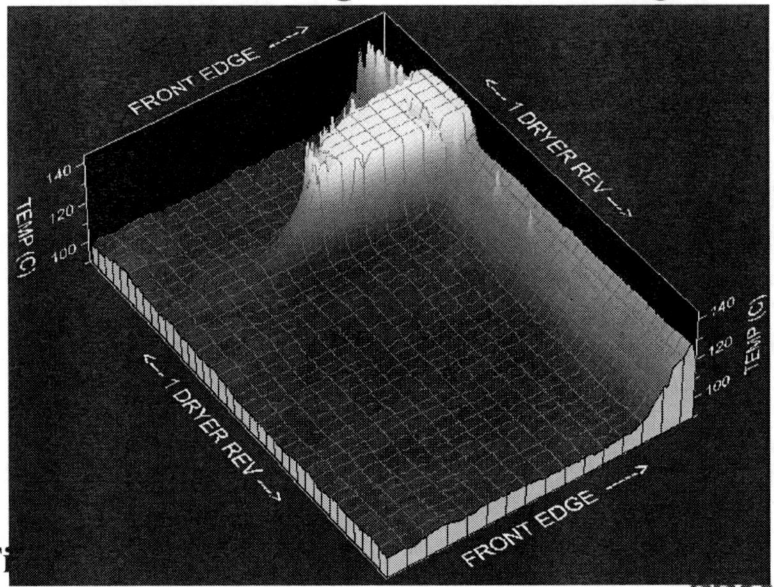

Ti

PROdry[SM]

An Example of Sheet Blistering Within the Web on the Yankee Dryer

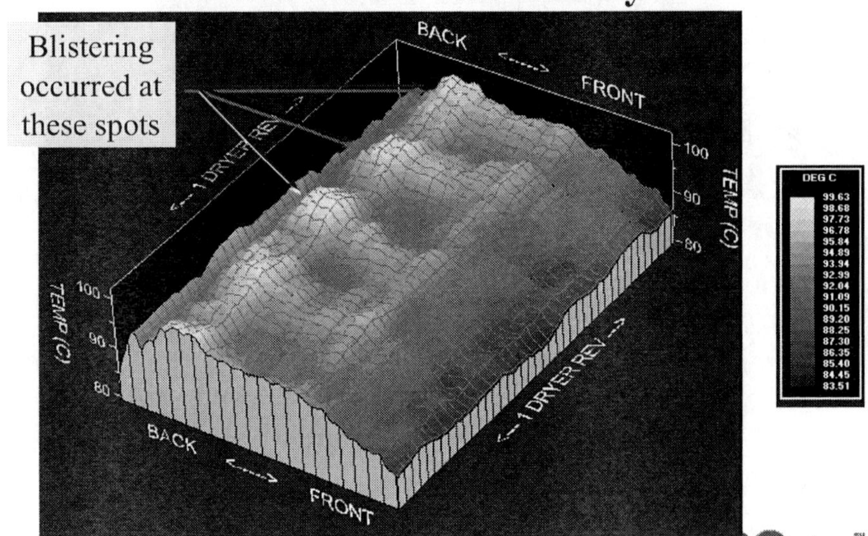

PROdry[SM]

Ways to Combat Sheet Blistering

1) Reduce sheet moisture coming to Yankee dryer
2) Reduce Yankee dryer surface temperature
3) Control sheet properties and other factors

PRO*dry*℠

Ways to Combat Sheet Blistering

1) Reduce sheet moisture coming to Yankee dryer
   - ✓ Higher solids to the pressure roll
     - Better wire drainage
     - Press improvements
   - ✓ Better pressure roll dewatering
     - Better nip integrity/crowns
     - Improved vacuum levels
     - More effective steambox usage
   - ✓ Better felt conditioning
     - Shower uniformity & volume
     - Uhle box dewatering & deckles

PRO*dry*℠

Formatting Issue - Best Available Paper/Graphic

Ways to Combat Sheet Blistering

2) Reduce Yankee dryer surface temperature

- ✓ External cooling of Yankee shell
  - Edge water sprays – CAUTION!
  - Increased hood infiltration on edges
  - Local area metalizing
- ✓ Improve Yankee dryer temperature profile
  - Manage blow-thru to reduce spots
- ✓ Adjust internal condensate removal system
  - Raise siphons
  - Profile siphons higher at edges
  - Check each siphon for clearance
- ✓ Shift drying to hoods

PROdry℠

Ways to Combat Sheet Blistering

3) Control sheet properties and other factors

- ✓ Increase uniformity of profiles
  - Basis weight profile (CD/MD)
  - Moisture profiles
  - Proper Yankee dryer crown
- ✓ Check sheet alignment over Yankee
  - Manage trim width and position
  - Alignment of sheet/Yankee/hood
- ✓ Increase sheet permeability
- ✓ Improve Yankee dryer chemical coating

PROdry℠

Beware!!!

Edge water showers used on a Yankee dryer can lead to high stresses in the corner and help promote crevice corrosion in the Head-to-Shell Joint. They should not be part of continuing operational practices.

PROdry℠

Chatter Image

PROdry℠

Formatting Issue - Best Available Paper/Graphic

Chatter Image

PRO*dry*℠

Factors Causing/Influencing Chatter

- Crepe Blade Set-up
- Crepe Blade Holder Mechanical Configuration
- Temperature Performance of Coating
- MD and CD Thermal Variation on the Dryer

PRO*dry*℠

Formatting Issue - Best Available Paper/Graphic

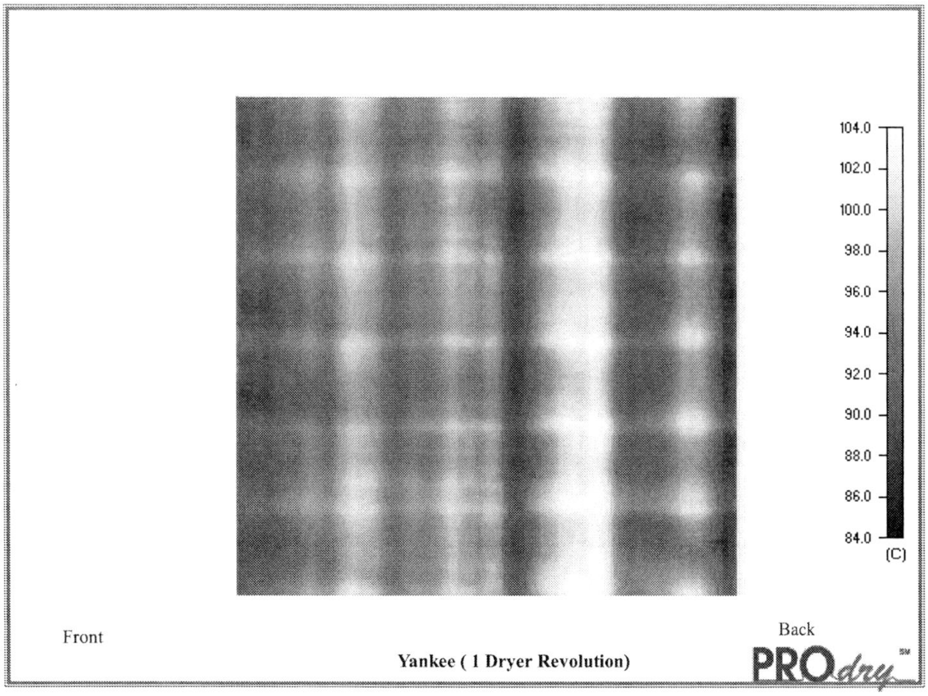

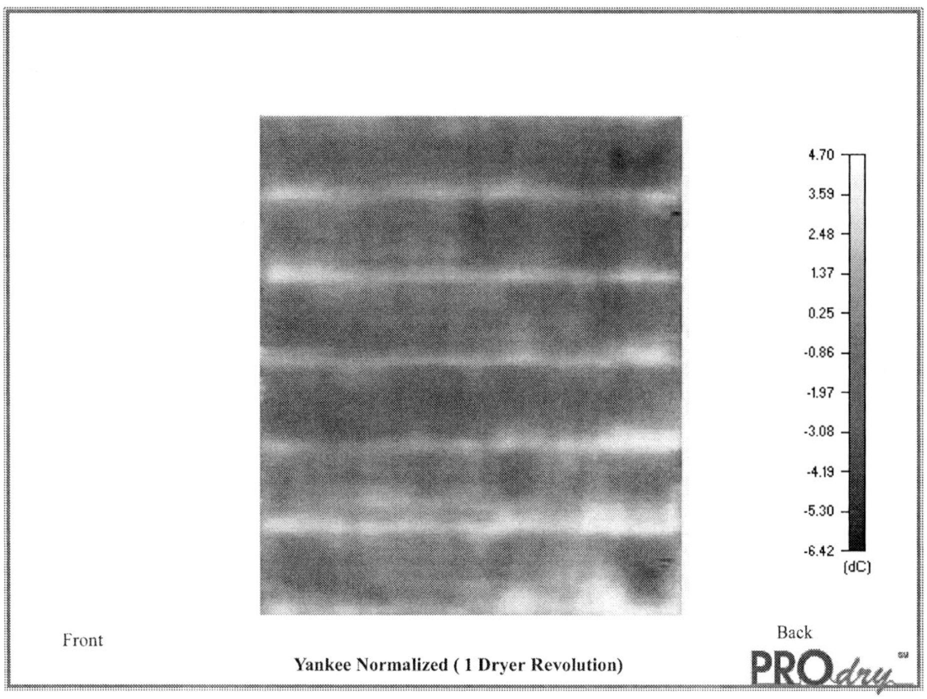

Slide 6

The "Out-of-Round" Phenomenon

- Characteristics of the Yankee During Out of Round Excursions:
  - "Sudden" onset of high vibration
  - Severe pressure roll bounce
  - High thermal variation in machine-direction
  - Blistering, blade plugging, sheet breaks
  - High vibration in ancillary equipment
  - High elastic deformation of the shell
  - Shell returns to "normal" after condensing load is removed.

PROdry℠

Slide 11

Normal Operation Temperature Distribution on a Smooth-Bore Yankee Dryer with Condensate Removal Scoops

The Yankee thermal distribution on this smooth-bore Yankee reflects the configuration and conditions inside the dryer.

The slightly warm CMD bands are locations of the full-width spoiler bars.

The relatively cool spots around the dryer reflect areas of missing spoiler bars and condensate riser leaks

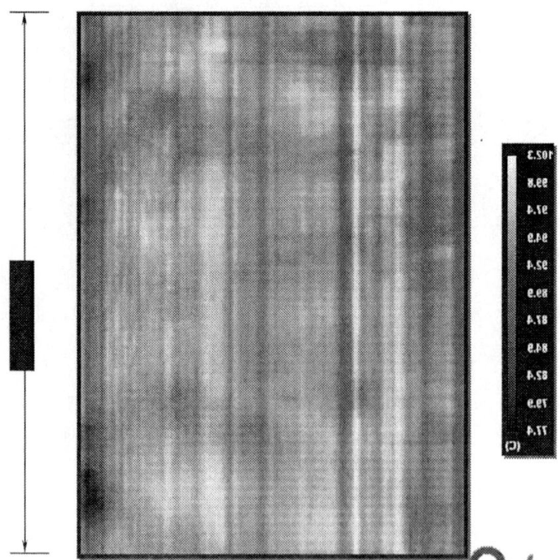

PROdry℠

Formatting Issue - Best Available Paper/Graphic

Slide 12

Evolution of Temperature Distribution of Yankee Experiencing an Out-of-Round Excursion

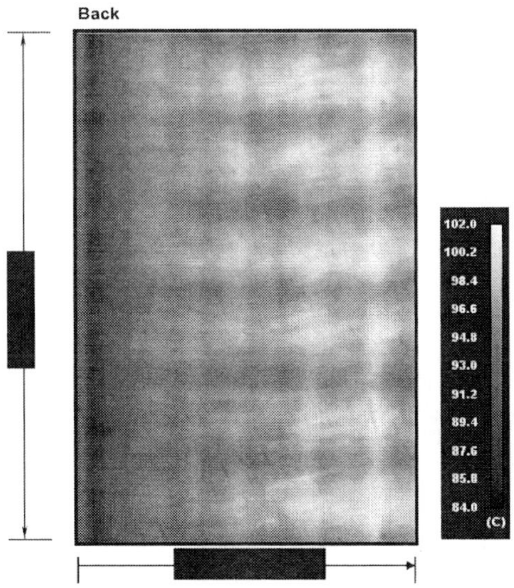

During Normal Operation, this 6-header Yankee dryer exhibits a typical thermal pattern with 6 warm and 6 cool bands in the machine-direction.

This Yankee dryer also shows a wet streak near the back edge of the sheet.

The average machine-direction temperature variation is about 7° C.

PROdry℠

Slide 13

Evolution of Temperature Distribution of Yankee Experiencing an Out-of-Round Excursion

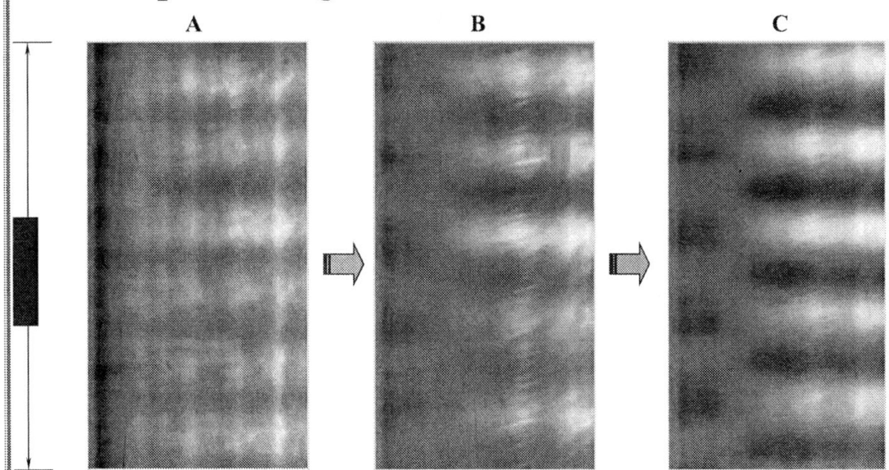

A	B	C
Range = 5.8 C	**Range = 12.7 C**	**Range = 19.2 C**
P to P = 6.5 C	**P to P = 20.0 C**	**P to P = 24.1 C**

PROdry℠

Slide 17

Temperature Distribution of Sheet on 6-Header Yankee Dryer During OOR Excursion

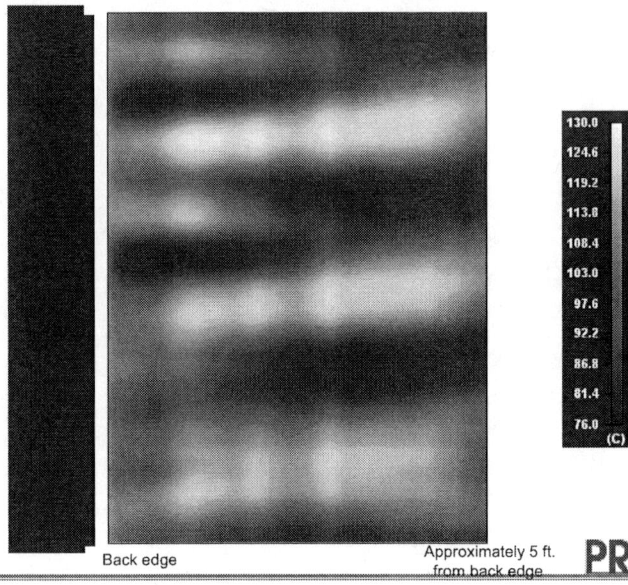

PROdry

Yankee Thermal Profile Optimization

- Evaluate mechanical condition of internal components
- Evaluate configuration of internal components
- Optimize profile of sheet before dryer
- Optimize drying protocol using high speed thermal imaging.

PROdry

Formatting Issue - Best Available Paper/Graphic

Thank you

PROdry℠

PROdry Technology, Inc.

Productivity · Reliability · Optimization

"Setting the standard in papermaking diagnostics"

44

PROdry℠

CLEARWATER PAPER

Identifying and Correcting
Yankee Condensate System Problems

Alan Coons and Bill Anderson

Problem:

Moisture variations were increasing, and drying efficiencies were gradually but measurably decreasing, over a two year period.

Problems with chatter were also increasing.

Annual internal inspections did not identify the cause.

Lewiston #3 Tissue Machine

Lewiston #2 Tissue Machine

Two Tissue Machines Analyzed and Repaired

- Two similar Yankees
- 5.5 meter diameter x 5.9 meter face
- Bathroom Tissue and Household Towel grades
- Machine speeds of 1600 to 2000 meters/minute

System Resistance Example - Plugging

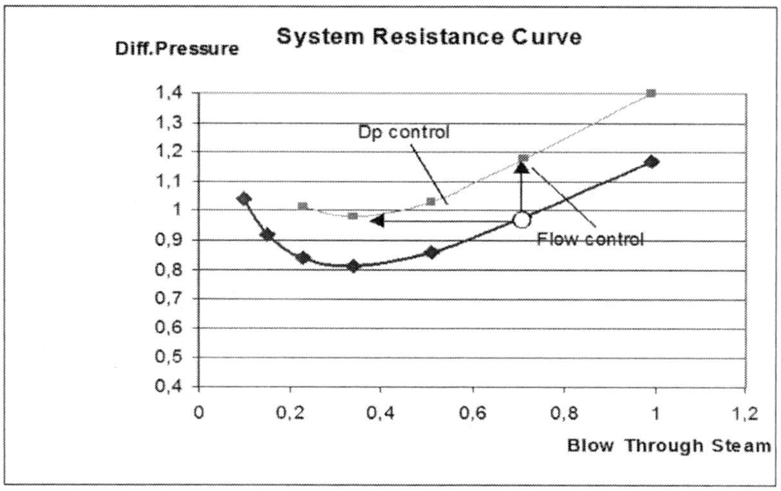

Example curves – courtesy of Metso

System Resistance Example - Leaking

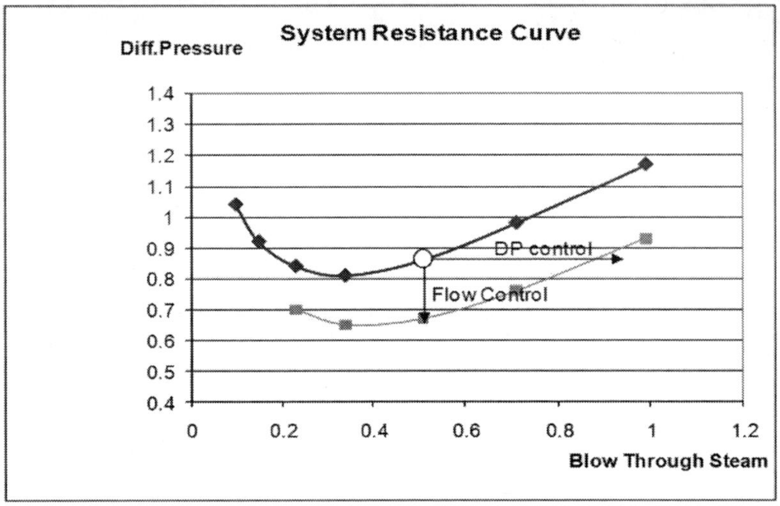

Example curves – courtesy of Metso

Troubleshooting Yankee Steam / Condensate Systems With Hi-Speed Moisture Sensor

Objective:

To identify "hot / cool" spots in the shell by using the on-line gauging (scanner) system. These changes in shell temperature show up as a moisture variation at the paper machine reel.

Requirements:

• Moisture sensor needs to be capable of responding to a change in water content at a high frequency. The sensor used in our system is at about 180Hz before it starts to "roll" off.

• Data acquisition at a rate that will supply enough data, at the resolution that is statistically sound, for the number of headers (condensate pick up points) in the Yankee rotating at speed. We set the sample rate to 20ms, i.e. 50Hz = 50 data points per second.

• FFT – Our example is from the on-line gauging system. The data can be acquired from a separate data acquisition system and processed off-line.

• Yankee Speed converted into Hz.

Fixed Point - Center of profile before replacement of header.

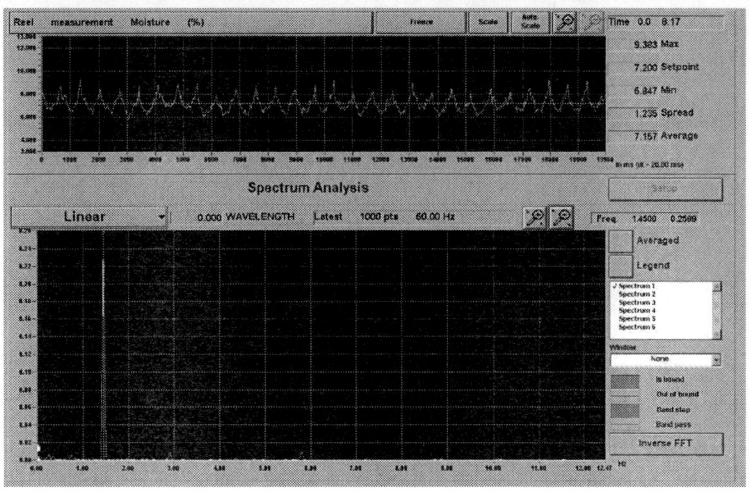

Notice the MAX to MIN values in the 20 second high speed trend.
Moisture variation = 9.383 – 5.847 = 3.536%

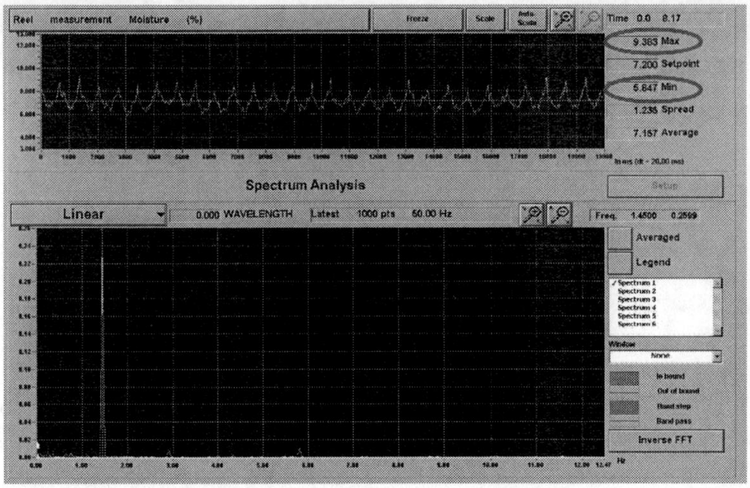

The dominant frequency is 1.45 Hertz
This matched our Yankee frequency
Frequency = Yankee Speed (meters/second) / (Diameter * Pi) = Revolutions per Second

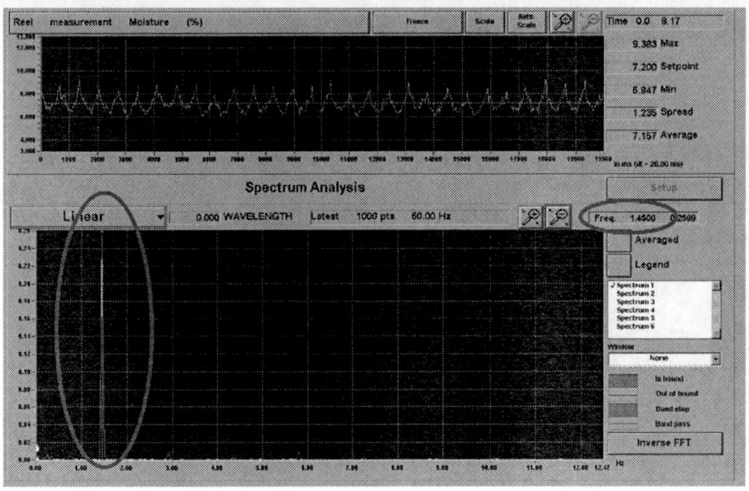

Example of the change in power

Moisture is indicated by "Power". Chart indicates moisture variation in the machine direction, shown at cross-machine machine locations, before and after replacement of the Yankee condensate headers .

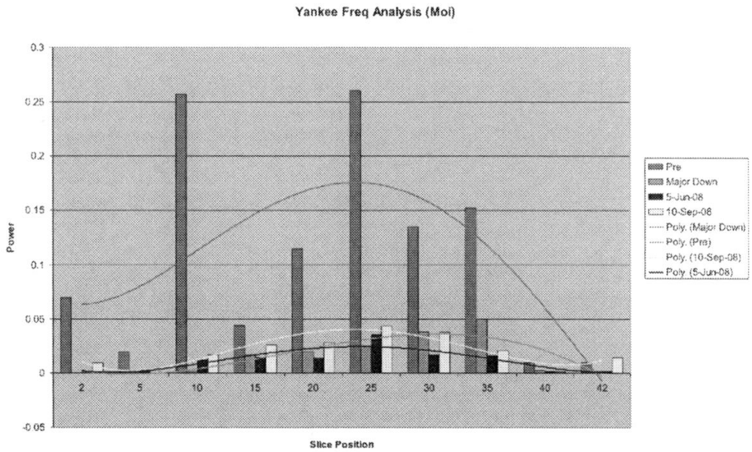

Infra-Red analysis also showed temperature variations

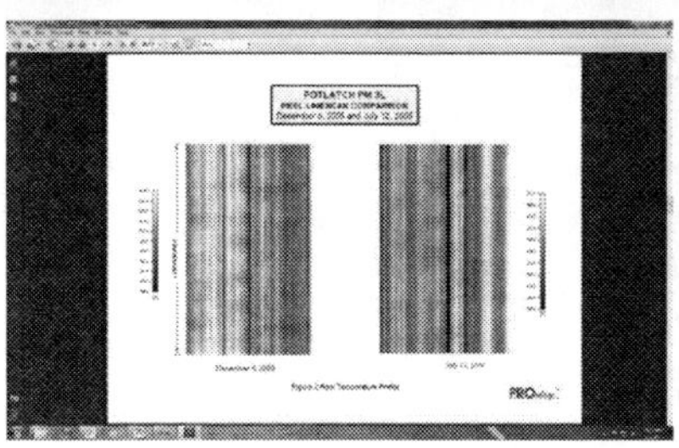

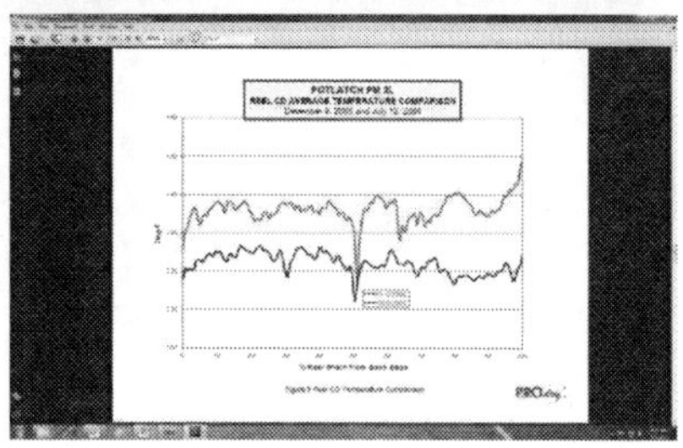

ProDry Technology, Inc.

July 12, 2006

Results of the survey indicated several significant findings:

- Generally, the machine's thermal profiles have slightly deteriorated from the last survey.
- There are two wet streaks visible in this survey that were not apparent in the older survey.
- The header near bolt 80 should be monitored as this one has appeared to have changed since the last survey. It is not significant enough to cause runnability problems but should be monitored.

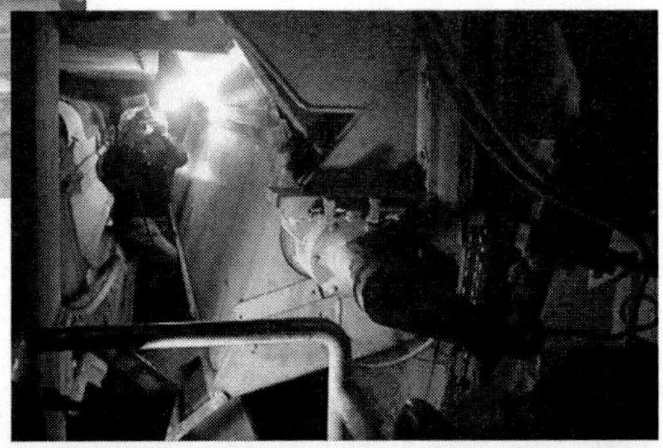

Both machines were experiencing more chatter problems as moisture and temperature variations increased.
When chatter marks developed they were repaired to extend production life between Yankee grinds.

#2L Yankee

The normal checks for plugged siphons used a shallow pool of water, enough to cover siphon ends.

Pipe connections were sprayed with soapy water to detect leaks.

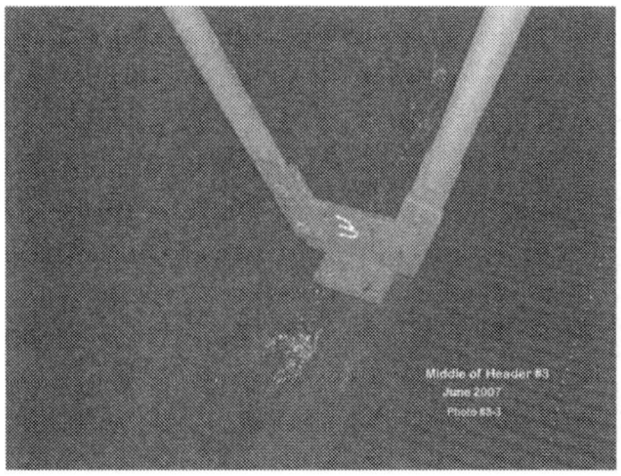

When more water was used and the headers were submerged, many leaks were found at siphon "cassette" connections.

Leaks at ends of headers were also found.

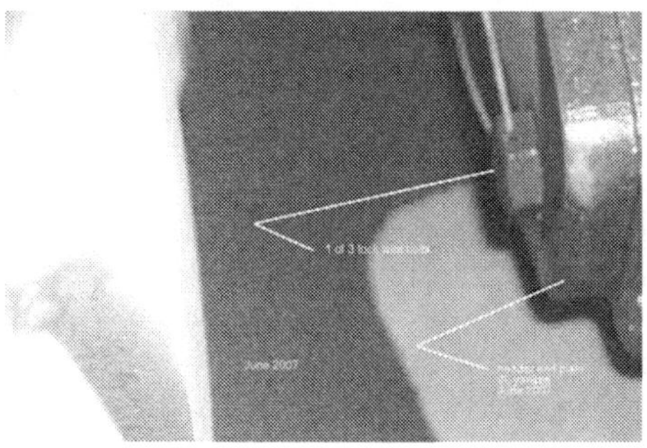

Plates on ends of headers leaked.

#2L Yankee

Leaking end plates were so close to head bolts they could not be removed & repaired.

Holes were cut in tops of headers and the joint bottom was welded internally, but this repair was not successful.

To eliminate condensate header leaks all headers and siphons were replaced.

New headers have welded ends and individually mounted siphon tubes.

Leaks at head flange connection

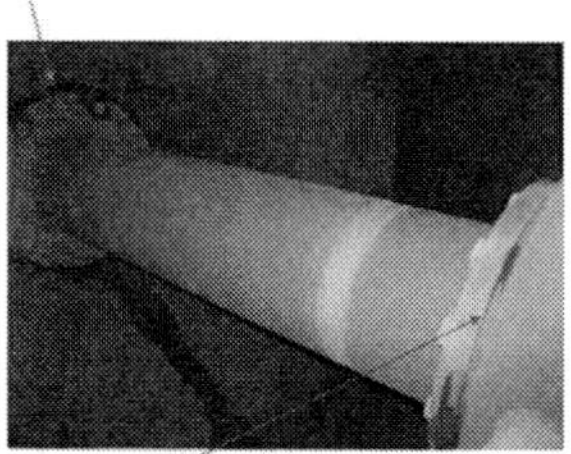

Leaks at collector flange

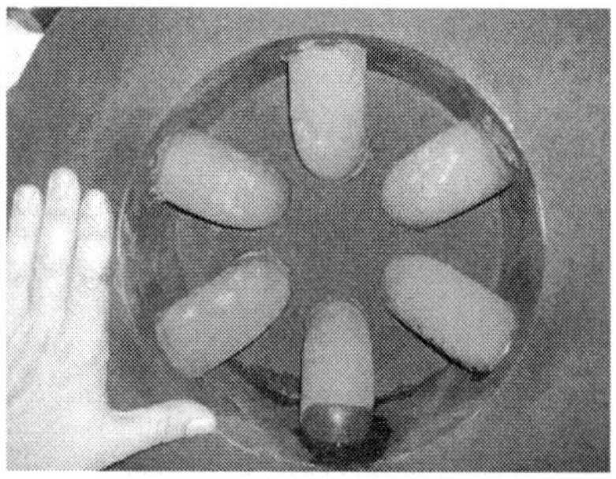

Leaks at riser elbows in collector

Additional locations prone to leaks

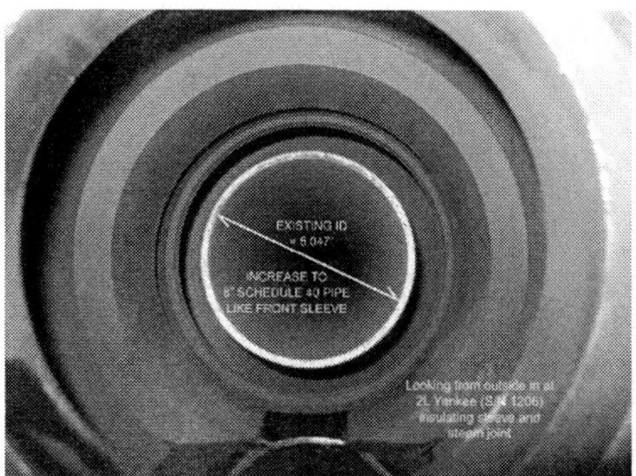

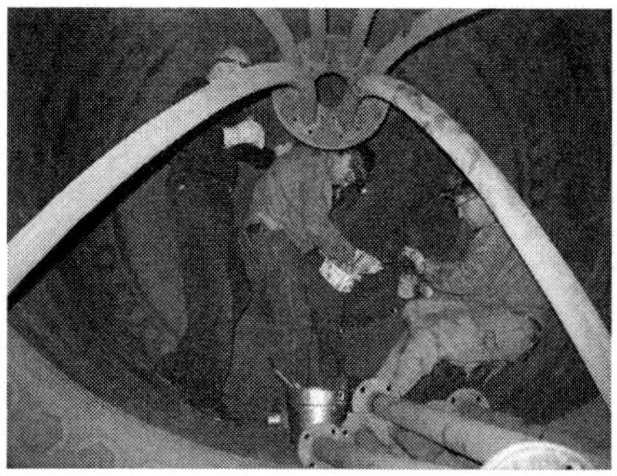

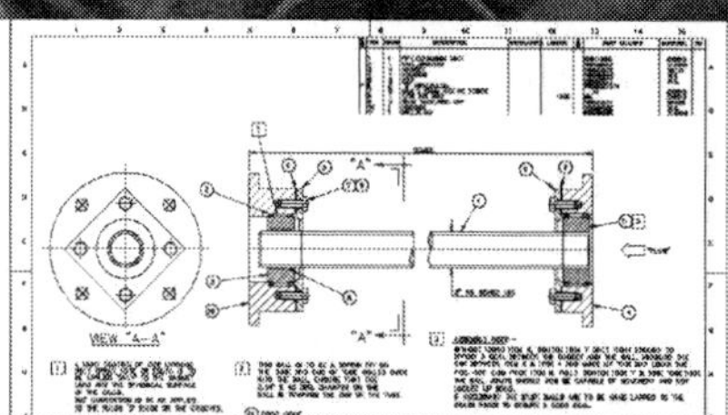

This Yankee was supplied with a 130 mm diameter insulating sleeve.

Proper size is 200 mm inside diameter.

One more possible restriction to condensate removal

Conclusion

• Analysis of Yankee condensate systems using on-machine scanners and infra-red analysis, in addition to trends of Yankee drive loads, steam flows, and steam pressures, can quantify Yankee drying problems.

•Internal inspections can pinpoint causes.

• Improved Yankee performance including the potential for reducing blade chatter can be achieved.

CLEARWATER PAPER

Thank you.

Gary Marzullo, Msquared

Sjaak Melkert, Msquared

Troubleshooting Yankee Dryer Condensate Removal Systems

After diagnosing more than 125 Yankee cylinder s under both condensing and non-condensing loads, the Engineers at Msquared have learned a lot about condensate system operating problems.

They will share their insights on common problems, troubleshooting techniques, and the results of their optimisation efforts

---

Gary Marzullo, Msquared:
After receiving his Chemical Engineering from the University of Washington, Gary entered the US Marine Corp as a Platoon Commander. After military service his tissue career started with Crown Zellerbach and then Georgia Pacific with positions in Technical Management. He then entered the Yankee Dryer Services industry by first joining Bender and then Voith in key Technical Management positions. Finally, Gary joined Sjaak Melkert in the formation of Msquared, the premier Yankee Dryer consulting organisation.

TissueWorld Yankee Dryer Workshop, Nice 2011

Company: msquared GmbH
Dülmener Strasse 89
48163 Münster
Germany
info@msquared.eu
www.msquared.eu
+49(0)2536 346 905

Authors: Gary B. Marzullo
Sjaak Melkert
Bertrand Niemeijer

0.0 Introduction:

msquared GmbH, a company of industry professionals, is dedicated to offer independent and unbiased diagnostic services and troubleshooting on tissue machines. Specialty repairs are also provided.

The Yankee dryer is the heart of the tissue machine and virtually any anomaly or inconsistency in the tissue manufacturing process will be manifested on the Yankee as either the shape or surface temperature variation.

By accurately measuring the shape and temperature profile of the Yankee under condensing load (during papermaking) and under conditions without a sheet (grinding conditions), a wealth of information now becomes available to determine the root cause of most Yankee operational anomalies.

After diagnosing more than 150 dryers over the last five years, it has become evident that the operation of the condensate removal system and the press nip uniformity are the paramount factors for optimizing the Yankee's operating efficiency.

Figure 0.1 Typical OTR installation FS and BS is shown on the brackets, behind the creping doctor.

1.0 Summary:

This paper and presentation highlights the various main types of Yankee dryer condensate removal systems employed throughout the industry and the effect of these systems in the operation of the Yankee dryer. Practical examples will show typical "footprints" of various condensate removal systems showing in both shape and temperature profiles of the dryer surface.

A uniform nip can only be achieved having a uniform shape of the Yankee surface, in both cross-machine (CD) and machine direction (MD). Consistent creping will only be achieved having a uniform temperature of the Yankee surface, in both cross-machine (CD) and machine direction. A correctly sized steam and condensate system, engineered with the correct control logic, and well maintained, combined with routine Yankee diagnostics to detect potential anomalies at an early stage does have a significant effect on the performance of the dryer, and thus the overall efficiency of the tissue machine.

Today, chatter development in the dryer surface is the most common Yankee surface anomaly driving down the performance of the tissue machine. Root causes for chatter development are manifold and rather complex but as chatter is always an effect related to the "stick-slip" behavior of the blade(s) tip against Yankee surface (and coating), dryer surface inconsistencies in both shape and temperature profile have a major influence.

Figure 1.1 Finely spaced surface chatter

2.0 Yankee Measurements

The shape profile measured of the Yankee under condensing load (making paper) and the temperature profile measured directly from Yankee surface provides valuable information about the operational performance of the Tissue machine, and in particular, the functionality of the condensate removal system.

The inverse of the temperature profile measured directly from the Yankee surface during production always matches the sheet moisture profile measured by the on-line scanner.

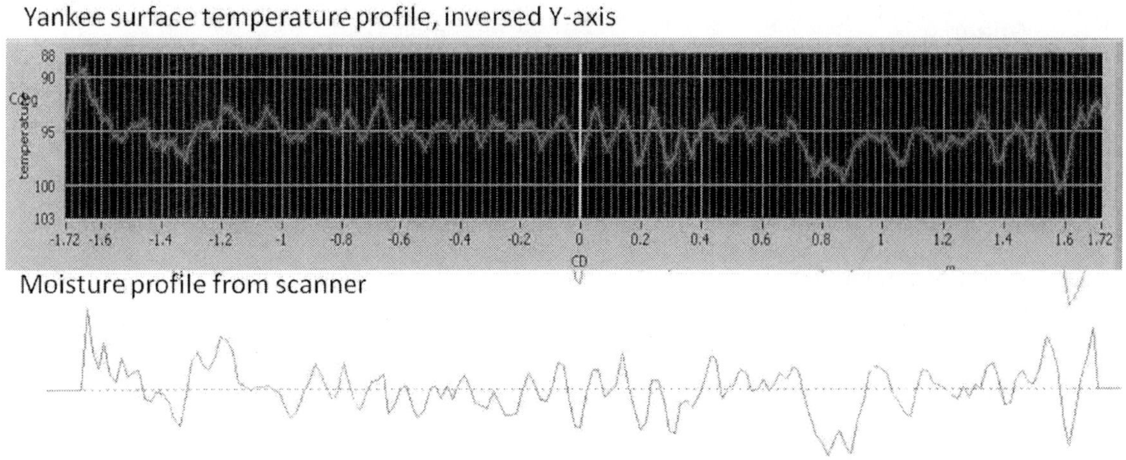

Figure 2.1 Correlation between the measured Yankee surface temperature and moisture profile achieved

Scanners are designed to measure cross machine profiles of the sheet. Fast machine directional variations in MD moisture are normally not visible using on-line moisture scanners, but do become very visible in the measured Yankee surface temperature profiles. Most of the machine directional variations (read also moisture variations) measured is the result of variances in the performance of condensate removal systems.

Although all types of condensate removal systems shown do, in principle, work, the examples given in this presentation are indicative for machines operating with poorly maintained systems or systems now operating and pushed beyond their original OEM design limits.

It is not **msquared**'s intention to favor one system over the other but to show the importance of operating the machine with a well functioning condensate removal system.

Furthermore, good and consistent CD & MD temperatures (and shape) are crucial for developing a uniform organic coating, and thus a uniform and consistent sheet crepe.

A good operating Yankee system will show maximum variations in machine direction (MD) of $\Delta T<2^oC$ and in cross machine direction $\Delta T<4$ to 5^oC.

3.0 Condensate Removal Systems:

Table 3.0.1 presents an overview of the most common systems employed. Other, less common, configurations are also in running.

System	Single point pickup	Full CD width pickup
Siphon Shoes	2-4 shoes	
Scoops	2-4 scopes; edge	
Radial Straw Headers		Normally 4-8 headers
Inclined Straw Headers		Up to 9 headers
Outrigger Straw Headers		Normally 6 headers; 12 pick-up points

Table 3.0.1 Overview condensate removal systems. In addition, on some MG cylinders full width headers without straws (series of bores) and full width scopes are also employed.

Scoop and Siphon systems are used in smooth bore dryers, and Header/Soda Straw systems are mainly in ribbed dryers.

3.1 Siphon Shoes

On tissue machines, siphon shoes are typically installed in somewhat older and slower machines on lower condensing rates. Various numbers of shoes and configurations are in use, either installed in the dryer center, on the dryer edges or also CD&MD spaced.

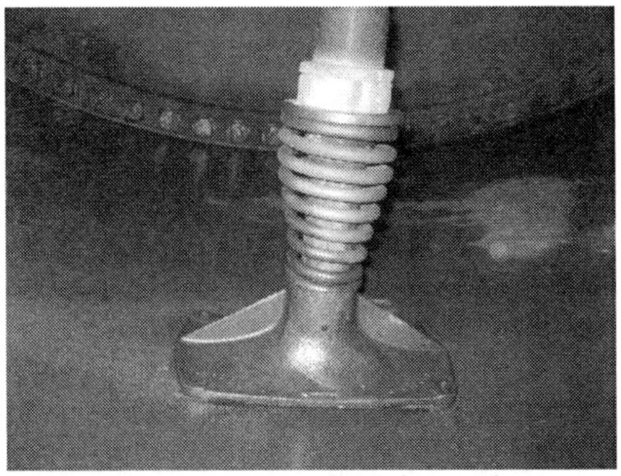

Figure 3.1.1 Typical Siphon Shoe

On various tissue machines measured, when operated with an increased condensing load, the shoe(s) start to mark quite strongly in surface temperature which will have a subsequent and direct effect on MD sheet moisture.

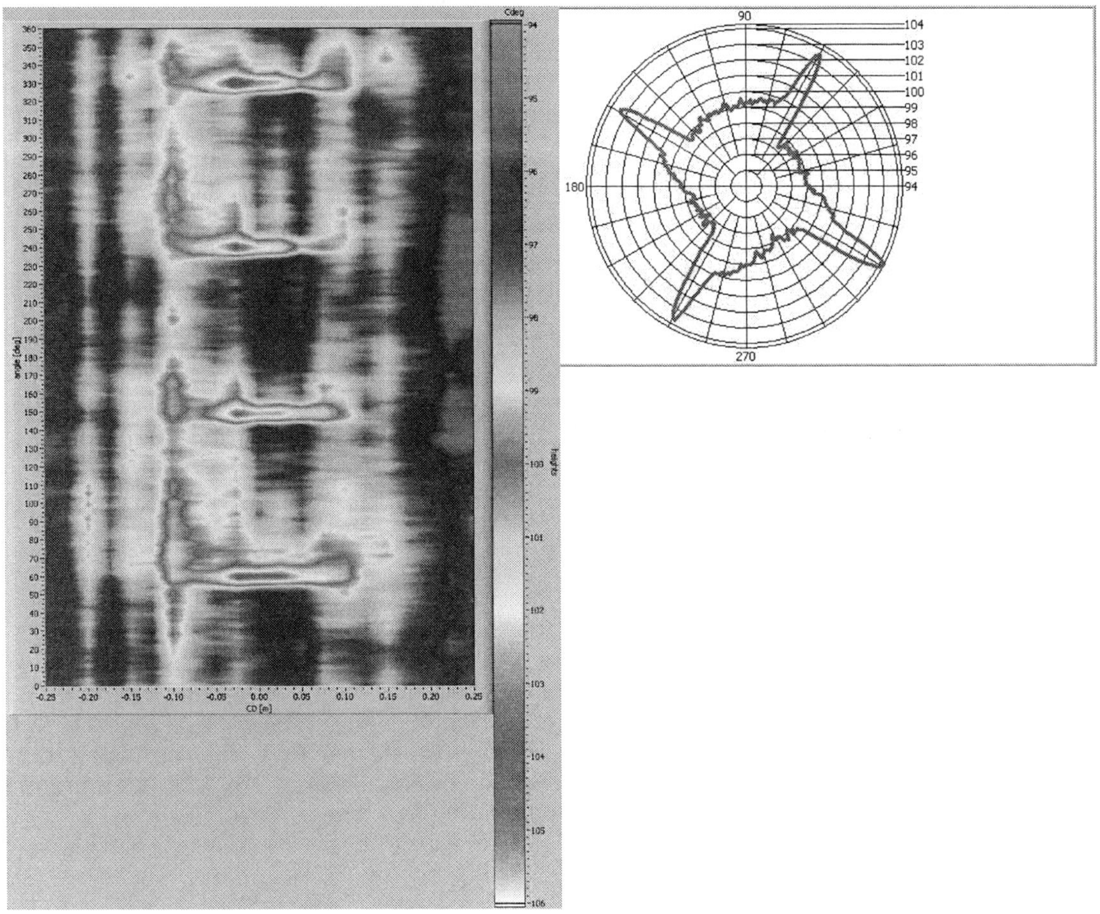

Figure 3.1.2 Temperature profile ±250 mm view around the centerline of a Yankee with four (4) center positioned shoes.

Although the overall Yankee surface temperature is good and consistent. The shoe positions are strongly marking, and are, in this particular machine, 8-9 °C warmer than the surrounding areas.

Siphon shoes have, for a long time, successfully been used in many and all kind of drying cylinders. On two known occasions, malfunctioning or improper functioning of the shoe did create significant damage to internal surface of a Yankee.

The shoes are normally spring loaded against the inner bore of the cylinder and the clearance between the shoe and Yankee (or dryer) surface is set using spacers. If, due to any malfunctioning reason, this gap is reduced, the erosive powers of the bi-phase flow of "steam and condensate" can cause damage to the inner bore of the Yankee.

Figure 3.1.3 steam erosion underneath a siphon shoe

This particular dryer had to be further de-rated as damaged area was large and the depth of the damage was significant affecting the RST (Cylinder Root Shell Thickness). Cross machine headers were installed.

Proper maintenance and regular Yankee diagnostics as preventive maintenance tool would have caught the anomaly at a very early stage and corrective action could have been initiated.

3.2 Scoops

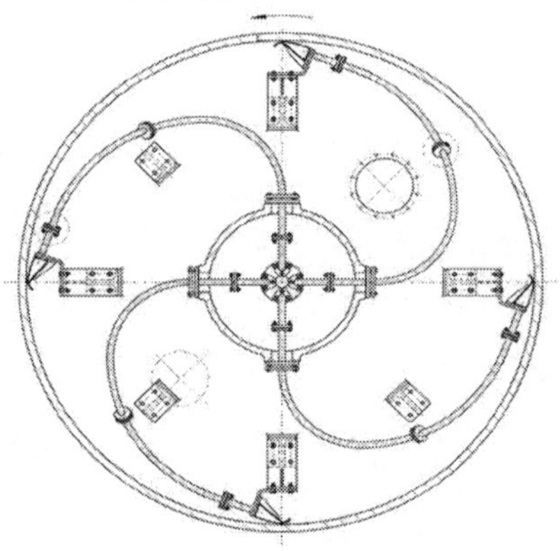

Figure 3.2.1 Scoops

Dryers operating with scoops normally show a good and consistent temperature profile. Without further discussing the thermal efficiency of these systems, this surface temperature uniformity is likely due to a relative large layer of condensate rimming (above rimming speed) in the dryer, equalizing any induced variances.

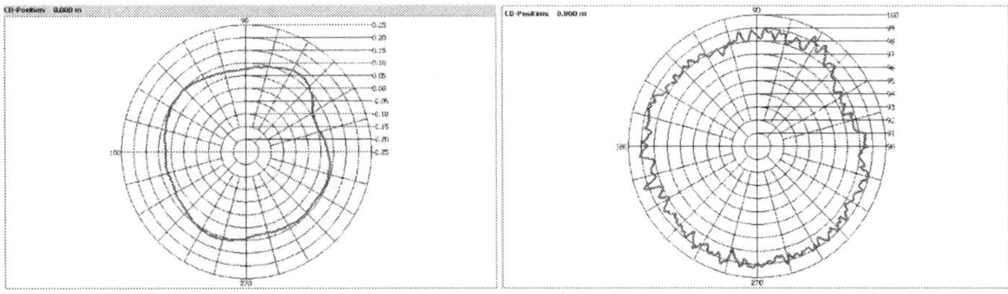

Figure 3.2.2 Centerline shape run-out and temperature run-out of a scooped cylinder

3.3 Cross machine headers, radial straws.

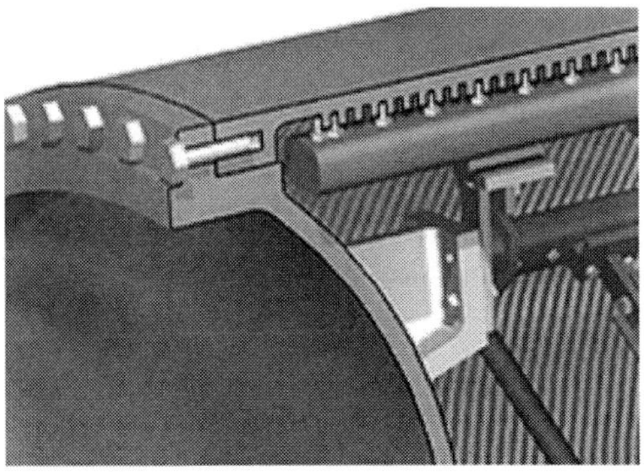

Figure 3.3.1 Headers with radial straws

This type of condensate removal systems is widely used, and various OEM's equip their dryers with this system.

The headers (normally 4 up to 8) are supported from the Center Trunnion, sometimes with additional support brackets from the FS & BS heads or, mainly on older systems, supported against the shell ID.

Straw distance (clearance between straw tip and shell) is normally set to approximately 5-7 mm, with increased straw clearance at the corner taper shoulder (outboard of the rib area) of the shell.

Temperature variations measured on Yankees equipped with this type of system are normally small, but with the individual headers nearly always visible.

Especially on systems which are only supported from the center trunnion, a strong "footprint" can sometimes become visible measured both in the Yankee shape and Yankee temperature profiles under condensing loads. (Shape variations are the result of temperature variations).
It is suspected that, although all straw distances are correctly set in stationary conditions, the straw clearances cross machine can significantly vary under condensing load due to shell deflection, and any header distortion.

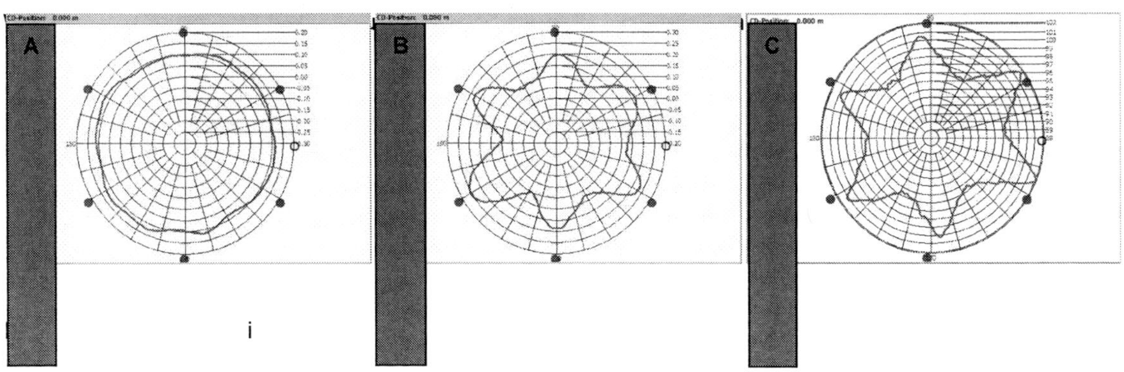

gure 3.3.2 Centerl ne polar pl of Yankee shape and temperatures

Plot A: is the centerline shape without a condensing load, dryer at production speed and steam pressure but without presses loaded and no sheet. Plot B: is the centerline shape during papermaking. Plot C: is the centerline temperature during papermaking

It is quite obvious that the six-pointed-star matches the header positions and distorts the Yankee in MD shape and temperature.

Figure 3.3.3 below shows the 3D temperature profile of this dryer in a "normalized" view. In this normalized view, CD variations are mathematically eliminated to better show the MD variations measured.

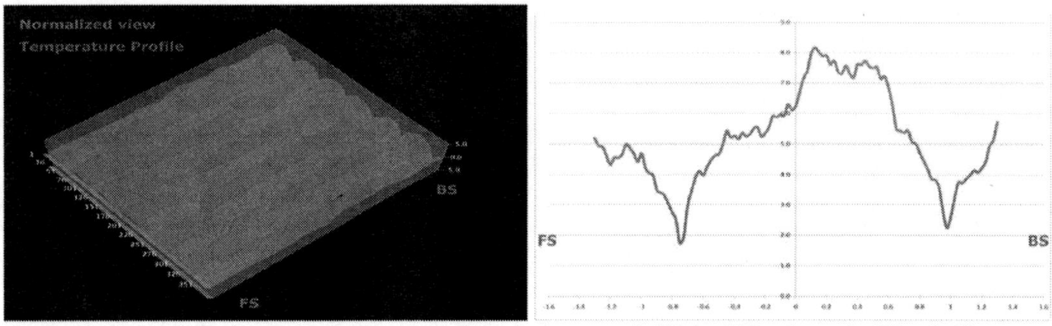

Figure 3.3.3 3D Temperature normalized view and ΔT in MD at each CD station measured

This dryer measures;"hot" beneath the headers at the center section and "colder" underneath the headers toward the edges, indicative for insufficient and non-uniform condensate removal.

The maximum MD ΔT =8.2ºC, which is considered to be a high value.

3.4 Cross machine headers; inclined straws

Figure 3.4.1 Headers with inclined straws

Many and some of the fastest running Yankees are equipped with this type system.

The condensate evacuation cross machine is normally very good, but in machine direction it quite often leaves a strong thermal and shape imprint.

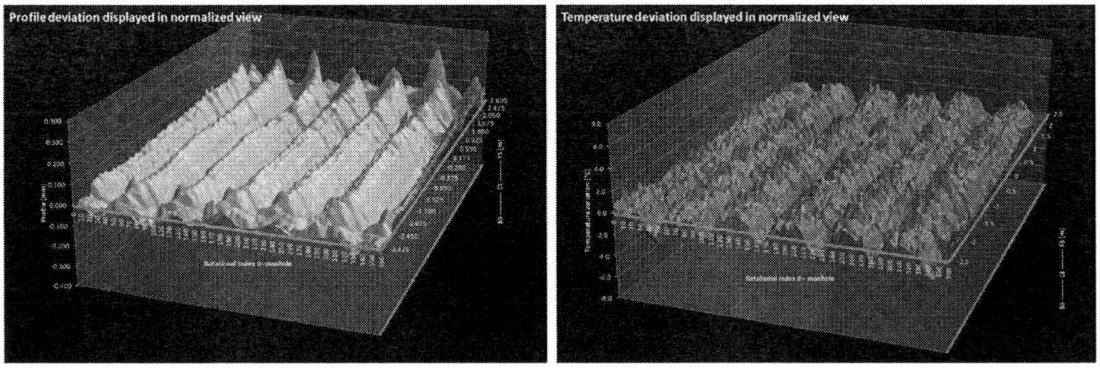

Figure 3.4.2 Shape and temperature variations

Figure 3.4.2 shows the six (6) times per revolution ripple in both shape and temperature. Neither shape nor temperature variations are large, but are very distinct.

By measuring the radial movement of the press roll while using the Yankee as rotational index the effects become very clear.

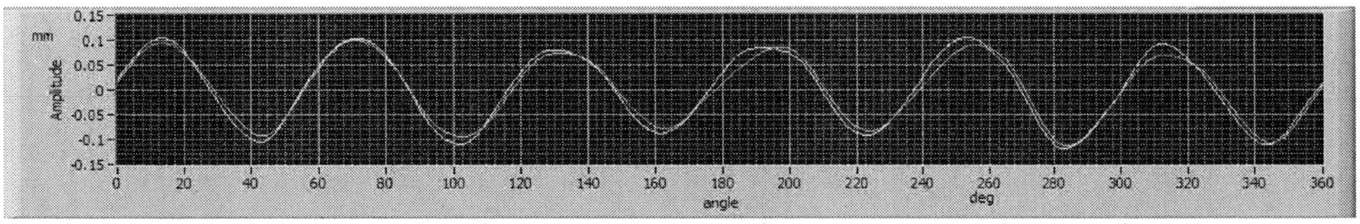

Figure 3.4.3 Radial movement of suction press in relation to one Yankee revolution

To optimize a machine for higher speeds and higher condensing loads, it is advisable that the distortions that do create machine vibrations be eliminated.

3.5 <u>Cross machine headers; outrigger straws.</u>

Figure 3.5.1 Headers with outrigger straws

Although originally designed by one OEM supplier, many Yankees from various suppliers are today equipped with this design condensate removal system.

The long straws are kept in position (and at straw distance from the groove bottom) with a slider block, spring loaded with a 2 kg force, against the top of the Yankee's rib. With this design, the straw distance becomes independent from any shell deformation and or header distortion and will keep a constant clearance.

Dryers measured with this system normally show very little MD temperature variations (and subsequently very little MD shape variations). Sometimes, little differences between the performances of leading straws versus trailing straws are noted but also these variations are normally quite small.

To obtain and maintain uniform CD and MD evacuation, the systems are more maintenance intensive, as the longer straws have a higher tendency to block in steam systems that are not optimal. Obviously good conditioning of the steam is furthermore crucial.

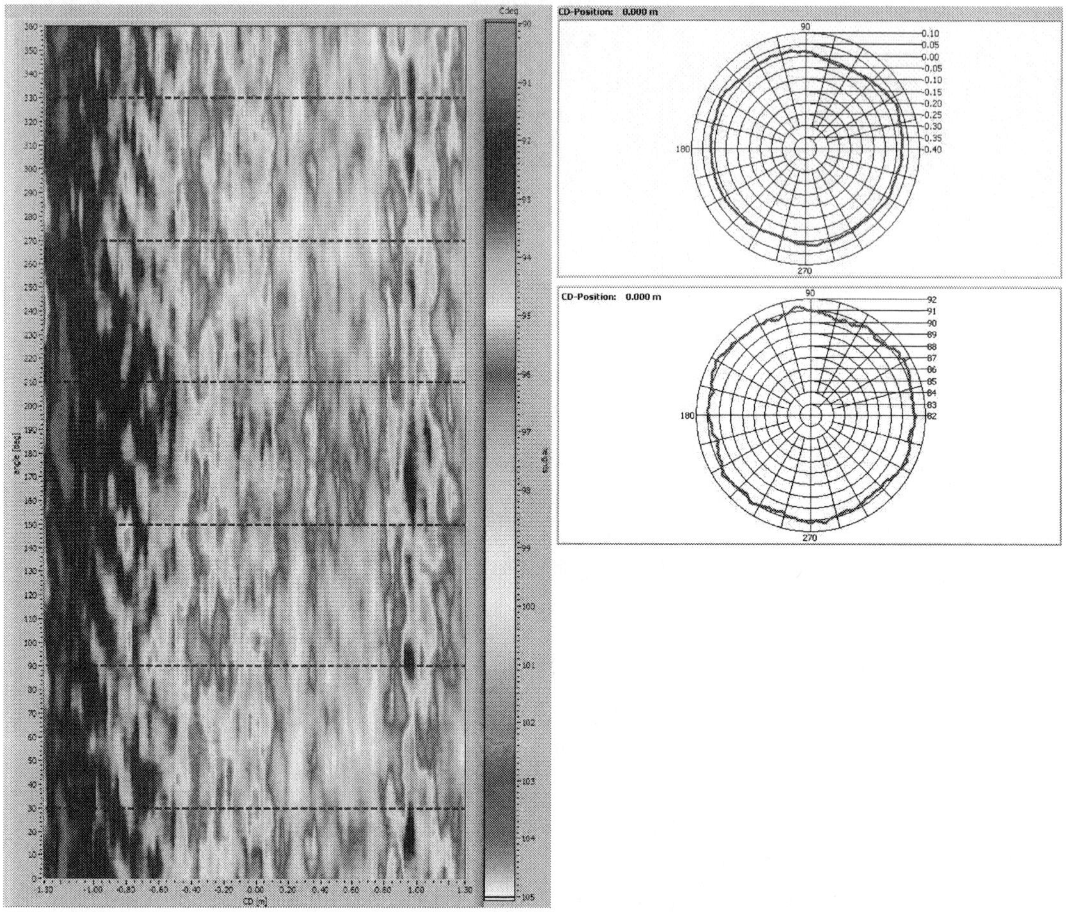

Figure 3.5.2 2D temperature profile and typical centerline shape/temperature run-out

Some CD variation in Yankee surface temperatures is noted, requiring further press nip optimization, but very little MD temperature variations and only little of the typical marking of the condensate headers is visible.

4.0 Other issues with condensate removal

4.1 OPTIMIZING THE BLOW THROUGH: A PRACTICAL EXAMPLE

The objective of the study was to determine the magnitude and shape of any Yankee deformation and the temperature distribution under operating conditions at various differential pressures. Previous measurements indicated that the thermocompressor was sized incorrectly for the operating values used in this Yankee's steam and condensate system. After metallization, a new, resized thermocompressor was installed.

Baseline operating conditions measurements were first performed each day to establish the conditions in effect since startup from the installation of the new thermocompressor. The machine was being operated at a ΔP of 1.05 bar and a steam pressure of 6.05 bar. The Yankee had been ground and metalized in April 2010 using the crown calculated for 6.0 bar pressure. It is important to make a change, then to take away that change to insure that the machine returns to "normal" reflecting the fact that the change made to the DP resulted in the changed noted on the Yankee.

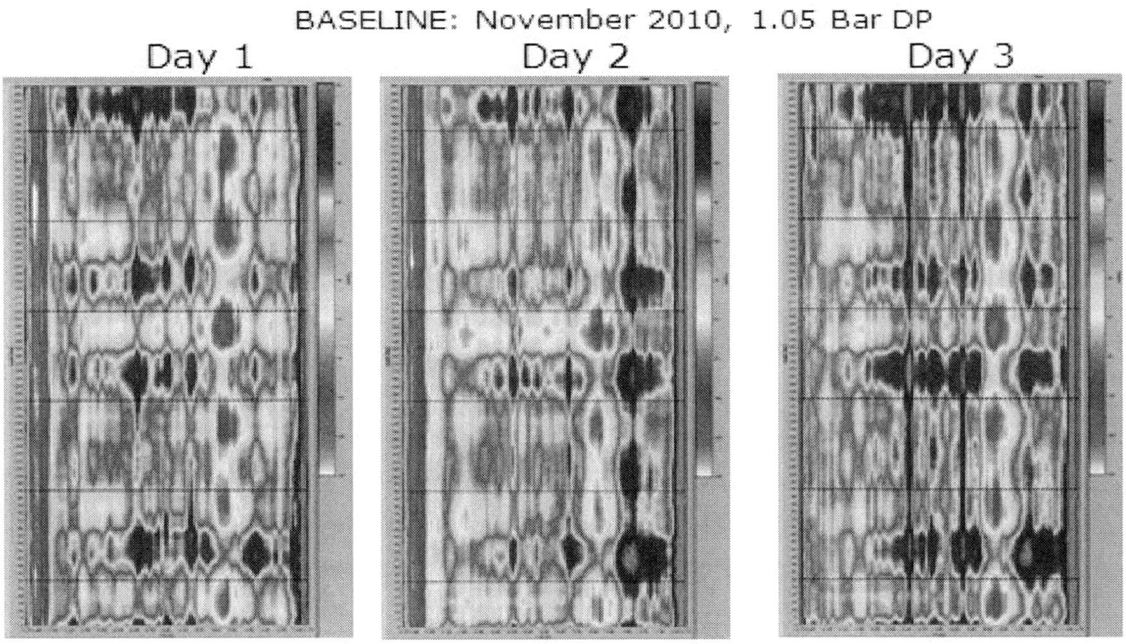

Figure 4.1.1 Baseline temperature measurements taken prior to making changes to the ΔP for each day's trial measurements. Each graph has the same temperature scaling of 85^0 to 105^0 C. Note that the machine returned to the same thermal profile at 1.05 bar ΔP profile after each trial period.

The software allows for the direct comparison of different measurements taken under different conditions. Displayed below is the **<u>CHANGE</u> IN TEMPERATURE** ON THE YANKEE SURFACE from the ΔP operated at 0.85 bar to a ΔP of 1.2 bar.

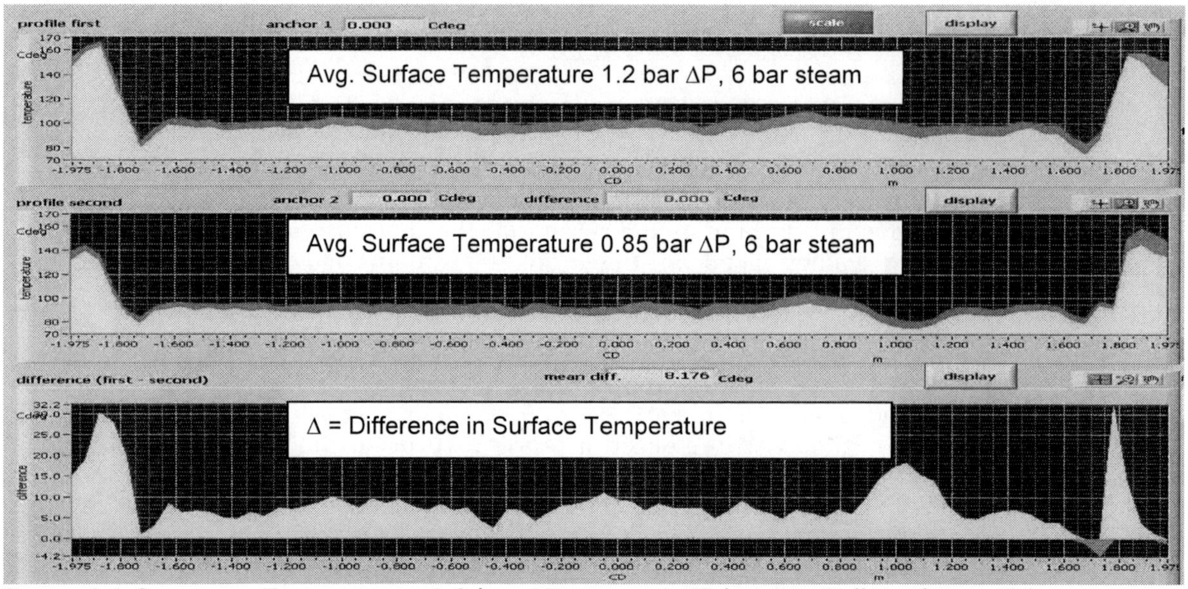

Figure 4.1.2 Temperature 1.2 bar ΔP verses 0.85 bar ΔP, Full Yankee width.

This graph displays the DIFFERENCE between operations at 6.0 bar steam pressure at 0.85 bar ΔP, and 6.0 bar pressure at 1.2 bar ΔP.

Note also the significant increase in temperature at the area between +1.0 and +1.2 m toward the drive side of centerline (on the right) where the cool band is present at the lower ΔP. It is apparent that this area has begun to have the condensate more uniformly removed.

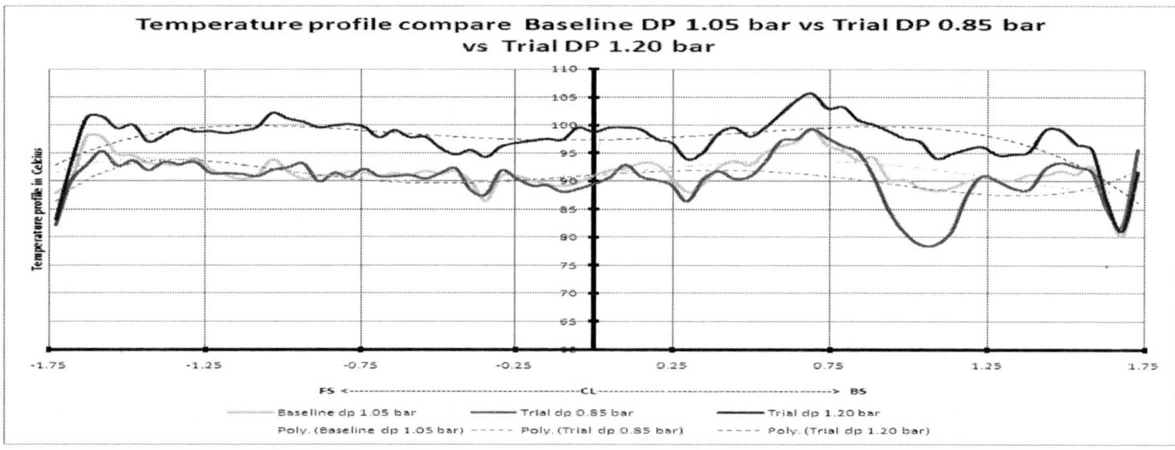

Figure 4.1.3 Temperature profiles at the various ΔP pressures operated for this investigation.

Shown above is the average temperature profile in the area beneath the sheet at the three differential pressures. Note the significant increase in temperature in the area between +1.0 and +1.25 m from centerline because of increasing the ΔP.

The significant observations regarding the surface temperature between the lowest ranges of ΔP investigated of 0.85 bar at the higher ΔP of 1.2 bar are:

- The cold band in the area between 1.0 and 1.2 m toward the drive side of centerline (the right side of the graph) disappeared.
- The entire surface is more uniform in temperature at the higher ΔP, and the average surface temperature has increase by approximately 6^0 C.
- This system has an apparent "threshold" of 1.2 bar ΔP to achieve uniform evacuation.
- The average surface temperature has increased by approximately +9^0 C at the lower ΔP.
- The makeup steam consumption was reduced by approximately 10% at the higher ΔP.

5.0 Concluding remarks

A well designed and, within its design limits, operated condensate removal system, including the auxiliary hardware and properly adjusted control logic is crucial to obtain a uniform temperature profile.

A uniform CD AND MD temperature profile will significantly increase machine efficiency and machine reliability and will further minimize effects for initiating and developing surface chatter in the Yankee surface.

Routine Yankee diagnostic (OTR-TNG and IRSense) and routine internal Yankee inspections can, at an early stage, recognize potential issues with condensate removal and is a valuable tool for further optimization of the Yankee as heart of the tissue machine.

AUTHOR INDEX